Science Fictions

Science Fictions

How FRAUD, BIAS, NEGLIGENCE, and HYPE Undermine the Search for Truth

STUART RITCHIE

Metropolitan Books

Henry Holt and Company   New York

Metropolitan Books
Henry Holt and Company
Publishers since 1866
120 Broadway
New York, New York 10271
www.henryholt.com

Metropolitan Books® and m® are registered trademarks of
Macmillan Publishing Group, LLC.

Library of Congress Cataloging-in-Publication data is available.

ISBN: 978-1-250-22269-5

Our books may be purchased in bulk for promotional, educational, or
business use. Please contact your local bookseller or the Macmillan Corporate
and Premium Sales Department at (800) 221-7945, extension 5442, or by
e-mail at MacmillanSpecialMarkets@macmillan.com.

First Edition 2020

Printed in the United States of America

3 5 7 9 10 8 6 4 2

For Katharine

Now *that* is scientific fact. There's no real *evidence* for it, but it is scientific fact.

Brass Eye

Contents

Preface

It is the peculiar and perpetual error of the human understanding
to be more moved and excited by affirmatives than by negatives.
Francis Bacon, *Novum Organum* (1620)

January 31, 2011 was the day the world found out that undergraduate students have psychic powers.

A new scientific paper had hit the headlines: a set of laboratory experiments on over 1,000 people had found evidence
for psychic precognition – the ability to see into the future
using extrasensory perception. This wasn't the work of some
unknown crackpot: the paper was written by a top psychology
professor, Daryl Bem, from the Ivy League's Cornell University.
And it didn't appear in an obscure outlet – it was published
in one of the most highly regarded, mainstream, peer-reviewed
psychology journals.[1] Science seemed to have given its official
approval to a phenomenon that hitherto had been considered
completely impossible.

At the time, I was a PhD student, studying psychology at the
University of Edinburgh. I dutifully read Bem's paper. Here's how
one of the experiments worked. Undergraduate students looked at
a computer screen, where two images of curtains would appear.
They were told that there was another picture behind one of the

curtains, and that they had to click whichever they thought it was. Since they had no other information, they could only guess. After they'd chosen, the curtain disappeared and they saw whether they'd been correct. This was repeated thirty-six times, then the experiment was over. The results were quietly stunning. When a picture of some neutral, boring object like a chair was behind one of the curtains, the outcome was almost perfectly random: the students chose correctly 49.8 per cent of the time, essentially fifty-fifty. However – and here's where it gets strange – when one of the pictures was pornographic, the students tended to choose it slightly more often than chance: 53.1 per cent of the time, to be exact. This met the threshold for 'statistical significance'. In his paper, Bem suggested that some unconscious, evolved, psychic sexual desire had ever-so-slightly nudged the students towards the erotic picture even before it had appeared on screen.[2]

Some of Bem's other experiments were less explicit, but no less puzzling. In one of them, a list of forty unrelated words appeared on the screen, one at a time. Afterwards came a surprise memory test, where the students had to type in as many of the words as they could remember. At that point, the computer randomly selected twenty of the words and showed them to the students again. Then the experiment ended. Bem reported that, during the memory test, the students were more likely to remember the twenty words *they were about to see again*, even though they couldn't have known – except by psychic intuition – which ones they were going to be shown. This would be a bit like studying for an exam, sitting the exam, then studying again afterwards, and that post-exam study somehow winding its way back in time to improve your grade. Unless the laws of physics had suddenly been repealed, time is supposed to run in only one direction; causes are supposed to come before, not after, their effects. But with the publication of Bem's paper, these bizarre results were now a part of scientific literature.

Crucially, Bem's experiments were extremely simple, requiring nothing more complicated than a desktop computer. If Bem was right, any researcher could produce evidence for the paranormal just by following his experimental instructions – even a PhD student with next to no resources. That is what I was, so that is exactly what I did. I got in touch with two other psychologists who were also sceptical of the results, Richard Wiseman of the University of Hertfordshire and Chris French of Goldsmiths, University of London. We agreed to re-run Bem's word-list experiment three times, once at each of our respective universities. After a few weeks of recruiting participants, waiting for them to complete the memory test and then dealing with their looks of bewilderment as we explained afterwards what we'd been looking for, we had the results. They showed … nothing. Our undergraduates weren't psychic: there was no difference in their recall of the words presented after the test. Perhaps the laws of physics were safe after all.

We duly wrote up our results and sent the resulting paper off to the same scientific journal that had published Bem's study, the *Journal of Personality and Social Psychology*. Almost immediately the door was slammed in our faces. The editor rejected the paper within a few days, explaining to us that they had a policy of never publishing studies that repeated a previous experiment, whether or not those studies found the same results as the original.[3]

Were we wrong to feel aggrieved? The journal had published a paper that had made some extremely bold claims – claims that, if true, weren't just interesting to psychologists, but would completely revolutionise science. The results had made their way into the public domain and received significant publicity in the popular media, including an appearance by Bem on the late-night talk show *The Colbert Report* where the host coined the memorable phrase 'time-travelling porn'.[4] Yet the editors wouldn't even consider publishing a replication study that called the findings into question.[5]

Meanwhile, another case was unfolding that also raised alarming questions about the current state of scientific practice. *Science*, widely considered one of the world's most prestigious scientific journals (second only to *Nature*), had published a paper by Diederik Stapel, a social psychologist at Tilburg University in the Netherlands. The paper, entitled 'Coping with Chaos', described several studies performed in the lab and on the street, finding that people showed more prejudice – and endorsed more racial stereotypes – when in a messier or dirtier environment.[6] This, and some of Stapel's dozens of other papers, hit the headlines across the world. 'Chaos Promotes Stereotyping', wrote *Nature*'s news service; 'Where There's Rubbish There's Racism', alliterated the *Sydney Morning Herald*.[7] The results exemplified a type of social psychology research that produced easy-to-grasp findings with, as Stapel himself wrote, 'clear policy implications': in this case, to 'diagnose environmental disorder early and intervene immediately'.[8]

The problem was that none of it was real. Some of Stapel's colleagues became suspicious after they noticed the results of his experiments were a little *too* perfect. Not only that, but whereas senior academics are normally extremely busy and rely on their students to do such menial tasks as collecting data, Stapel had apparently gone out and collected all the data himself. After the colleagues brought these concerns to the university in September 2011, Stapel was suspended from his professorship. Multiple investigations followed.[9]

In a confessional autobiography he wrote subsequently, Stapel admitted that instead of collecting the data for his studies, he would sit alone in his office or at his kitchen table late into the night, typing the numbers he required for his imaginary results into a spreadsheet, making them all up from scratch. 'I did some things that were terrible, maybe even disgusting,' he wrote. 'I faked research data and invented studies that had never happened. I worked alone, knowing exactly what I was doing … I didn't feel anything: no disgust, no shame, no regrets.'[10] His scientific fraud was surprisingly elaborate. 'I invented entire schools where I'd done my research, teachers with whom I'd discussed the experiments, lectures that I'd given, social-studies lessons that I'd contributed to, gifts that I'd handed out as thanks for people's participation.'[11]

Stapel described printing off the blank worksheets he'd ostensibly be giving to his participants, showing them to his colleagues and students, announcing he was heading off to run the study ... then dumping the sheets into the recycling when nobody was looking. It couldn't last. The findings of the investigations were clear; he was fired not long after his suspension. Since then, no fewer than fifty-eight of his studies have been retracted – struck off the scientific record – due to their fake data.

The Bem and Stapel cases – where esteemed professors published seemingly impossible results (in Bem's case) and outright fraudulent ones (in Stapel's) – sent a jolt through psychology research, and through science more generally. How could prestigious scientific journals have allowed their publication? How many other studies had been published that couldn't be trusted? It turned out that these cases were perfect examples of much wider problems with the way we do science.

In both cases, the central issue had to do with *replication*. For a scientific finding to be worth taking seriously, it can't be something that occurred because of random chance, or a glitch in the equipment, or because the scientist was cheating or dissembling. It has to have really happened. And if it did, then in principle I should be able to go out and find broadly the same results as yours. In many ways, that's the essence of science, and something that sets it apart from other ways of knowing about the world: if it won't replicate, then it's hard to describe what you've done as scientific at all.

What was concerning, then, wasn't so much that Bem's experiments were unreliable or that Stapel's were a figment of his imagination: some missteps and spurious results will always be with us (and so, alas, will fraudsters).[12] What was truly problematic was how the scientific community had handled both situations. Our attempted replication of Bem's experiment was unceremoniously rejected from the journal that published the original; in the case of Stapel, nobody had ever even *tried* to replicate his findings. In other words, the community had demonstrated that it was content to take the dramatic claims in these studies at face value, without checking how durable the results really were. And if there are no double-checks on the replicability of results, how do we know they aren't just flukes or fakes?

Perhaps Bem himself best summed up many scientists' attitudes to replication, in an interview some years after his infamous study. 'I'm all for rigor,' he said, 'but I don't have the patience for it … If you looked at all my past experiments, they were always rhetorical devices. I gathered data to show how my point would be made. I used data as a point of persuasion, and I never really worried about, "Will this replicate or will this not?"'[13]

Worrying about whether results will replicate or not isn't optional. It's the basic spirit of science; a spirit that's supposed to be made manifest in the system of peer review and journal publication, which acts as a bulwark against false findings, mistaken experiments and dodgy data. As this book will show, though, that system is badly broken. Important knowledge, discovered by scientists but not deemed interesting enough to publish, is being altered or hidden, distorting the scientific record and damaging our medicine, technology, educational interventions and government policies. Huge resources, poured into science in the expectation of a useful return, are being wasted on research that's utterly uninformative. Entirely avoidable errors and slip-ups routinely make it past the Maginot Line of peer review. Books, media reports and our heads are being filled with 'facts' that are either incorrect, exaggerated, or drastically misleading. And in the very worst cases, particularly where medical science is concerned, people are dying.

Other books feature scientists taking the fight to a rogue's gallery of pseudoscientists: creationists, homeopaths, flat-Earthers, astrologers, and their ilk, who misunderstand and abuse science – usually unwittingly, sometimes maliciously, and always irresponsibly.[14] This book is different. It reveals a deep corruption within science itself: a corruption that affects the very culture in which research is practised and published. Science, the discipline in which we should find the harshest scepticism, the most pin-sharp rationality and the hardest-headed empiricism, has become home to a dizzying array of incompetence, delusion, lies and self-deception. In the process, the central purpose of science – to find our way ever closer to the truth – is being undermined.

The book begins by showing, in Part I, that doing science involves much more than just running experiments or testing hypotheses. Science is inherently a *social* thing, where you have to convince other people – other scientists – of what you've found. And since science is also a *human* thing, we know that any scientists will be prone to human characteristics, such as irrationality, biases, lapses in attention, in-group favouritism and outright cheating to get what they want. To enable scientists to convince one another while trying to transcend the inherent limitations of human nature, science has evolved a system of checks and balances that – in theory – sorts the scientific wheat from the chaff. This process of scrutiny and validation, which leads to the supposed gold-standard of publication in a peer-reviewed scientific journal, is described in Chapter 1. But Chapter 2 shows that the process must have gone terribly wrong: there are numerous published findings across many different areas of science that can't be replicated and whose truth is very much in doubt.

Then, in Part II, we'll ask why. We'll discover that our publication system, far from neutralising or overriding all the human problems, allows them to leave their mark on the scientific record – and does so precisely because it *believes* itself to be objective and unbiased. A peculiar complacency, a strange arrogance, has taken hold, where the mere existence of the peer-review system seems to have stopped us from recognising its flaws. Peer-reviewed papers are supposedly as near as one can get to an objective factual account of how the world works. But in our tour through many dozens of those papers, we'll discover that peer review can't be relied upon to ensure scientists are honest (Chapter 3), detached (Chapter 4), scrupulous (Chapter 5), or sober (Chapter 6) about their results.

Part III digs deeper into scientific practice. Chapter 7 shows that it's not just that the system fails to deal with all the kinds of malpractice we've discussed. In fact, the way academic research is currently set up *incentivises* these problems, encouraging researchers to obsess about prestige, fame, funding and reputation at the expense of rigorous, reliable results. Finally, after we've diagnosed the problem, Chapter 8 describes a set of often-radical reforms to scientific practice that could help reorient it towards its original purpose: discovering facts about the world.

To make the case about the frailties of scientific research, throughout the book I'll draw on cautionary tales from a wide variety of scientific fields. Partly because I'm a psychologist, there'll be a preponderance of examples from that subject.[15] My background isn't the only reason there's so much psychology in the book: it's also because after the Bem and Stapel affairs (among many others), psychologists have begun to engage in some intense soul-searching. More than perhaps any other field, we've begun to recognise our deep-seated flaws and to develop systematic ways to address them – ways that are beginning to be adopted across many different disciplines of science.

The first step in fixing our broken scientific system is learning to spot, and correct, the mistakes that can lead it astray. And the only way to do this is with more science. Throughout the book, I'll draw on *meta-science*: a relatively new kind of scientific research that focuses on scientific research itself. If science is the process of exposing and eliminating errors, meta-science represents that process aimed inwards.

Much can be learned from mistakes. On one of his albums, the musician Todd Rundgren has a spoken-word introduction encouraging the listener to play a little game he calls 'Sounds of the Studio'. Rundgren describes all the missteps that can be made when recording music: hums, hisses, pops on the microphone whenever someone sings a word with a 'p' in it, choppy editing, and so on. He suggests that the reader listen for these mistakes in the songs that follow, and on other records. And just as a better understanding of recording studio slip-ups can give you a new insight into how music is made, learning about how science goes wrong can tell you a lot about the process by which we arrive at our knowledge.

Discovering the serious problems with the way we do science will be disconcerting. How many intriguing results that you've read about in the news and popular science books, or seen in documentaries – discoveries you've been excited enough to share with friends, or that made you rethink how the world works – are based on weak research that can't be replicated? How many times has your doctor prescribed you a drug or other treatment that rests on flawed evidence? How many times have you changed your diet, your purchasing habits, or some other aspect of your lifestyle on the basis of a scientific study, only for the evidence to be completely overturned by a new study a few months later? How many times have politicians made laws or policies that directly impact people's lives, citing science that won't stand up to scrutiny? In each case, the answer is: a lot more than you'd like to think.

It's naïve to hope that every single scientific study will be true – that is, a report of ironclad facts that will never be overturned in future research. The world is far too messy a place for that. All we can hope for is that our scientific studies are trustworthy – that they honestly report what occurred in the research. If the much-vaunted peer-review process can't justify that trust, science loses one of its most basic and most desirable qualities, along with its ability to do what it does best: revolutionise our world with a steady progression of new discoveries, technologies, treatments and cures.

I come to praise science, not to bury it; this book is anything but an attack on science itself, or on its methods. Rather, it is a defence of those methods, and of scientific principles more generally, against the way science is currently practised. What makes all the disasters we'll encounter so disturbing is the importance of science: by allowing it to become so tarnished, and its progress to be so badly stalled, we're in danger of ruining one of the greatest accomplishments of our species.

But the damage isn't irreparable. In principle, if not in practice, science still has the potential to be the robust and reliable system of knowledge we need it to be. As we explore the litany of scientific failures in the book, the positive thought to hold onto – the fragile scrap of hope and reassurance that emerges from the Pandora's box of fraud, bias, negligence and hype that we'll prise open in what follows – is that nearly all of these problems have been uncovered *by other scientists*. The clever meta-scientific ideas that have been proposed to combat these problems and clear up the mess that has been created have come, in substantial part, from within the scientific community. Even if it's been deeply buried in many fields, the self-critical spirit that animates genuine science remains.

And that's just as well, because as we're about to find out, there really is *quite* a mess.

PART I
OUGHT AND IS

1

How Science Works

Science is a social construct.

Before that statement makes you toss the book across the room, let me explain what I mean. I don't mean it in the sense used by extreme relativists, post-modernists, anti-science crusaders, and others who suggest that there's no real world out there, that science is only one not-particularly-special way of knowing about it, or even that science is just one 'myth' among many that we could choose to believe.[1] Science has cured diseases, mapped the brain, forecasted the climate, and split the atom; it's the best method we have of figuring out how the universe works and of bending it to our will. It is, in other words, our best way of moving towards the truth. Of course, we might never fully get there – a glance at history shows how hubristic it is to claim any facts as absolute or unchanging. For ratcheting our way towards better knowledge about the world, though, the methods of science are as good as it gets.

But we can't make progress with those methods alone. It's not enough to make a solitary observation in your lab; you must also convince other scientists that you've discovered something real. This is where the social part comes in. Philosophers have long discussed how important it is for scientists to *show* their fellow researchers how they came to their conclusions. John Stuart Mill puts it this way:

> In natural philosophy, there is always some other explanation possible of the same facts; some geocentric theory instead of heliocentric, some phlogiston instead of oxygen; and it has to be shown why that other theory cannot be the true one: and until this is shown, and until we know how it is shown, we do not understand the grounds of our opinion.[2]

And so, scientists work together in teams, travel the world to give lectures and conference speeches, debate each other in seminars, form scientific societies to share research and, perhaps most importantly, publish their results in peer-reviewed journals. These social aspects aren't just a perk of the job, nor mere camaraderie. They're the process of science in action: an ongoing march of collective scrutiny, questioning, revision, refinement and *consensus*. Although it might sound paradoxical at first, the subjective process of science is what provides it with its unmatched degree of objectivity.[3]

It's in this sense, then, that science is a social construct. Any claim about the world can only be described as scientific knowledge after it's been through this communal process, which is designed to sieve out errors and faults and allow other scientists to say whether they judge a new finding to be reliable, robust and important. That each discovery has to run such a gauntlet imbues the eventual products of the scientific process – the published, peer-reviewed studies – with a great deal of power in society. This is no mere cant, rhetoric, or opinion, we say: this is *science*.

Science's social nature does come with weaknesses, however. Because scientists focus so much on trying to persuade their peers, which is the way they get those studies through peer review and onward to publication, it's all too easy for them to disregard the real object of science: getting us closer to the truth. And because scientists are human beings, the ways that they try to persuade each other aren't always fully rational or objective.[4] If we don't take great care, our scientific process can become permeated by very human flaws.

This book is about how we haven't taken enough care of our precious scientific process. It's about how we ended up with a scientific system that doesn't just overlook our human foibles, but amplifies them. In recent years, it's become increasingly, painfully obvious that peer review is far from the guarantee of accuracy and reliability it's cracked up to be, while the system of publication that's supposed to be a crucial strength of science has become its Achilles' heel.

To understand how the scientific publication system has gone so wrong, though, we first need to know how it's supposed to work when it goes right.

Let's imagine you want to do some science. The first step is to read the scientific literature. This consists of a vast library of journals, the specialist magazines that are the main outlets for new scientific knowledge. The idea of a periodical where scientists could share their work dates back to 1665, when Henry Oldenburg of the UK's Royal Society published the first issue of, to give it its full title, *Philosophical Transactions: Giving Some Accompt of the Present Undertakings, Studies, and Labours of the Ingenious in Many Considerable Parts of the World.*[5] The intention was that those ingenious scientists could send in letters describing their exploits, for the perusal of other interested readers. Before that, scientists either laboured alone in the courts of wealthy rulers or for private patrons or guilds (where their science was often seen as more akin to a parlour trick than an effort to discover the truth), published standalone books, or formed letter-writing circles with like-minded peers. Indeed, this latter kind of correspondence club is where institutions like the Royal Society originated.[6]

The initial issues of Oldenburg's journal were more like a newsletter, with descriptions of recent experiments and discoveries. For example, Volume 1, Issue 1 described the first ever observation of what was probably the Great Red Spot of Jupiter, by the natural philosopher and polymath Robert Hooke. The entire entry read:

> The Ingenious Mr. Hook did, some months since, intimate to a friend of his, that he had, with an excellent twelve foot Telescope, observed, some days before, he than spoke of it, (*videl.* on the ninth of *May*, 1664, about 9 of the clock at night) a small Spot in the biggest of the 3 obscurer *Belts* of *Jupiter*, and that, observing it from time to time, he found, that within 2 hours after, the said Spot had moved from East to West, about half the length of the Diameter of *Jupiter*.[7]

The journal still exists to this day, with the somewhat easier-to-remember title of *Philosophical Transactions of the Royal Society*.[8] As time went on, the brief news items were replaced with longer articles containing detailed descriptions of experiments and studies. It's now part of a global ecosystem of over 30,000 journals, ranging from the very general (like the highly prestigious journals *Nature* and *Science*, which aim to publish the world's most noteworthy research from any scientific field) to the very specific (like the *American Journal of Potato Research*, which is only interested in papers about one tuberous topic in particular).[9] Some journals, like *Philosophical Transactions*, are still run by scientific societies, but most are owned by commercial outfits such as Elsevier, Wiley and Springer Nature.[10] A recent advancement is that scientific journals are all online, allowing anyone who can afford to pay the publisher's subscription fees – or have their university library do so on their behalf – to have the world's scientific knowledge at their fingertips.[11]

After reading the journals relevant to your field, you might alight on a research question. Maybe there's a scientific theory that makes a prediction – an hypothesis – that you can test in some clever way; maybe there's a gap in our existing knowledge that you know just how to plug; maybe you've had a spark of inspiration and have come up with an experiment that tests something entirely new. Before you can do any of this, though, you'll normally need some money to fund the study: for instance, to buy new equipment or materials, to recruit participants, or to pay the salaries of the scientists you'll hire to do the legwork. Unless you happen to be, say, a pharmaceutical company that can afford to run its own laboratories, the main way to get that all-important funding is to apply for a grant. This might come from your government, a business, an endowment fund, a non-profit, a charity, or even a wealthy individual. You might apply to the National Institutes of Health or the National Science Foundation (both of which are taxpayer-funded agencies in the United States), or to a science-funding charity like the Wellcome Trust or the Bill & Melinda Gates Foundation.[12]

Funding is by no means assured, and any scientist will tell you that one of the most gruelling parts of the job is trying to get their latest research ideas funded, with failure grindingly common. This grasping for cash has important knock-on effects on the science itself, and we'll return to them later in the book. But for now, let's imagine you're successful in securing a grant. You can then get to work. Collecting the data might involve smashing particles together in an underground collider, finding fossils in the rocks of the Canadian Arctic, setting up the precise environment for bacterial growth in a petri dish, organising hundreds of people to come to a lab and fill in questionnaires, or running a complex computer model; it can take days, months, decades.

Once the data are in, you'll normally have a set of numbers that you, or a more mathematically minded colleague, can analyse using some variety of statistics (another minefield to which we'll return). Then you need to write it all up in the form of a scientific paper. The typical paper starts with an Introduction, where you summarise what's known on the topic and what your study adds. There follows a Method section, where you describe exactly what you did – in enough detail so that anyone could, in theory, run exactly the same experiment again. You'll then move on to a Results section, where you present the numbers, tables, graphs and statistical analyses that document your findings, and you'll end with a Discussion section where you speculate wildly – er, I mean, provide thoughtful, informed consideration – about what it all means. You'll top the whole thing with an Abstract: a brief statement, usually of around 150 words, that summarises the whole study and its results. The Abstract is always available for anyone to read, even if the full paper is behind the journal's subscription paywall, so you'll want to use it to make your results sound compelling. Papers come in all lengths and sizes, and sometimes mix up the above order, but in general your paper will end up along these lines.[13]

When the paper is ready, you enter the world of scientific journals, and the competition for publication. Until recently, submitting a paper to a journal meant printing out several hard copies and mailing them to the editor, but nowadays everything is handled online – though many journals still use such archaic, buggy websites that you might as well send your paper by carrier pigeon. The journal's editor, often a senior academic, will read the paper (or, let's be honest, probably just the Abstract) and decide whether it might be worth publishing. Most journals, especially the highly prestigious ones, pride themselves on their exclusivity and thus their low acceptance rate (*Science*, for example, accepts less than 7 per cent of submissions), so the majority of papers will be bounced back to the authors at this point, in what's called a 'desk rejection'.[14] This is the initial step in quality control: a sorting by the editor of the papers into those that match the theme of the journal and have potential in terms of their scientific interest or quality, and those that aren't worth a second look. For the fraction of articles that do take the editor's fancy, now comes the moment of peer review. The editor will find two or three scientists who are experts in your field of research and ask them whether they'd like to evaluate your manuscript. They'll probably decline because they're too busy, so the editor will keep going down the list of possible reviewers until a few agree. And so begins the nail-biting wait to see if your work will receive their endorsement.

Most people, including scientists, assume peer review has always been a crucial feature of scientific publication, but its history is more complicated. Although in the seventeenth century the Royal Society tended to ask some of its members whether they thought a paper was interesting enough to publish in *Philosophical Transactions*, requiring them to provide a written evaluation of each study wasn't tried until at least 1831.[15] Even then, the formal peer review system we know today didn't become universal until well into the twentieth century (as you can tell from a letter Albert Einstein sent in 1936 to the editors of *Physical Review*, huffily announcing that he was withdrawing his paper from consideration at their journal because they had dared to send it to another physicist for comment).[16] It took until the 1970s for all journals to adopt the modern model of sending out submissions to independent experts for peer review, giving them the gatekeeping role they have today.[17]

Peer reviewers are usually anonymous, which is both a blessing and a curse: a blessing because it allows them to speak their minds without concern about repercussions from the scientists whose work they're criticising (a junior scientist can be truly honest about the flaws of a big-name professor's work), but a curse because, well, it allows them to speak their minds without concern about repercussions from the scientists whose work they're criticising. The following are genuine excerpts from peer reviews:

- 'Some papers are a pleasure to read. This is not one of them.'
- 'The results are as weak as a wet noodle.'
- 'The manuscript makes three claims: The first we've known for years, the second for decades, the third for centuries.'

- 'I am afraid this manuscript may contribute not so much towards the field's advancement as much as towards its eventual demise.'
- 'Did you have a seizure while writing this sentence? Because I feel like I had one while reading it.'[18]

If the reviewers' evaluations look like this, the editor will probably reject your paper. At that point you might want to give up, or start the whole process again by submitting to a different journal, and if that fails a different one, and if that fails a different one, and so on – it's not uncommon for papers to go through half a dozen or more journals, usually of ever-lower prestige, before they get accepted for publication. If, on the other hand, the reviewers are more impressed, you might get the opportunity to revise your paper to respond to their critiques – perhaps running new analyses or new experiments, or rewriting certain sections – and submit it to the editor again. The back-and-forth revising process can go through multiple rounds, and often takes months. Eventually, if the reviewers are satisfied, the editor gives the go-ahead and the paper is published. If the journal still prints hard copies, you'll get to see your precious work in print; otherwise, you'll have to settle for the thrill of seeing it on the journal's official website. That's it. You've made your mark on the scientific literature, and you have a publication that you can add to your CV and that can be cited by other researchers. Congratulations – take the rest of the day off.

The above summary is all too brief and general, but essentially every scientific field follows that process in some form. We might ask ourselves whether, after being put through the mangle of peer review, the eventual publication still provides a faithful representation of what was done in the study. We'll get to that in later chapters. For now, we need to consider something else. What ensures that the participants in the process just described – the researcher who submits the paper, the editor at the journal, the peers who review it – all conduct themselves with the honesty and integrity that trustworthy science requires? There's no law requiring that everyone acts fairly and rationally when evaluating science, so what's needed is a shared ethos, a set of values that aligns the scientists' behaviour.[19] The best-known attempt to write down these unwritten rules is that of the sociologist Robert Merton.

In 1942, Merton set out four scientific values, now known as the 'Mertonian Norms'. None of them have snappy names, but all of them are good aspirations for scientists. First, *universalism*: scientific knowledge is scientific knowledge, no matter who comes up with it – so long as their methods for finding that knowledge are sound. The race, sex, age, gender, sexuality, income, social background, nationality, popularity, or any other status of a scientist should have no bearing on how their factual claims are assessed. You also can't judge someone's research based on what a pleasant or unpleasant person they are – which should come as a relief for some of my more disagreeable colleagues. Second, and relatedly, *disinterestedness*: scientists aren't in it for the money, for political or ideological reasons, or to enhance their own ego or reputation (or the reputation of their university, country, or anything else). They're in it to advance our understanding of the universe by discovering things and making things – full stop.[20] As Charles Darwin once wrote, a scientist 'ought to have no wishes, no affections, – a mere heart of stone.'[21]

The next two norms remind us of the social nature of science. The third is *communality*: scientists should share knowledge with each other.[22] This principle underlies the whole idea of publishing your results in a journal for others to see – we're all in this together; we have to know the details of other scientists' work so that we can assess and build on it.[23] Lastly, there's *organised scepticism*: nothing is sacred, and a scientific claim should never be accepted at face value. We should suspend judgement on any given finding until we've properly checked all the data and methodology. The most obvious embodiment of the norm of organised scepticism is peer review itself.

*

It looks good in theory: by following the four Mertonian Norms, we should end up with a scientific literature we can trust – the shoulders of giants, as in Newton's famous phrase, on which we stand to see farther. Of course, those giants often had it wrong: just to take the two examples mentioned above by John Stuart Mill, we used to believe that the Sun orbited the Earth, and that flammable objects were full of a special element called phlogiston that was released when they burned.[24] But we eventually consigned these theories to the scrapheap as better data came in. Indeed, it's a virtue for a scientist to change their mind. The biologist Richard Dawkins recounts his experience of 'a respected elder statesman of the Zoology Department at Oxford' who for years had:

> passionately believed, and taught, that the Golgi Apparatus (a microscopic feature of the interior of cells) was not real: an artefact, an illusion. Every Monday afternoon it was the custom for the whole department to listen to a research talk by a visiting lecturer. One Monday, the visitor was an American cell biologist who presented completely convincing evidence that the Golgi Apparatus was real. At the end of the lecture, the old man strode to the front of the hall, shook the American by the hand and said – with passion – "My dear fellow, I wish to thank you. I have been wrong these fifteen years." We clapped our hands red ... In practice, not all scientists would [say that]. But all scientists pay lip service to it as an ideal – unlike, say, politicians who would probably condemn it as flip-flopping. The memory of the incident I have described still brings a lump to my throat.[25]

This is what people mean when they talk about science being 'self-correcting'. Eventually, even if it takes many years or decades, older, incorrect ideas are overturned by data (or sometimes, as was rather morbidly noted by the physicist Max Planck, by all their stubborn proponents dying and leaving science to the next generation).[26] Again, that's the theory. In practice, though, the publication system described earlier in this chapter sits awkwardly with the Mertonian Norms, in many ways obstructing the process of self-correction. The specifics of this contradiction – between the competition for grants and clamour for prestigious publications on the one hand, and the open, dispassionate, sceptical appraisal of science on the other – will become increasingly clear as we progress through the book.

For now, though, notice what it was that changed the mind of Dawkins's elder statesman: 'completely convincing evidence'. There's little point in trying to correct and update our scientific theories with data if the data themselves aren't convincing – or worse, aren't even accurate. This brings us back to the idea we discussed in the Preface: for results to warrant our trust, they need to be replicable. As the philosopher of science Sir Karl Popper puts it:

> Only when certain events recur in accordance with rules or regularities, as is the case with repeatable experiments, can our observations be tested – in principle – by anyone. We do not take even our own observations quite seriously, or accept them as scientific observations, until we have repeated and tested them. Only by such repetitions can we convince ourselves that we are not dealing with a mere isolated 'coincidence'.[27]

It's not as if this is a revolutionary idea – or one that was new to Popper, writing in the 1950s. If we return to the early days of *Philosophical Transactions* in the seventeenth century, we find the co-founder of the Royal Society, the chemist Robert Boyle, going to extraordinary lengths to ensure the replicability of his findings. He would repeatedly demonstrate his experiments, which used his famous air pump to show various properties of air and the vacuum to groups of observers, before having them sign sworn testimony that they'd witnessed the phenomena in question.[28] He would ensure that his writings were detailed enough 'that the person I addressed them to might, without mistake, and with as little trouble as possible, be able to repeat such unusual experiments.'[29] And despite the great difficulty of building the complex apparatus, he encouraged and assisted other natural philosophers to replicate his air-pump experiments in different parts of Britain and Europe.[30]

Replication, then, has long been a key part of how science is supposed to work – and incidentally, it's another of its social aspects, with results only being taken seriously after they've been corroborated by multiple observers. But somewhere along the way, between Boyle and modern academia, a great many scientists forgot about the importance of replication. In the collision of our Mertonian ideals with the realities of the scientific publication system – not to mention the realities of human nature – the ideals have proven the more fragile, leaving us with a scientific literature full of untrustworthy, unreliable, unreplicable studies that often do more to confuse than enlighten.

In the next chapter, we'll see just *how* untrustworthy, unreliable and unreplicable the scientific literature has become.

2

The Replication Crisis

Vaulting ambition, which o'erleaps itself ...
William Shakespeare, *Macbeth*, 1.7.27

Published and *true* are not synonyms.
Brian Nosek, Jeffrey Spies and Matt Motyl

Easily the most popular psychology book of the past decade is Daniel Kahneman's *Thinking, Fast and Slow*. There are few better guides to the human mind than Kahneman: he won the Nobel Prize in Economics in 2002 for his work on human (ir)rationality and has published dozens of clever experiments showing the limits of our reasoning ability. *Thinking, Fast and Slow* was a publishing sensation, eventually accruing sales in the millions, and it still sells well to this day. There's a reason for that: it's a vivid and beautifully explained tour through all the mistakes and biases in human thinking.[1]

Among the many topics in the book, Kahneman covered work on what psychologists call 'priming'. Some kinds of priming have to do with language. It's established, for example, that if I show you a series of words one at a time on a computer screen and ask you to press a button whenever the word *spoon* appears, you'll press it ever-so-slightly faster if the preceding word was *fork* (or something else cutlery-related) than if it was *tree* (or something else unrelated to eating utensils). Seeing the word *fork* psychologically 'primes' you to be quicker to react to a word that's akin to it in meaning.[2]

What Kahneman described, though, was more surprising. He covered social-psychology research that implied that priming certain *concepts* – usually unconsciously – could dramatically alter our behaviour. One example is known as the 'Macbeth effect'. In a study published in *Science* in 2006, researchers found that being asked to copy down a story about unethical deeds made participants more likely to want to buy soap, and asking them to recall something unethical they'd done made it more likely that they'd pick up an antiseptic wipe upon leaving the lab ('Out, damned spot!'). This went way beyond priming *words*: it implied that the brain worked in a far more joined-up way than we'd imagined, with strong connections being made between concepts that might've seemed only loosely linked. In this case, it seemed to be evidence for some deep-seated overlap between the concepts of morality and cleanliness. The authors argued that it might even tell us why handwashing is part of so many religious rituals across the world.[3]

Kahneman also reviewed research on 'money priming'. In another *Science* paper, also from 2006, social psychologists had found that subtly reminding people about money – for instance, having them sit at a desk where there happened to be a computer showing a screensaver of floating banknotes – made them feel and behave as if they were more self-sufficient, and made them care less about others.[4] Being exposed to money priming, the authors said, made participants prefer 'to play alone, work alone, and put more physical distance between themselves and a new acquaintance'.[5] Indeed, when asked to arrange the room's seating for a face-to-face conversation with a stranger, the money-primed participants, compared to those who'd seen a blank screen, set their chairs almost 30 centimetres further apart. Quite an impact, you might think, for a simple screensaver. This was a pattern in the most prominent priming studies: very subtle primes appeared to cause impressive changes in the way people behaved.

Kahneman concluded that these kinds of priming studies 'threaten our self-image as conscious and autonomous authors of our judgments and our choices'. He had little doubt about their soundness. 'Disbelief is not an option,' he wrote. 'The results are not made up, nor are they statistical flukes. You have no choice but to accept that the major conclusions of these studies are true. More important, you must accept that they are true about *you*.'[6]

But perhaps Kahneman shouldn't have placed such complete trust in these priming effects, despite their being published in one of the most renowned scientific journals. As it turns out, along with the discovery of Diederik Stapel's fraud and the publication of Daryl Bem's weird psychic results, it was a priming study – or, rather, the attempted replication of one – that was among the initial spurs of what has become known as 'the replication crisis'.[7]

In that original priming study, the researchers asked participants to find the odd one out from a jumbled list of words, the rest of which could be rearranged to form a sentence. For half of the participants, the odd-words-out were random and neutral; for the other half, these words had to do with elderly people. These included, for example, *old*, *grey*, *wise*, *knits*, and *Florida* – the last of which is known in the US for having a large population of retirees. Having completed the task, the participants were free to leave – but unbeknownst to them, the experimenters were timing how quickly they walked down the corridor to leave the building. Demonstrating again the mental connection between concepts and actions, the participants who had been primed with elderly-related words *walked more slowly out of the lab* than those in the control group.[8]

Published in 1996, this study has now been cited over 5,000 times by other researchers and has become a staple of psychology textbooks – I remember being taught about it myself when I was a student. In 2012, though, an independent group tried running exactly the same experiment again, but with a bigger sample size and better technology. They found no differences in walking speed. They proposed that the original study might have come up with its results because the research assistants, who timed the participants with stopwatches, knew which participants were expected to behave in which way, possibly influencing their timing. Measuring the participants' walking speed with infrared beams, as was done in the replication study, appeared to nullify the supposed priming.[9] Within a few years, other labs tried to replicate both the Macbeth effect and the money-priming effect, also in much larger, more representative samples.[10] These efforts also conspicuously failed. There's no reason to think, to use Kahneman's terms, that the various priming results were 'made up'; we have to assume they were arrived at in good faith. But 'statistical flukes'? Perhaps exactly that.

Other priming studies haven't done much better. One claimed that participants who were primed with 'distance' – by having them plot two points far apart on a piece of graph paper – were more likely to feel 'distant' from their friends and relatives; it failed to replicate in 2012.[11] Another study claimed that when written moral dilemmas were printed with a surrounding checkerboard pattern, participants made more polarised judgements, because the pattern made them think of the concept 'black and white'; this failed to replicate in 2018.[12] On a similar topic, a line of research that claimed that you can make people more morally judgemental by priming their disgust was thrown into doubt by a review in 2015.[13]

To give Kahneman his due, he later admitted that he'd made a mistake in overemphasising the scientific certainty of priming effects. 'The experimental evidence for the ideas I presented in that chapter was significantly weaker than I believed when I wrote it,' he commented six years after the publication of *Thinking, Fast and Slow*. 'This was simply an error: I knew all I needed to know to moderate my enthusiasm … but I did not think it through.'[14] But the damage had already been done: millions of people had been informed by a Nobel Laureate that they had 'no choice' but to believe in those studies.

Priming isn't the only psychological effect to have been given an audience in the millions. Harvard psychologist Amy Cuddy rocketed to fame in 2012 with a TED talk advocating 'power posing'. She recommended that just before you enter a stressful situation, such as an interview, you should find two minutes in a private place (such as a bathroom stall) to stand in an open, expansive posture: for example, with your legs apart and your hands on your hips. This powerful posture would give you a psychological – and hormonal – boost. An experiment by Cuddy and her colleagues in 2010 had found that, compared to those who were asked to sit with arms folded or slouched forward, people who were made to power-pose not only felt more powerful, but had higher risk tolerance in a betting game and had increased levels of testosterone and decreased levels of the stress hormone cortisol.[15]

Cuddy's message that people who used the two-minute power pose could 'significantly change the outcomes of their life' struck a chord: hers became the second-most-watched TED talk ever, with over 73.5 million views in total.[16] It was followed in 2015 by Cuddy's *New York Times*-bestselling self-help book, *Presence*, whose publisher informed us that it presented 'enthralling science' that could 'liberate [us] from fear in high-pressure moments'.[17] Provoking quite some degree of mockery, the UK's Conservative Party seemed to take Cuddy's message to heart, with a spate of photos appearing that same year of their politicians adopting the wide-legged stance at various conferences and speeches.[18] Alas, also in 2015, when another team of scientists tried to replicate the power-posing effects, they found that while power-posers did report feeling more powerful, the study 'failed to confirm an effect of power posing on testosterone, cortisol, and financial risk'.[19]

The critical spotlight that was activated in the replication crisis has also been aimed at older pieces of psychology research, with similarly worrying results. Perhaps the most famous psychology study of all time is the 1971 Stanford Prison Experiment, where psychologist Philip Zimbardo split a group of young men into mock 'guards' and 'prisoners', and had them stay for a week in a simulated prison in the basement of the Stanford University psychology department. Disturbingly quickly, according to Zimbardo, the 'guards' began to punish the 'prisoners', abusing them so sadistically that Zimbardo had to end the experiment early.[20] Along with Stanley Milgram's 1960s studies of obedience, which found many participants willing to administer intense electric shocks to hapless 'learners' (the shocks and the learners were both fake, but the participants didn't know it), Zimbardo's experiment is held up as one of the prime pieces of evidence for the power of the situation over human behaviour.[21] Put a good person into a bad situation, the story goes, and things might get very bad, very fast. The Stanford Prison Experiment is taught to essentially every undergraduate psychology student on the planet and on its basis Zimbardo became among the most well-known and respected modern psychologists. He used the findings of his experiment, for example, to testify as an expert witness at the trials of the US military guards at Abu Ghraib prison in Iraq, arguing that the situation the guards were in, and the roles they were made to take on, were the reasons for their shocking abuse and torture of their prisoners.[22]

Although its implications have always been controversial, only recently have we begun to see just how poor a study the Stanford Prison Experiment was.[23] In 2019, the researcher and film director Thibault Le Texier published a paper entitled 'Debunking the Stanford Prison Experiment', where he produced never-before-seen transcripts of tapes of Zimbardo intervening directly in the experiment, giving his 'guards' very precise instructions on how to behave – going as far as to suggest specific ways of dehumanising the prisoners, like denying them the use of toilets.[24] Clearly, this heavily stage-managed production was far from an organic example of what happens when ordinary humans are assigned specific social roles. Despite the enormous attention it's received over the years, the 'results' of the Stanford Prison Experiment, such as they are, are scientifically meaningless.[25]

As you might expect, the confluence of failed replications (like the priming studies) and bizarre results (like Bem's paranormal discoveries), along with revelations of misrepresentation (like Zimbardo's experiment) and fraud (like Stapel's fake data) spooked psychologists. Just how many of the studies in their field, they wondered, could be trusted? To get an idea of how bad things were, they started banding together to run large-scale replications of prominent studies across multiple different labs. The highest profile of these involved a large consortium of scientists who chose 100 studies from three top psychology journals and tried to replicate them. The results, published in *Science* in 2015, made bitter reading: in the end, only 39 per cent of the studies were judged to have replicated successfully.[26] Another one of these efforts, in 2018, tried to replicate twenty-one social-science papers that had been published in the world's top two general science journals, *Nature* and *Science*.[27] This time, the replication rate was 62 per cent. Further collaborations that looked at a variety of different kinds of psychological phenomena found rates of 77 per cent, 54 per cent, and 38 per cent.[28] Almost all of the replications, even where successful, found that the original studies had exaggerated the size of their effects. Overall, the replication crisis seems, with a snap of its fingers, to have wiped about half of all psychology research off the map.[29]

Maybe it's not quite that bad, for two reasons. First, we would expect some results that really are solid to fail to replicate sometimes, merely due to bad luck.[30] Second, some replications might have failed due to their being run with slight changes to the methodology from the original (though if a result is fragile enough that it disappears after minor modifications to the experiment, one might wonder how useful or meaningful it really is).[31] For these reasons, it's sometimes tricky to decide whether a finding is 'replicable' or not based just on one or two replication attempts. What's more, the replication rate seems to differ across different areas of psychology: for example, in the 2015 *Science* paper, cognitive psychology (studies of memory, perception, language, and so on) did better than social psychology (which includes the sorts of metaphor-priming studies we saw above).[32]

In general, though, the effect on psychology has been devastating. This wasn't just a case of fluffy, flashy research like priming and power posing being debunked: a great deal of far more 'serious' psychological research (like the Stanford Prison Experiment, and much else besides) was also thrown into doubt. And neither was it a matter of digging up some irrelevant antiques and performatively showing that they were bad – like when Pope Stephen VI, in the year 897, exhumed the corpse of one of his predecessors, Pope Formosus, and put it on trial (it was found guilty). The studies that failed to replicate continued to be routinely cited both by scientists and other writers: entire lines of research, and bestselling popular books, were being built on their foundation. 'Crisis' seems to be an apt description.

We might try to tell ourselves that there's something unique about psychology as a discipline that caused its replication crisis. Psychologists have the unenviable job of trying to understand highly variable and highly complicated human beings, with all their different personalities and backgrounds and experiences and moods and quirks. The things they study, like thought, emotion, attention, ability, and perception, are usually intangible – difficult, if not impossible, to pin down in a lab experiment – and in social psychology, they have to study how all those complicated humans interact with one another. Could the sheer complexity of the task make findings in psychology particularly untrustworthy, compared to other sciences?

There is something to this argument: many studies in psychology barely scratch the surface of the phenomena they're interested in, while other 'harder' sciences – say, physics – have better-developed theories and more precise and genuinely objective measurements. But it's not as if psychology is alone in having problems with replicability – although no other sciences have yet investigated their replication rates as systematically and in as much detail, there are glimmers of the same kinds of problems across very many different fields:

- In economics, a 2016 replication survey of eighteen microeconomic studies (not too different from psychology research, with people coming into the lab and taking part in experiments on their economic behaviour) had a replication rate of only 61 per cent.[33]
- In neuroscience, a study in 2018 found that standard studies in functional brain imaging, where the brain's activity is recorded using MRI while the person is completing some kind of task (or just lying in the scanner), were likely only 'modestly replicable'.[34] The world of functional brain-imaging was also rocked by a paper which revealed that a default setting in a software package commonly used to analyse imaging data had a statistical error. It led to a vast number of accidental, uncorrected false-positive results, and it might have compromised around 10 per cent of all studies that had ever been published on the topic.[35]
- In evolutionary biology and ecology, a series of classic findings, repeated in textbooks and taught to generations of students, have fallen to replication attempts and critical reviews. For instance, the famous 'domestication syndrome', where Russian foxes that were selected for tameness started to take on the physical characteristics of domesticated species (like floppy ears and wider faces), turns out to have been hugely exaggerated, with most of the 'domesticated' traits existing before the selection even took place.[36] And much of what we thought we knew about sexual selection in birds has been shot down by better evidence. For instance, despite what we thought we knew, putting a red band on male finches' legs probably *doesn't* make them super-attractive to females; male sparrows with larger patches of black plumage on their

throats (their so-called 'bib') probably *don't* have higher dominance in the flock; and the evidence that female blue tits are more attracted to particular plumage colours in males is inconclusive.[37]

- In marine biology, a massive replication attempt in 2020 found that the effects of ocean acidification (one of the consequences of climate change) on fish behaviour were non-existent.[38] It thus failed to replicate several highly publicised studies from the previous decade that had apparently shown that more acidic conditions caused fish to become disoriented, and in some cases to swim *towards*, rather than away from, the chemical cues produced by predators.

- In organic chemistry, the journal *Organic Syntheses*, which operates an unusual policy where an editorial board member attempts to replicate in their own lab the results of every paper they receive as a submission, reported rejecting 7.5 per cent of submissions because of replication failures.[39]

There are countless other examples: almost every case I'll describe in this book involves a scientific 'finding' that, upon closer scrutiny, turned out to be either less solid than it seemed, or to be completely untrue. But more worryingly still, these examples are drawn just from the studies that have received that all-important scrutiny. *These are just the ones we know about.* How many other results, we must ask ourselves, would prove unreplicable if anyone happened to make the attempt?

One reason that we live in such uncertainty is that, as we learned from the Preface, hardly anyone runs replication studies. Though we don't have the numbers for most fields, scans of the literature from certain subjects draw a bleak conclusion. In economics, a miserable 0.1 per cent of all articles published were attempted replications of prior results; in psychology, the number was better, but still nowhere near good, with an attempted replication rate of just over 1 per cent.[40] If everyone is constantly marching onwards to new findings without stopping to check if our previous knowledge is robust, is the above list of replication failures that much of a surprise?

Here's something that's perhaps even more alarming. You'd think that if you obtained the exact same dataset as was used in a published study, you'd be able to derive the exact same results that the study reported. Unfortunately, in many subjects, researchers have had terrible difficulty with this seemingly straightforward task. This is a problem sometimes described as a question of *reproducibility*, as opposed to *replicability* (the latter term being usually reserved to mean studies that ask the same questions of different data). How is it possible that some results can't even be reproduced? Sometimes it's due to errors in the original study. Other times, the original scientists weren't clear enough with reporting their analysis: they took twists and turns with the statistics that weren't declared in their scientific paper, and thus their exact steps can't be retraced by independent researchers. When new scientists run the statistics in their own way, the results come out differently. Those studies are like a cookbook including mouth-watering photographs of meals but providing only the patchiest details of the ingredients and recipe needed to produce them.

In macroeconomics (research on, for example, tax policies and how they affect countries' economic growth), a re-analysis of sixty-seven studies could only reproduce the results from twenty-two of them using the same datasets, and the level of success improved only modestly after the researchers appealed to the original authors for help.[41] In geoscience, researchers had at least minor problems getting the same results in thirty-seven out of thirty-nine different studies they surveyed.[42] And when machine-learning researchers analysed a set of papers about 'recommendation algorithms' – the kind of computer programs used by websites such as Amazon or Netflix to suggest what you might want to buy or watch next, based on what people like you have chosen in the past – they could reproduce only seven out of the eighteen studies on the topic that had been recently presented at prestigious computer science conferences.[43] These papers are the real-life version of the classic Sidney Harris cartoon.

"I THINK YOU SHOULD BE MORE EXPLICIT HERE IN STEP TWO."

You might wonder why some of the above points matter. Although we've seen poor replicability in some weightier areas, such as economic theory, how much difference can it make to our lives if a bunch of academics end up disagreeing over whether power posing works, or whether alpha-male sparrows have more black patches of feathers? There are two responses to make. The first is that there's a wider principle at stake here: science is crucial to our society, and we mustn't let any of it be compromised by low-quality, unreplicable studies. If we let standards slip in any part of science, we risk tarnishing the reputation of the scientific enterprise more generally. The second response is to look at a scientific field we haven't yet considered, where the immediate consequences of a lack of replicability are indisputable. That field is, of course, medical research.

Around the time that the replication crisis was brewing in psychology, scientists at the biotechnology company Amgen attempted to replicate fifty-three landmark 'preclinical' cancer studies that had been published in top scientific journals (preclinical studies are those done at an early stage of drug development, perhaps in mice or in human cells *in vitro*).[44] A mere six of the replication attempts (that is, just 11 per cent) were successful. The results from similar attempts at another firm, Bayer, weren't much more impressive, at around 20 per cent.[45] This lack of a firm underpinning in preclinical research might be among the reasons why the results from trials of cancer drugs are so often disappointing – by one estimation, only 3.4 per cent of such drugs make it all the way from initial preclinical studies to use in humans.[46]

Just like in psychology, these revelations made cancer researchers wonder about the wider state of their field. In 2013 they formed an organised, collaborative attempt to replicate fifty-one important preclinical cancer studies in independent labs.[47] Those studies included claims that a particular type of bacterium might be linked to tumour growth in colorectal cancer, and that some mutations found in leukaemia were related to the activity of a specific enzyme.[48] But before the replicators could even begin, they hit a snag. In every single one of the original papers, for every single one of the experiments reported, there wasn't enough information provided for researchers to know how to re-run the experiment.[49] Technical aspects of the studies – such as the specific densities of cells that were used, or other aspects of the measurements and analyses – simply weren't included. Replication attempts ran aground, prompting voluminous correspondence with the original scientists, who often had to dig out their old lab books and contact former members of their groups who'd moved on to other jobs, to find the specific details of their studies.[50] Some were reluctant to collaborate: 45 per cent were rated by the replicators as either 'minimally' or 'not at all' helpful.[51] Perhaps they were worried that the replicators might not be competent, or that failures to replicate their results could mean their future work wouldn't get funded.[52]

Later, a more comprehensive study took a random sample of 268 biomedical papers, including clinical trials, and found that all but *one* of them failed to report their full protocol. Meaning that, once again, you'd need additional details beyond the paper even to *try* to replicate the study.[53] Another analysis found that 54 per cent of biomedical studies didn't even fully describe what kind of animals, chemicals or cells they used in their experiment.[54] Let's take a moment to think about how odd this is. If a paper only provides a superficial description of a study, with necessary details only appearing after months of emailing with the authors (or possibly being lost forever), what was the point of writing it in the first place? Going back at least to Robert Boyle in the seventeenth century, recall that the original, fundamental motivation for scientists reporting all the specifics of their studies was so that others could scrutinise, and try to replicate, their research. The papers here failed that elementary test, just as the journals that published them also failed to perform their basic critical function.

For the cancer research replication project, all the troubles with trying to replicate the studies, combined with some financial problems, meant that the researchers had to cut steadily down the number of intended to-be-replicated studies, going from fifty to only eighteen.[55] Fourteen of these have been reported at the time of this writing, and they're a mixed bag, with five clearly replicating crucial results from the original studies (including the enzyme-leukaemia claim), four replicating parts of them, three that were clear failed replications (including the bacteria-colorectal cancer link), and two where the results couldn't even be interpreted.[56] Replicating, it's fair to say, ain't easy.

Replication problems in medicine haven't only affected laboratory based, preclinical research: they can have direct effects on the treatments doctors give to their patients. Common treatments, it turns out, are often based on low-quality research; instead of being solidly grounded in evidence, accepted medical wisdom is instead regularly contradicted by new studies. This phenomenon occurs so frequently that the medical scientists Vinay Prasad and Adam Cifu have a name for it: 'medical reversal'.[57]

One particularly striking medical reversal relates to 'anaesthesia awareness', the low-key name for the nightmarish (but thankfully rare) phenomenon of waking up during an operation, sometimes feeling the excruciating pain of the incision, and being unable to move, speak, or do anything about it. Studies in the 1990s had supported the use of a device called a 'bispectral index monitor', essentially an electrode attached to the scalp to reassure the surgeons that the patient was truly unconscious. These studies hardened into common knowledge: half of US operating theatres had a bispectral index monitor by 2007, and an estimated 40 million operations across the world used one.[58] But it turned out that these initial studies weren't up to scratch. When a bigger, better-quality study was conducted in 2008, the monitor was found to be useless: 'anaesthesia awareness occurred even when [bispectral index monitor] values ... were within the target ranges.'[59]

In 2019, Prasad, Cifu and their colleagues performed a review of over 3,000 papers in three top medical journals and found no fewer than 396 studies that overturned the consensus on a medical practice.[60] Here are just a few examples:

- *Childbirth*. Some previous studies had indicated that having a planned caesarean section was the safest option for babies when a mother was giving birth to twins, and as a result this became standard practice (at least in North America). But a large randomised trial in 2013 showed that it made no difference to the health of the children.[61]
- *Allergy*. Peanut allergies can be deadly, and if a parent has one, their children are at higher risk of developing one too. For many years, the guidelines for at-risk babies, based on

previous research, were to avoid giving them peanuts until they were at least three years old and for breastfeeding mothers to avoid peanuts as well. It turns out this advice was exactly backwards: a high-quality randomised trial in 2015 showed that only around 2 per cent of at-risk children who ate peanuts early in life developed an allergy to them by age five, compared to the almost 14 per cent of at-risk children who had avoided peanuts.[62]

- *Heart attack*. Some small trials found that cooling a cardiac-arrest patient down by a few degrees meant that they'd be more likely to survive. Advice based on this finding began to be included in guidelines for paramedics. But a big study in 2014 showed that it made no difference to survival rates and may even have increased the likelihood of a second heart attack while the patient was being transported to the hospital.[63]

- *Stroke*. Research had suggested that the best thing to do in the days after someone has a stroke is to get them moving as soon as possible: have them sit up in bed, stand, walk around if they can. This concept of 'early mobilisation' appears in many widely used hospital guidelines. A large-scale randomised trial in 2015, however, showed that early mobilisation in fact led to *poorer* outcomes for the stroke patients.[64] Similarly, a 2016 study showed that the accepted practice of giving stroke patients a platelet transfusion (a procedure that replenishes the blood cells involved in clotting, which in theory helps to prevent further bleeding) actually made things worse.[65]

It's understandable that doctors, and those who write the guidelines for medical treatments, sometimes find themselves relying on low-quality evidence. Often the alternative is no evidence at all, and their job is to help patients who need treatment *right now*. And it's inevitable that advances in technology, methodology, and funding allow scientists to do better research today than was possible a few years ago: that's normal scientific progress. But scientists have let doctors and patients down by creating such a constant state of flux in the medical literature, running and publishing poor-quality studies that even students in undergraduate classes on research design would recognise as inadequate. Even at the time many of these original studies were published, we knew how to do better – and yet we didn't.

The extent of the uncertainty in medical science can fully be appreciated if we take a look across the whole literature. One way of doing so is to consult all the many comprehensive reviews published by the Cochrane Collaboration, a highly reputable charity that systematically assesses the quality of medical treatments. Of those, a startling 45 per cent conclude that there's insufficient evidence to decide whether the treatment in question works or not.[66]

How many patients have had their hopes raised, have suffered, or have even died because their doctors have used worthless or harmful treatments that only *seemed* like they were backed by science? And human misery aside, think of the money being squandered. If you assume that only half of preclinical research is replicable – a reasonable, though of course debatable, assumption – the amount spent every year on low-quality studies that can't be replicated has been calculated at $28 billion in the US alone (that includes pharmaceutical company investment, government grants, and funds from other sources).[67] Other estimates are substantially higher.[68] Even if the replicability situation is much better than 50 per cent, there's still an eye-watering amount of good money being thrown after bad research. Worse, the calculation only covered preclinical research; even more will be wasted 'downstream' on later-stage studies, such as drug trials in humans, that try to build on those unreliable foundations. And it only included the cost of the research itself – it didn't include the additional waste involved in rolling out an ineffective treatment, like the bispectral index monitor for anaesthesia awareness, for use with millions of patients.

*

Put all these failures and reversals together, and it's no surprise that so many scientists feel so nervous about the level of replicability in their field. A 2016 survey of over 1,500 researchers – though admittedly not a properly representative one, since it just involved those who filled in a questionnaire on the website of the journal *Nature* – found that 52 per cent thought there was a 'significant crisis' of replicability. A further 38 per cent believed there was, in a somewhat peculiar turn of phrase, at least a 'slight crisis'.[69] Nearly 90 per cent of chemists said that they'd had the experience of failing to replicate another researcher's result; nearly 80 per cent of biologists said the same, as did almost 70 per cent of physicists, engineers, and medical scientists. Only a slightly lower percentage of scientists said they'd had trouble in replicating their *own* results. Because the study wasn't a proper poll, and scientists who were already worried about replicability were probably more likely to complete it, these numbers will be somewhat exaggerated. But they do tell us that there are widespread concerns about just how much of the scientific literature, including even the research we've done ourselves, we can trust.

We should have seen this coming. A 2005 article by the meta-scientist John Ioannidis carried the dramatic title 'Why Most Published Research Findings Are False', and its mathematical model concluded just that: once you consider the many ways that scientific studies can go wrong, any given claim in a scientific paper is more likely to be false than true.[70] The article may have attracted a great deal of attention and discussion, being cited over 800 times in the first five years after its publication, but in terms of scientists starting to make the necessary changes to improve the quality of research, its Cassandra-like warnings fell on deaf ears.[71] Only since the revelations of the replication crisis – beginning with the publication of Bem's parapsychological claims and the exposure of Stapel's fraud in 2011, along with the psychological priming and cancer-reproducibility failures around that same time – has there been a widespread acknowledgement of our problems, and a stark realisation that they go to the very heart of how science is practised today.[72]

So how have we reached the point where an incendiary title like 'Why Most Published Research Findings Are False' seems not like an absurd overstatement but a reasonable proposition? We'll now look at the many ways that science can – and does – go wrong.

PART II
FAULTS AND FLAWS

3
Fraud

Though we may not desire to detect fraud, we must not, on that account, endeavor to be insensible of it ... If men dread knaves, they also despise fools.

Norman MacDonald, *Maxims and Moral Reflections* (1827)

Some of the most genuinely affecting moments on the internet are videos of people with illnesses or disabilities having their lives changed in an instant by a new piece of technology. Babies given a cochlear implant, noticing to their surprise and delight that they can hear for the first time; children born with cataracts getting the surgery they need to be able to see; soldiers who lost their legs in battle taking their first steps with new prosthetic limbs.[1] These videos go viral on social media, and for good reason: not only are they heart-warming, but they also serve to remind us of the power of science, at its best, to enhance our health and our lives.

But here's a story of how even scientific effects as pure as these can be perverted and corrupted. A story of how patients who thought they were receiving a cutting-edge, transformative medical treatment ended up victims of one of the worst scientific frauds we've seen this century. To make things worse, it wasn't a scam run by some alternative-medicine mountebank recruiting desperate patients online. It was a fraud perpetrated in the halls of the world's most distinguished medical schools and in the pages of one of the world's most respected scientific journals. It'll show us that even the most outrageous fraudsters can sometimes be hiding in plain sight.

When the trachea (the windpipe) is badly damaged by disease or injury, surgeons can no longer reattach the broken ends of the tube, and providing a new trachea is the only way of saving the patient.[2] As with any large organ transplant, transplanting a trachea is extremely difficult: not only is it hard to find potential donors (who obviously have to be dead), but if the donor is genetically different to the recipient, the trachea is usually rejected by the recipient's immune system. So instead of using transplants, surgeons have for decades tried implanting their patients with artificial tracheas, made from an amazing variety of materials: plastic, stainless steel, collagen, even glass. Yet the attempts almost always failed: the synthetic windpipes moved around, became obstructed and attracted infection. By the beginning of the twenty-first century, it was the medical consensus that artificial tracheas just weren't a viable option.[3]

Enter Paolo Macchiarini, an Italian surgeon who in 2008 had published a blockbuster paper in the top medical journal the *Lancet* on his successful transplant of a trachea.[4] Macchiarini's new notion was to 'seed' the donor trachea with a sample of the recipient's stem cells – cells that can divide endlessly, repairing and replacing other cells in the body – before it was implanted. After some time in a specially designed incubator, the stem cells had 'colonised' the donor trachea and so seemed to prevent its rejection during the later transplant. This was a big step. But the holy grail was still the creation of completely artificial tracheas, ones that wouldn't require donors at all. Could Macchiarini's idea – making foreign objects more acceptable to the body by having an outer layer of compatible cells grown around them – finally reach that goal?

Just a few years later, the answer appeared to be 'yes'. Since his 2008 paper Macchiarini's reputation as a surgical genius had blossomed, and in 2010 he'd been recruited – at the recommendation of fourteen professors who already worked there – by Sweden's Karolinska Institute, where he became a visiting professor, and its associated Karolinska Hospital, where he was made a lead surgeon. The Karolinska Institute isn't just the number-one university in a country with a lot of great universities; it's the home of the Nobel Prize in Physiology or Medicine. It made perfect sense that Macchiarini, a surgeon who was revolutionising regenerative medicine with his ingenious stem cell techniques, would be employed by such an august institution.

In July 2011, the Institute excitedly announced that the next step had been taken: Macchiarini had just successfully transplanted, 'for the first time in history', a completely synthetic carbon-silicon trachea, seeded with stem cells, into a cancer patient at the Karolinska Hospital.[5] In November that year, the scientific paper reporting the details of the operation was published, by which time Macchiarini had performed a similar operation on another Karolinska patient.[6] The paper, again published in the Lancet, described the 'solid evidence' of the transplantation's success. During 2012, Macchiarini conducted artificial trachea operations on another three patients, one at the Institute and two at his secondary base of operations in Krasnodar, Russia. Two more operations in Russia would happen across the next two years, and Macchiarini spread the good news by publishing even more scientific papers.[7]

One of these papers, a 2014 report in the journal Biomaterials, bursting with pretty electron-microscope photos of the 'electrospun tracheal scaffold', admitted rather tersely that the first patient had experienced some difficulties, before continuing on briskly to describe the wonders of the new technology. The authors had omitted a devastating detail: the patient in question had died – seven weeks before the paper was even accepted for publication.[8] The patient from the second operation had died even earlier, just three months after his procedure.[9] The third Karolinska patient would die in 2017 after several failed follow-up surgeries.[10] The Russian patients fared little better.

The first of them, a ballet dancer from St Petersburg named Julia Tuulik, described her tragic state to a journalist:

> Everything is very, very bad with me. I have spent over six months in the hospital in Krasnodar. I have undergone over thirty surgeries under general anaesthesia. Three weeks after the first operation a purulent fistula [a hole leaking pus] opened, and my neck has since rotted. I weigh 47kg. I can barely walk. I have trouble breathing, and now I have no voice. And it smells so strongly ... that people back away.[11]

Tuulik died in 2014, two years after her operation.[12] Most heartbreakingly, she hadn't even been in a life-threatening situation before the operation.[13] One other Russian patient died in what was described as a 'bicycle accident', another died in uncertain circumstances the year after the operation, and another survived but only after the synthetic trachea had been removed.[14] Macchiarini also operated on a Canadian-South Korean toddler at a hospital in Peoria, Illinois in the US in 2013, amid substantial media attention. She died just a few months later.[15]

A group of doctors at the Karolinska Hospital who had looked after Macchiarini's patients post-operation couldn't reconcile the patients' terrible condition with the glowing outcomes reported in the scientific papers. They got together and complained to the heads of the Karolinska Institute. Instead of surprise and concern, they were met with stonewalling and attempts to have them silenced. The Institute even reported the doctors to the police, alleging that by looking through the patients' medical records they had violated their privacy (these charges were rapidly dismissed).[16] Eventually, however, the Institute high-ups bowed to the pressure, bringing in an independent researcher – a professor from nearby Uppsala University – to investigate the claims.

His report, which appeared in May 2015 and ran to 20,000 words, couldn't have been clearer: Macchiarini was 'guilty of scientific misconduct' on multiple counts.[17] Across seven papers,

Macchiarini had: falsely claimed that patients' conditions had improved, when the necessary examinations hadn't even taken place; misstated the follow-up periods to make it seem like the patients had been healthier for longer; failed to report that the patients had experienced severe complications and sometimes had to undergo further operations; failed to get the correct ethical permission to run what was essentially medical experimentation on human subjects; and falsified data in a lab study where he'd replaced the tracheas of rats.[18]

At this point, one would think the story would be over. But after Macchiarini responded to the charges laid out in the independent investigation, the Karolinska Institute decided to run their own, internal inquiry. In August 2015, based on information that wasn't made public, it concluded that, in actual fact, no misconduct had taken place.[19] The next week, the *Lancet* published an editorial to celebrate the fact that 'Paolo Macchiarini is not guilty of scientific misconduct'.[20] Macchiarini had apparently just been exonerated – and by two of the biggest institutions in the medical world.

Then, in January 2016, came two developments that beggared belief, and couldn't be ignored or covered up. First, an article appeared in *Vanity Fair* magazine giving a lengthy account of Macchiarini's romance with the NBC News producer Benita Alexander, to whom he proposed in 2014.[21] He claimed to her that he was the personal doctor to Pope Francis, and that he'd be inviting the pontiff as well as some of the world's biggest celebrities – including the Obamas, Russell Crowe and Elton John – to their wedding.[22] When *Vanity Fair* contacted the Vatican, however, they were informed that the Pope didn't have any doctors by the name of Macchiarini. Not only that, but it was discovered that during his entire relationship with Alexander, Macchiarini was married to someone else – and had two young children.[23]

The second development was the broadcast of a three-part Swedish TV documentary, *The Experiment*, in which the filmmakers showed, in harrowing detail, how the lives of

Macchiarini's patients had been ruined, and sometimes lost, due to his gross incompetence. The documentary included footage of a bronchoscopy (where a tiny camera is sent down the trachea) on the first of the Karolinska patients, which revealed a scarred, blocked and even perforated windpipe – a far cry from the 'almost normal airway' Macchiarini had described in his paper in the *Lancet*.[24]

The Karolinska Institute was shaken into setting up a completely new investigation. This time, heads rolled. The university's vice-chancellor, who had supported Macchiarini from the start, resigned. So did the dean of research; then the chair of the university board; then a member of the Nobel Committee who had pushed for Macchiarini's appointment.[25] In March 2016, Macchiarini was at last removed from his post.[26] It had been seven years since the first botched synthetic trachea operation.

After their staunch defence of just a few years earlier, the *Lancet* retracted Macchiarini's synthetic trachea papers.[27] The university confirmed in mid-2018 that it had also found scientific misconduct in several of his other papers, which it listed online; all have now been retracted.[28] After his dismissal, Macchiarini retreated to Russia and continued his 'research' – though now he had moved on from the trachea to the oesophagus (the food pipe).[29] Thankfully, in his most recent published paper, no patients were involved; instead, he and his new colleagues tested the compatibility of their plastic oesophagus with cells taken from dead baboons.[30] His current status is unclear: the Russian government pulled his funding in mid-2017, and he's unlikely to be able to perform any more operations.[31] And there may yet be some form of justice for his patients: in December 2018, the Swedish Prosecution Authority announced that it was resuming an investigation against Macchiarini for two cases of manslaughter.[32]

What led to such a drawn-out process, where even the horrific, painful deaths of multiple patients hadn't moved the university to fire, or even censure, the man responsible? Indeed, what led to the university lashing out against the very people who blew the whistle on the falsified results?[33] Macchiarini clearly used his 'breakthrough' operations to burnish his reputation and fame. But he was also an asset to the Karolinska Institute and its planned international expansion. It has been suggested that the university's association with the superstar surgeon helped to grease the wheels of its new regenerative medicine centre that was being set up in Hong Kong.[34] Sheer panic and embarrassment were likely also a factor in their wanting to cover up the case: a university with as high a standing as the Karolinska would hardly be keen to admit to themselves, let alone the public, that they'd let a lethally dangerous fraudster loose on vulnerable patients.

Few episodes of scientific fraud have effects on people's lives that are as chillingly direct as the case of the artificial tracheas. Few scientific fraudsters are as outrageous and flamboyant as Macchiarini. Yet there are some wider lessons to learn from his story. The first is how much of science, despite its built-in organised scepticism, comes down to trust: trust that the studies really occurred as reported, that the numbers really are what came out of the statistical analysis, and, in this case, that the patients really did recover in the way that was claimed. Fraud shows just how badly that trust can be exploited. The second lesson is that the same scepticism we level at studies and at people also needs to be levelled at institutions. There will always be blackguards whose craving for fame and success overrides all other concerns, but we should be able to trust famed scientific institutions like the Karolinska Institute and the *Lancet* to do their utmost to prevent them from having an effect on science – and to expose and punish them whenever they arise. Alas, the reputation-motivated desire of these institutions to employ and publish glamorous scientists can result in a wilful blindness to the activities of fraudsters, and sometimes even extends to shielding them from the consequences of their actions.

We can go even further. The fact that the scientific community so proudly cherishes an image of itself as objective and scrupulously honest – a system where fraud is complete anathema – might, perversely, be what prevents it from spotting the bad actors in its midst. The very idea of villains like Macchiarini existing in science is so abhorrent that many adopt a see-no-evil attitude, overlooking even the most glaring signs of scientific misconduct. Others are in denial about the prevalence and the effects of fraud. But as we'll see in this chapter, fraud in science is not the vanishingly rare scenario that we desperately hope it to be. In fact, it's a distressingly common one.

One of the best-known, and most absurd, scientific fraud cases of the twentieth century also concerned transplants – in this case, skin grafts. While working at the prestigious Sloan-Kettering Cancer Institute in New York City in 1974, the dermatologist William Summerlin presaged Macchiarini by claiming to have solved the transplant-rejection problem that we encountered above. Using a disarmingly straightforward new technique in which the donor skin was incubated and marinated in special nutrients prior to the operation, Summerlin had apparently grafted a section of the skin of a black mouse onto a white one, with no immune rejection. Except he hadn't. On the way to show the head of his lab his exciting new findings, he'd coloured in a patch of the white mouse's fur with a black felt-tip pen, a deception later revealed by a lab technician who, smelling a rat (or perhaps, in this case, a mouse), proceeded to use alcohol to rub off the ink. There never were any successful grafts on the mice, and Summerlin was quickly fired.[35]

Summerlin is hardly alone among scientists in indulging an illicit artistic urge. It's commonly seen in the figures that illustrate scientific papers. With computer graphics, it's never been easier to crop, duplicate, touch up, splice, recolour, or otherwise alter scientific images to make them show whatever you want. Of course, producing fraudulent photographs was eminently possible well before the Photoshop era (just ask Commissar Nikolai Yezhov, who was famously 'disappeared' from a photo with Joseph Stalin after he fell out of favour with the Soviet leader). In 1961, *Science* magazine apologised for publishing an article by Indian veterinary researchers who claimed to have found the parasite *toxoplasma gondii* for the first time in chicken eggs (a potential health risk since the parasite can cause toxoplasmosis, a dangerous disease for those with weakened immune systems).[36] The evidence of the presence of the parasite – microscope photos of its cysts in an egg – was, it turned out, fake. What the researchers claimed were two different cysts were in fact the same photo that had been zoomed out and flipped horizontally, a duplication that's crashingly obvious in retrospect, but that the peer reviewers missed. After the foul play was discovered, the researchers were soon forced to retire, or suspended from their positions.[37]

You might think that only the laziest scientific frauds would use duplicated images in their papers, making their deceit visible to the vigilant naked eye. But image duplication comes up again and again, and has been a central feature of some of the most prominent fraud cases of recent decades. In 2004, the South Korean biologist Woo-Suk Hwang announced in a *Science* paper that he'd successfully cloned human embryos. The next year, in the same journal, he reported that he'd produced, from those embryos, the first cloned human stem cell lines. What gives stem cells their potential, beyond the fact that they can keep multiplying indefinitely, is that they're 'pluripotent', meaning they can be transformed in the lab, Swiss-Army-knife-like, into many different types of tissues (neurons, liver cells, blood cells, and so on). Cloned cell lines – eleven of them had been produced for Hwang's paper – might have allowed the production of personalised stem cell treatments, and thus the repair of damaged tissues and the regeneration of injured or diseased organs. As with the trachea transplants, this would mean that the personalised stem cells came from that same individual, and that their immune system would be less likely to reject any treatments that used them. In yet another breakthrough that same year, Hwang's team at Seoul National University introduced the world to the first ever cloned dog, an Afghan hound named Snuppy.[38]

It's hard to overstate how famous these achievements made Hwang in South Korea. He was venerated in the media.[39] Posters of his face, with statements like 'Hope of the World – Dream of Korea', appeared in the streets and on public transport.[40] The Korean post office issued a special postage stamp in 2005 celebrating his work, showing (perhaps rather prematurely) a series of silhouettes of a person standing up from their wheelchair, leaping into the air, and hugging a loved one.[41] The Korean government, who named Hwang 'the Supreme Scientist', poured enormous sums into his research. Hundreds of women reached out to Hwang to donate their eggs for his studies.[42]

You can guess what happened next. Further inspection of the paper in *Science* revealed that two of the images, purportedly showing Hwang's cell lines for different patients, were identical (and just to be clear, there was effectively zero chance of two images of this sort being identical by mere coincidence). There was also overlap in two more, which came from parts of the same photo but had been passed off as entirely different.[43] If these were the only problems, it could easily have been an oversight – someone might have mixed up or mislabelled the photos. Far from it. Whistleblowers from Hwang's lab revealed that only two cell lines had been created, not eleven, and neither were from cloned embryos.[44] The rest of the cell photos had been doctored or deliberately mislabelled under Hwang's instructions. The entire research project had been a charade.

Even before the concerns about image manipulation were raised, Hwang had been in hot water for obtaining eggs from donors who were neither fully informed of how their eggs would be used, nor of the potential dangers of having them extracted. Hwang had also pressured the female members of his lab to donate their own eggs for experimentation.[45] The revelations of misconduct kept coming: Hwang had siphoned off some of his research funding into a network of bank accounts that he controlled – and although he claimed the money was still spent on scientific apparatus, an investigation revealed that the 'apparatus' included a new car for his wife and donations to supportive politicians.[46]

The glowing media coverage of Hwang had been so intense that not even this egregious scientific fraud was enough to put off his admirers. Protestors lined the streets outside the offices of media organisations who ran negative stories and filled up their online forums with thousands of irate pro-Hwang posts.[47] Nevertheless, the authorities had to react. Hwang was fired from his university, then criminally prosecuted, although he managed to avoid jail with a two-year suspended sentence.[48] These days he's still working on cloning, but at a less prestigious university, and his work receives a tiny fraction of the attention it once captured. Incidentally, amongst all Hwang's counterfeit accomplishments, Snuppy was real: DNA tests revealed he was a genuine clone of another Afghan hound named Tai. Snuppy died in 2015, but he's still around, in a sense: four further clones of him were born two years later.[49]

Hwang was easily the most famous scientist in his country, and one of the world's most prominent biologists. Given all the attention he received in that position, why did he think he could get away with such blatant, careless fraud? The answer speaks not just to his own (lack of) character, but to something broken about the scientific system. As we noted above, the system is largely built on trust: everyone basically assumes ethical behaviour on the part of everyone else. Unfortunately, that's exactly the sort of environment where fraud can thrive – where fakers, like parasites, can free-ride off the collective goodwill of the community. The fact that Hwang's acts were so shameless only showcases how gullible reviewers and editors – the very people we rely upon to be rigorously sceptical – can be when faced with such exciting, 'ground-breaking' results.

Aside from microscope photos of cells, one of the most common targets for image-based fraud in biology is blotting. Blotting is a technique with many variants that molecular biologists use to reverse-engineer the composition of the chemicals they produce or examine in their experiments. The original is the Southern blot, named after its inventor, the biochemist Edwin Southern.[50] Southern blots detect DNA sequences: using radioactive tagging, the blotting produces the images you've perhaps seen accompanying news articles on genetics, with semi-rectangular blurs of varying size organised into vertical ladders or 'lanes'.[51] Other subtypes of blotting can detect different chemicals, and each one has a name that plays on the original: northern blots, western blots, and so on.[52] Biological experiments often hinge on a particular blot coming out a particular way: for example, western blots can be used to diagnose some diseases by picking up on the production of proteins that indicate the presence of certain bacteria or viruses. Scientists often proudly display the blots as figures in their papers, with the blot providing the crucial evidence that some chemical has been detected in the experiment. This is where the tampering comes in.

A decade after Hwang's 'discoveries', in 2014, scientists at Japan's RIKEN institute published two papers in *Nature* reporting new results on induced pluripotent stem cells.[53] Unlike the stem cells that were part of Hwang-gate, *induced* pluripotent cells can be produced from mature adult cells, reducing the need to use cells from embryos.[54] The trouble is that the standard process for creating this kind of stem cell, which won its inventors a Nobel Prize in 2012, is laborious and inefficient, taking several weeks and producing a lot of waste.[55] The RIKEN group, though, claimed to have found another way to produce the cells: a technique called STAP, for 'Stimulus-Triggered Acquisition of Pluripotency'. All you had to do, apparently, was bathe the adult cells in a weak acid (or provide another kind of mild stress, like physical pressure), and they would turn into pluripotent stem cells without all the hassle. The lead researcher, Haruko Obokata, assembled an array of impressive-looking evidence, illustrated with microscope images, graphs and blots showing DNA evidence that the adult cells had been reprogrammed into pluripotency.

This was revolutionary stuff and Obokata suddenly became a household name in Japan. Articles about her and her quirky lab arrangements (she had a pet turtle, her lab was decorated with characters from the Moomins, and instead of a white lab coat she wore a *kappogi*, a Japanese apron she'd been given by her grandmother) were splashed all over the press and she was held up as a shining example of an exceptional female scientist.[56] It didn't last long. Within days of publication, other scientists began noticing discrepancies in the illustrations in Obokata's papers, particularly in the four 'lanes' of her DNA blot. Each one was supposed to come from the same blot, but if you looked very closely, the background of one was darker than the others, with suspiciously sharp edges. It turned out that it had been spliced in from another, separate blot picture, and subtly resized to fit better with the other lanes.[57] There was no indication in the text that this had been done; hardly the action of a scientist working with painstaking transparency. After this was highlighted, more anomalies were discovered. Some of the colours in the photographs had been touched up in post-production, and Obokata had also engaged in the practice of image duplication: two of the supposedly different figures in her second paper were the same photo, only in one case it had, no surprises here, been flipped onto its side.

Meanwhile, and rather unusually, labs around the world had gone into overdrive to try and replicate the results. One of STAP's downfalls may have been that it was such a simple technique that it was easy for others to double-check. One cell biology professor created a website where researchers could send reports of in-progress replication attempts. He used a green font for positive or encouraging results, and red for failures; as the reports rolled in, almost all were in red.[58] Adding to this combined pressure from the image-checkers and the replicators, RIKEN set up an investigation, which confirmed that the images had been doctored. Obokata and her colleagues requested that *Nature* retract the papers, which they did by June 2014; she resigned her RIKEN job that December.[59]

A more detailed investigation revealed that Obokata's rap sheet went beyond the initial charge of fabricating images: she'd also included figures from older research that she rebranded as new, and faked data that showed how quickly the cells had grown. Any actual evidence of pluripotency was due to her allowing her samples to become contaminated with embryonic stem cells.[60]

The final part of the STAP story is dreadfully sad. Yoshiki Sasai, a brilliant stem cell biologist and co-author on the papers, who hadn't been involved in the fraud but who, according to the RIKEN report, bore 'grave responsibility' for not double-checking Obokata's results, committed suicide, hanging himself in the RIKEN institute building in August 2014.[61] He was fifty-two. In his suicide note, he mentioned the media furore that had been kicked off by the discovery of Obokata's deception.[62]

The stories of Hwang and Obokata are both unusual in one obvious respect: the extraordinary prominence of the fraudulent papers. These were publications in *Science* and in *Nature*, two of the world's leading journals. That such conspicuous fakes made it through these journals' vetting process is concerning enough, but their prestige meant that the papers immediately captured the world's attention – and its scrutiny. If this kind of fraud occurs at the very highest levels of science, it suggests that there's much more of it that flies under the radar, in less well-known journals. Which raises the question: how often *do* biologists fake the images in their papers? In 2016, the microbiologist Elisabeth Bik and her colleagues decided to find out.

They searched forty biology journals for papers that included western blots, eventually finding 20,621.[63] In a genuinely heroic set of analyses, Bik personally looked through every single one and checked for inappropriate duplication in its photographic images. What she found was enough to populate a gallery of dodgy scientific pictures several times over: not just basic duplications (see Figure 1, below, for an example), but duplications with Hwang-esque cropping, Obokata-esque splicing and resizing, and a whole host of other dishonest techniques. In all, 3.8

per cent of the published papers (around one in twenty-five) included a problematic image. In a later analysis of papers from just one cell-biology journal, Bik and her colleagues found an even higher percentage: 6.1 per cent.[64] Of these, many were just honest mistakes and the authors could issue a correction that solved the problem. However, around 10 per cent of the papers were retracted, implying something more nefarious was going on. If those numbers are representative of cell biology papers in general, Bik calculated that there are up to 35,000 papers in the literature that need to be retracted. There was at least some positive news: more prestigious journals appeared on average less likely to have published papers with image duplication. Perhaps most intriguing, though, were the results about repeat offenders: when Bik and her team found a paper with faked images, they checked to see if other publications by the same author also had image duplications. This was true just under 40 per cent of the time. Duplicating one image may be regarded as carelessness; duplicating two looks like fraud.

So far, our focus has mainly been on image-based deception, but that's hardly the only place fraud occurs. Perhaps a more effective place to commit – and to hide – scientific fraud is in the numbers: the rows and columns of figures that make up

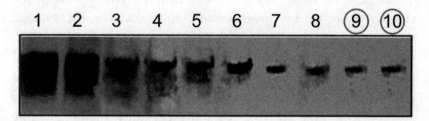

Figure 1. A duplicated western blot discovered by Bik and colleagues. The last two bands (columns 9 and 10) turn out to be identical, the image having been duplicated (and the right-hand one slightly resized), perhaps using software like Photoshop. The paper in question was later corrected. Image adapted from Bik et al. (2016) *mBio*.

a study's dataset. In the Preface, we met Diederik Stapel, who simply typed the results he wanted into his spreadsheets and passed them off as real. How often does data fraud of this kind occur? And how easy is it to spot?

Fortunately, just as it's a monumentally difficult task to forge a compelling Rembrandt or Vermeer (or a compelling western blot), it's not at all easy to fake a dataset convincingly. Data pulled out of thin air don't have the properties we'd expect of data collected in the real world.[65] Fundamentally, this is because no science is really an exact science: numbers are *noisy*. Every time you try to measure anything, you'll be slightly off from the true value, be it the economic performance of a country, the number of rare orangutans left in the world, the speed of a subatomic particle, or even something as simple as how tall someone is. With height, for instance, the person might be a bit slouched, your tape measure might slip by a fraction of an inch, or you might accidentally write down the wrong number. This is called *measurement error*, and it's hard to get around completely, even if there are ways to reduce it.[66]

Measurement error's equally annoying cousin is *sampling error*. As scientists we can rarely, if ever, examine every single instance of a phenomenon – no matter whether we're trying to study a set of cells, or exoplanets, or surgical operations, or financial transactions. Instead, we take samples, and try to generalise from them to the set as a whole (statisticians call the whole set the 'population', even if it's not a set of people). The trouble is, the characteristics of any given sample you take (say, the average height of all the people in your study) are never a precise match to what you really want to know (say, the average height of all the people in the country). Just through the random chance of who was included, every sample will have a marginally different average. And some samples, again just by chance, might be wildly different from the true average in the overall set.[67]

Both measurement error and sampling error are unpredictable, but they're predictably unpredictable. You can always expect data from different samples, measures or groups to have somewhat different characteristics – in terms of the averages, the highest and lowest scores, and practically everything else. So even though they're normally a nuisance, measurement error and sampling error can be useful as a means of spotting fraudulent data. If a dataset looks too neat, too tidily similar across different groups, something strange might be afoot. As the geneticist J. B. S. Haldane put it, 'man is an orderly animal' who 'finds it very hard to imitate the disorder of nature', and that goes for fraudsters as much as for the rest of us.[68]

This kind of reasoning is what caught out social psychologists Lawrence Sanna and Dirk Smeesters in 2011. Sanna published a study in which he claimed to find that people are more prosocial when standing at higher elevations; Smeesters claimed to show that seeing the colours red and blue affects how people think about celebrities.[69] The results in both papers looked impressive at first glance, easily confirming their proposed theories about human behaviour. But a closer look revealed something distinctly odd. The psychologist Uri Simonsohn showed that in the various groups in Sanna's experiment, the range of the data (the difference between the highest and lowest scores) was nearly identical, although the groups were otherwise quite different. Simonsohn calculated that the chances of this happening in real data were minuscule. It was the same for Smeesters, except it was the averages of his groups that were too similar; again, these similarities just weren't consistent with what would happen in real data, where error would have nudged the numbers further apart.[70] Once these problems, among others, were exposed, the offending papers were retracted, and both researchers resigned in disgrace.[71]

These kinds of statistical red flags are analogous to what makes your bank freeze your credit card after it's suddenly used to spend large sums on a tropical cruise: unusual activity that's out of line with normal expectations, and which might be due to fraud.[72] And there are a host of other features of fraudulent data that might cause readers to become suspicious when they dig into the details. The dataset might look a little too immaculate, for example, with too few missing datapoints, which come about for all sorts of reasons in real datasets: participants dropping out of the study or instruments failing, for example. Perhaps the distribution of numbers might not follow certain expected mathematical rules.[73] Or the effects might be vastly larger than seems plausible in the real world, and thus too good to be true.[74]

Some forgers are well aware of how tricky it is to make fake numbers look real, so have tried more creative ways to cover their tracks. The political scientist Michael LaCour, at the time a graduate student at the University of California, Los Angeles, published an eye-catching result in *Science* in 2014 from a large-scale door-to-door survey.[75] The data showed that being approached by a gay campaigner, as opposed to a straight one, had a big, positive long-term impact on people's opinions towards gay marriage. That is, meeting someone from the minority whose rights are in question made respondents much more likely to support those rights. It was an optimistic message, and it was immediately used by the (ultimately successful) campaign for the legalisation of same-sex marriage in Ireland's 2015 referendum.[76]

Two other political scientists, David Broockman and Joshua Kalla, were among the many researchers impressed with the findings, and wanted to try a similar study themselves. But upon examining LaCour's dataset, they found some very weird anomalies. They noticed that the distributions of the results, that is, the spread of high to low scores on the 'how much do you support same-sex marriage' scales, were suspiciously similar to those in another, older survey dataset they knew well, called the Cooperative Campaign Analysis Project (CCAP). In fact, the distributions were nearly identical. Data from the follow-up waves of the study, where participants were surveyed after the initial contact with the campaigner to see whether their attitudes had changed, were also strange: not a single one of the participants had moved further than a set amount from their original views. Once again, numbers are noisy: you'd expect to see far more to-ing and fro-ing over time in a big dataset like this.

It turned out that LaCour's dataset was a palimpsest: he'd taken the results from the CCAP survey, jiggled around the numbers by adding random noise, and pretended they'd come from his new study. The follow-up waves were just the same data with additional jiggling, while the many details LaCour provided in the paper about training the canvassers were entirely fictional. The survey had never taken place. His unlucky co-author, Donald Green, a respected professor of political science who wasn't involved in faking the data, later marvelled at the 'incredible mountain of fabrications with the most baroque and ornate ornamentation' that LaCour had shown him. 'There were stories, there were anecdotes ... graphs and charts. You'd think no one would do this except to explore a very real data set.'[77]

I distinctly remember the appearance of Broockman and Kalla's report about LaCour's study, in May 2015.[78] It was posted online just before I got on a plane from Edinburgh to a conference in San Francisco; by the time I arrived, thirteen hours later, the scientists I followed on social media had been talking about little else. The report was a demolition, exposing LaCour's data-tampering in minute detail. There was nowhere left to go: *Science* retracted the paper soon afterwards, and LaCour waved goodbye to a job offer he'd had – largely on the basis of his precocious publication in such a prestigious journal – from Princeton University.[79]

Given the lengths to which LaCour went to hide his fraud, it surely would have been easier for him just to do the study. This would certainly have avoided the downside of his career being ruined when proper scrutiny was applied to his dataset. But what LaCour, like Sanna, Smeesters, and Stapel before him, got from his data fabrication was *control*. The study fit the exact specifications required to convince *Science*'s peer reviewers it was worth publishing. It was what the publication system and the university job market appeared to demand: not a snapshot of a messy reality where results can be unclear and interpretations uncertain, but a clean, impactful result that could immediately translate to use in the real world.

Once again, it was not just the reviewers' desire for attractive and exciting findings that had been exploited, but also their trusting nature. It's unavoidable that *some* trust will be involved in the peer review process: reviewers can't be expected to double-check every single datapoint for signs of tampering. But the lesson of the data fraud stories is that the bar might be set rather too low for true organised scepticism to be taking place. For the sake of the science, it might be time for scientists to start trusting each other a little less.

We're now ready to ask the question: across all its different variations, how prevalent is scientific fraud? One way to estimate the scale of the problem is to look at the number of papers that have been retracted. Retraction is the ultimate inglorious end for a paper; it has even been called 'science's death penalty'.[80] After this capital punishment, retracted studies enter a kind of purgatory. They're not simply deleted, because to do so would just cause more confusion, especially if they've already been cited in other work. Instead, they're kept online permanently on the journal's website, with an indication that they're no longer considered legitimate. This often takes the form of the word RETRACTED printed diagonally across the pages of the document in enormous bold red type.

The best place for information about retractions is the website Retraction Watch, which chronicles every new retraction, contacting journals and authors for statements to find out what went wrong. In 2018 the website's owners, Ivan Oransky and Adam Marcus, launched a database that catalogues all of the over 18,000 retractions in the scientific literature since the 1970s – a treasure trove for those interested in the scandalous side of science. A glance at the section labelled 'Reason(s) for Retraction' hints at some of the torrid stories associated with the papers: 'Conflict of Interest'; 'Forged authorship'; 'Misconduct of author'; 'Sabotage of Materials'; 'Criminal Proceedings'.[81]

The Retraction Watch Database isn't a perfect list: some retractions might have been missed, since journals vary widely in the extent to which they acknowledge or highlight retracted articles. It's also important to note that a retraction doesn't necessarily mean fraud – many papers are retracted because the authors noticed a mistake and withdrew the paper themselves. Other retractions are more ambiguous: for instance, in early 2020 the Nobel Prize-winning chemical engineer Frances Arnold announced that her team were retracting a paper on enzymes from *Science* because the results wouldn't replicate and because 'careful examination of the first author's lab notebook … revealed missing contemporaneous entries and raw data for key experiments.'[82] Whether this implied mere error or something worse on behalf of that lead author, a student in Arnold's lab, is unclear. Arnold's admission was painfully candid: 'I apologise to all,' she tweeted. 'I was a bit busy when this was submitted, and did not do my job well.'[83]

Among retractions in general, honest mistakes only make up around 40 per cent or less of the total. The majority are due to some form of immoral behaviour, including fraud (around 20 per cent), duplicate publication and plagiarism.[84] The number of retractions is also increasing over time, though this might not imply an increase in fraud: rather, it might mean that journal editors are getting wiser to it, or that authors are more willing, like Arnold, to admit that they screwed up.[85]

In the same way that a small number of lawbreakers commit a disproportionate number of crimes in society, the Retraction Watch Database shows that just 2 per cent of individual scientists are responsible for 25 per cent of all retractions.[86] The worst repeat offenders are added to the Retraction Watch Leaderboard, a position on which is a sort of reverse Nobel Prize.[87] It features our acquaintance Diederik Stapel at number five, with fifty-eight retractions. The undisputed heavyweight champion of retractions, however, is currently the Japanese anaesthesiologist Yoshitaka Fujii, who invented data from non-existent drug trials and whose retracted papers number an astonishing 183. In 2000, a letter to the journal *Anaesthesia & Analgesia* described Fujii's reported data as 'incredibly nice!'.[88] As the curators of Retraction Watch have observed, this wasn't a compliment.[89] The letter's authors had noticed that the number of participants in Fujii's trials who reported headache as a side effect was exactly identical across the different groups in thirteen of his studies, and almost identical in a further eight. The data, just like in the cases we saw above, were far too uniform to be real. But absolutely nothing was done for over a decade, and all the while Fujii continued to publish fake papers in highly regarded anaesthesiology journals. It wasn't until 2012, when another analysis found yet more of his data to be extremely improbable, that a formal investigation ended his career.[90] In addition to documenting 172 papers that were confirmed to contain fake data (more have been discovered since, taking Fujii to his world-beating total), the investigators also listed the papers published by Fujii that they concluded *didn't* contain any fraud. There were three.[91]

So, if you combine the Leaderboard's heavy hitters with all the other scientists who retract articles for dishonest reasons, how many scientists actually commit fraud? The overall proportion of papers retracted – around 4 in 10,000 published studies, or 0.04 per cent – is reassuringly low. It also isn't very helpful, as we know on the one hand that some retractions aren't due to fraud, but on the other that some journals either don't catch false findings or don't bother retracting them. What happens if instead you simply ask scientists, anonymously, whether they've ever committed fraud?

The biggest study on this question to date pooled research from seven surveys, finding that 1.97 per cent of scientists admit to faking their data at least once.[92] As if one in fifty scientists admitting to being fraudsters wasn't alarming enough, consider that people are naturally loath to confess to fraud, even in an anonymous survey, so the real number is surely much higher. Indeed, when surveys asked scientists how many *other* researchers they know who have falsified data, the figure jumped up to

14.1 per cent (although of course, some of those polled might have been mistaken, paranoid, or exaggerating problems in their rivals' research).[93]

Who are these fraudsters? Can we put together an FBI-like 'profile' of the quintessential fraudster, to aid us in preventing further acts of data fabrication? In a review of fraud cases, the neuroscientist Charles Gross lamented the lack of solid evidence on who commits fraud. He did, however, have a go at describing the type of character who regularly appears in well-publicised fraud reports in the media. That person, he wrote, tends to be 'a bright and ambitious young man working in an elite institution in a rapidly moving and highly competitive branch of modern biology or medicine, where results have important theoretical, clinical, or financial implications.'[94] By this point in the chapter, it's a familiar picture: for instance, it fits Paolo Macchiarini almost perfectly.

Notably, Gross described the fraudster as a man. This is a clear pattern among the worst fraudsters: of the thirty-two scientists currently on the Retraction Watch Leaderboard, only one is a woman.[95] To know whether this tells us something important, we'd need to know what the base rate of men versus women was in each relevant field, and thus whether men were overrepresented. A study in 2013, focusing on the life sciences, took those baseline differences into account and found that men were indeed overrepresented as the subject of fraud reports from the US Office for Research Integrity.[96] A 2015 paper examining retractions and corrections across all scientific fields, meanwhile, found no gender differences, although it's not clear whether it considered the all-important baseline.[97]

After collecting their database of papers with duplicated western-blot images, Elisabeth Bik and her colleagues also checked to see if there were any characteristics that differentiated the problematic papers from others.[98] One thing that stood out was that image duplication was more likely to take place in some countries rather than others: India and China were overrepresented in the number of papers with duplicated images, while the US, the UK, Germany, Japan and Australia were underrepresented. The authors proposed that these differences were cultural: the looser rules and softer punishments for scientific misconduct in countries like India and China might be responsible for their producing a higher quantity of potentially fraudulent research.[99] This once again emphasises that the social milieu in which science is conducted can have serious effects on its quality.

Others have made related speculations. After citing research showing that a rather suspicious *one hundred per cent* of trials of acupuncture from scientists in China had positive results (even if acupuncture worked perfectly, we'd expect to see a few failures just by chance), the doctor and writer Steven Novella argued that the political circumstances in China might not be conducive to good science:

> There is also legitimate concern that totalitarian governments do not create an environment in which science can flourish. Science requires transparency, it requires valuing method over results, and it should be ideologically neutral. These are not concepts that flourish under a totalitarian regime. Also, the scientists who get promoted to positions of respect and power are likely to be those who please the regime, by proving, for example, that their cultural propaganda is real. So the selective pressures for advancement do not prioritize research integrity.[100]

Whatever its cause, Chinese scientists would seem to agree that there's a major problem. In one survey of Chinese biomedical researchers in the early 2010s, participants estimated that around 40 per cent of all biomedical articles published by their compatriots were affected by some kind of scientific misconduct; 71 per cent said that the authorities in China paid 'no or little attention' to misconduct cases.[101]

Apart from broad statements about their gender or their country of origin, though, we only have a very vague idea of the profile of the typical scientific fraudster. So, if we can't identify them by their demographics, would knowing their motives help?

Why do fraudsters commit such brazen acts, especially when they have so much to lose? A paper in 2014 found that scientists in the US who went on to be censured for misconduct by the Office for Research Integrity were ones who had been struggling for funds in the years beforehand. This might imply that a desperation for grant money is part of the story; however, the alternative explanation is that their funding dried up during the period that they were under investigation (that is, the misconduct caused the lack of funding rather than the other way around).[102]

Another motive may be the fraudster's pathologically mistaken views on what science is about. The immunologist and Nobelist Sir Peter Medawar has argued, perhaps counterintuitively, that scientists who commit fraud care *too much* about the truth, but that their idea of what's true has become disconnected from reality. 'I believe,' he wrote, 'that the most important incentive to scientific fraud is a passionate belief in the truth and significance of a theory or hypothesis which is disregarded or frankly not believed by the majority of scientists – colleagues who must accordingly be shocked into recognition of what the offending scientist believes to be a self-evident truth.'[103] The physicist David Goodstein agrees: 'Injecting falsehoods into the body of science is rarely, if ever, the purpose of those who perpetrate fraud,' he suggests. 'They almost always believe that they are injecting a truth into the scientific record ... but without going through all the trouble that the real scientific method demands.'[104]

Goodstein was, in part, thinking of a high-profile example from his own discipline. In 2001, the German condensed-matter physicist Jan Hendrik Schön, working at the famed Bell Labs in America, wowed the world by claiming to have invented a carbon-based transistor – the device that controls electronic signals by switching and amplifying current, and that forms the basis for practically any electronic circuit – where the switching action happened within a single molecule.[105] This was far smaller than any transistor that could be made from silicon, the standard material found in microchips, and it promised a dramatic change in the way circuits are built. We could finally begin to build molecular-scale circuits, which would constitute a huge advance for nanotechnology. A Stanford professor enthused that Schön's technique was 'particularly elegant in its simplicity', and he won numerous prizes from scientific societies.[106] The minuscule transistor and other rapid technological advances gave Schön a world-beating publication record: between 2000 and 2002, in addition to many papers in excellent specialist physics journals, he managed to publish nine articles in *Science* and seven in *Nature*. For most scientists, even one paper in either of these ultra-impressive journals would be the pinnacle of a career. There were whispers of a Nobel Prize.

The whispers, however, quickly turned suspicious. Other labs had major difficulty repeating Schön's experiments (which were, like the STAP technique for stem cells, straightforward enough that they could easily be double-checked). Then it was noticed that several of his papers, ones that claimed to be reporting entirely different experiments, showed exactly the same figure for their results.[107] Bell Labs started a detailed investigation, asking Schön for the raw data that supported his findings. But the dog had eaten his homework: Schön informed them that he'd deleted most of his data because his computer 'lacked sufficient memory'.[108] In the data he did provide, the investigation committee was still able to find clear evidence of misconduct. For example, in what was presented as a comparison of two different molecules' currents, he simply doubled all the data for the first molecule then pretended that this result was for the second molecule. When you divided it in half, the data for both molecules were identical down to the fifth decimal place.[109] The investigation found many other examples of duplication, manipulation and full-on fabrication. Once again, a hubristic fraudster had concocted data that were far too perfect to be real.

Particularly interesting for our hunt for the motives of fraudsters, though, is what comes at the end of the lengthy investigation report: a short reply from Schön himself. He pleads that, although he 'made mistakes' and 'realise[s] there is a lack of credibility' he also 'truly believe[s] that the reported scientific effects are real, exciting, and worth working for'.[110] We should of course be extremely wary about taking fraudsters at their word. But Schön's apparent faith in his theory, even after so much of his work had been expunged from the scientific literature (he's currently fifteenth on the Retraction Watch Leaderboard, with thirty-two retractions) speaks to the kind of delusion and self-deception that Medawar and Goodstein noted above.[111] It's impossible to know for sure, but Schön might genuinely have believed in his incredible transistors and seen his rule-breaking as a necessary evil in his quest to bring them to the world's attention.

In his book, Diederik Stapel expressed a similar sentiment, describing how he began to falsify his social-psychology studies after he started getting disappointing results in studies that should, to his mind, have succeeded:

> When the results are just not quite what you'd so badly hoped for; when you know that that hope is based on a thorough analysis of the literature; when this is your third experiment on this topic and the first two worked great; when you know that there are other people doing similar research elsewhere who are getting good results; then, surely, you're entitled to adjust the results just a little?[112]

In this way, fraudsters like Schön and Stapel represent the most extreme end of what we'll encounter again and again in this book: scientists taking liberties to give the 'truth', or what they *want* to be the truth, a helping hand.[113]

Whatever their reasons, scientific fraudsters do grievous and disproportionate damage to science and thereby to one of our most precious human institutions. First, consider the waste of time. Investigating a fraud case can take weeks, months or even years of intense work, especially because the misconduct often metastasizes beyond a single study into many more, each of which needs to be examined with forensic care. Most of the people who do this work aren't specifically employed to do it: they're often busy researchers themselves, who put their own studies on hold to investigate fraud for the greater good. The process doesn't stop with just the gathering of evidence, either. As whistleblowers, data sleuths and anyone else who's contacted a scientific journal or university with allegations of impropriety will tell you, getting even a demonstrably fraudulent paper retracted is a glacial process – and that's if you aren't simply ignored or fobbed off by the authorities in the first place.

To the waste of time, of course, we can add the waste of money. The loss from direct theft, such as the amount of money Woo-Suk Hwang embezzled from his research grants, is dwarfed by the loss of all the money spent on wild-goose-chase research attempting to build on results that were never real in the first place. The obesity researcher Eric Poehlman, for example, the first scientific fraudster to be jailed in America, wasted millions of dollars of taxpayer money in grants from the US government in producing a decade's worth of useless, fabricated data.[114] And how many perfectly innocent scientists have wasted their own research grant money trying to follow or replicate his work, or that of other fraudsters?

Aside from the waste, fraud has a terribly demoralising effect on scientists. As we've seen, one reason that so many frauds manage to infiltrate the literature is that, in general, scientists are open-minded and trusting. The norm for peer reviewers is to be sceptical of how results are interpreted, but the thought that the data are fake usually couldn't be further from their minds. The sheer prevalence of fraud, though, means that we all need to add a depressing option to our repertoire of reactions to questionable-looking papers: someone might be lying to us. Nor is it just other people's papers that require this extra vigilance: fraud can happen on any scientist's own doorstep. Because papers are rarely authored by lone researchers, a fraudulent co-author can sometimes tarnish the reputation of entire teams of innocent colleagues. In many cases the perpetrator is a junior lab member who drags their senior co-authors' names through the mud, as in the case of Michael LaCour's fake gay-marriage canvassing study. Sometimes it goes the other way, with established scientists recklessly jeopardising the careers of their subordinates (the report into Diederik Stapel's fraud noted, for example, that no fewer than ten of his students' PhD theses were reliant on his faked data).[115] And we already saw the ultimate cost of reputational damage in the case of Yoshiki Sasai, who took his own life after finding himself involved in the STAP stem-cell scandal.

Fraud also pollutes the scientific literature. Although retraction noticeably reduces the number of citations a paper gets, often it's not enough: retracted papers still regularly get cited by oblivious scientists.[116] In 2015, researchers followed up the case of the fraudulent anaesthesiologist Scott Reuben, who had twenty papers retracted by 2009 (he's now at number 27 on the Leaderboard, with twenty-four retractions). In the subsequent five years, these papers garnered 274 citations, and in only about a quarter of cases did the citing scientists appear to realise that they had been retracted.[117] An analysis of several other retracted papers found that 83 per cent of post-retraction citations were positive and didn't mention the retraction – these zombie papers were still shambling around the scientific literature, with hardly anyone noticing they were dead.[118] Some journals clearly do a worse job than others at flagging that a paper has been retracted.[119] Whatever its cause, the fact that retracted papers still regularly get cited, and not just to be criticised, means that a great many scientific studies are relying on completely false information. An act of fraud affects far more of the literature than just the original paper presenting the fake results.

The damage isn't just done to science: the effects of fraud bleed out far beyond the journals where the fabrications are published. Practitioners who rely on research, like doctors, can be misled into using treatments or techniques that either don't work or are actively dangerous. An example of the latter is the anaesthesiologist Joachim Boldt, current occupier of the second spot on the Retraction Watch Leaderboard.[120] Boldt fabricated data on hydroxyethyl starch, a chemical sometimes used during trauma surgery as a blood volume expander (the idea was that it could prevent shock after blood loss by helping the remaining blood to circulate). Boldt's faked results made it look as if hydroxyethyl starch was safe for this purpose, a verdict bolstered by the fact that a 'meta-analysis' – a review study that pools together all the previous papers on the subject – reached the same conclusion. This was only true, however, because Boldt's fraud hadn't yet been revealed; the meta-analysis included his fake results as part of its review. When Boldt's deception became known, and his papers were excluded from the meta-analysis, the results changed dramatically: patients who'd been given hydroxyethyl starch were, in fact, more likely to die.[121] Boldt's fraud had distorted the entire field of research, endangering patients whose surgeons, perfectly understandably, took the results at face value.[122]

Among the very worst scientific fraud cases was one that not only misled scientists and doctors, but also had an enormous impact on the public perception of a vitally important medical treatment. It caused so much fear and confusion that even though it occurred more than two decades ago, we still live with its noxious effects today. I'm referring, of course, to the infamous 1998 study on vaccines published in the *Lancet* by the British doctor Andrew Wakefield.[123] On the basis of a sample of twelve children, Wakefield and his co-authors claimed that the combined Measles, Mumps & Rubella (MMR) vaccine was linked to autism. The theory was that measles virus left in the system from the MMR vaccine was a cause of both gut- and brain-related symptoms (Wakefield called the condition, which he was describing for the first time, 'autistic enterocolitis').[124] In interviews and a press conference after the paper's publication, Wakefield repeatedly stated that the MMR should be split up into three individual vaccines, because the combination was 'too much for the immune system of some children to handle'.[125]

By now, most people know that Wakefield's findings have been discredited. Since 1998, there have been several large-scale, rigorous studies showing no relation between the MMR vaccine (or any other vaccine) and autism spectrum disorder.[126] It's also been shown that combination vaccines are just as safe as individual ones.[127] What many *aren't* aware of, though, is that the Wakefield paper, far from being an honest mistake or an understandable dead-end in a tentative line of research, was fraudulent right from the beginning.[128]

After the study's publication and the attendant contro-versy, the investigative journalist Brian Deer began to dig into Wakefield's data and, crucially, his motivations. In a series of stunning articles in *The BMJ* (formerly the *British Medical Journal*), Deer described how Wakefield misrepresented or altered the medical details of every single one of the twelve children included in his paper.[129] He simply invented the 'fact' that all the children showed their first autism-related symptoms soon after receiving the MMR whereas in reality, some had records of symptoms beforehand, others only had symptoms many months afterwards, and some never even received a diag-nosis of autism at all.[130] As for the motivation, Deer showed, Wakefield had two major financial interests in the research turning out the way it did.[131] First, he was being retained, on a substantial fee, by a lawyer who had plans to sue the mak-ers of the vaccines on behalf of the parents of children with autism.[132] Indeed, an anti-vaccine pressure group linked to this lawyer was how Wakefield recruited the patients for his study. Second, the year before the study's publication, he had applied for a patent for his own single measles vaccine and would thus have profited had his research frightened people away from the combined MMR.[133] Inexcusably, neither of these interests were disclosed in the paper: it merely noted that the work was funded by 'Special Trustees', and that the parents of the children 'provide[d] the impetus for these studies'.[134]

When Deer first contacted the *Lancet* with his concerns, in 2004, he encountered fierce resistance from the editors of the journal (which, we might note, foreshadowed the same thing happening at the same journal in the case of Macchiarini around a decade later).[135] As a result of Deer's investigation, Wakefield's paper was finally retracted in 2010, having by this point been a part of the official scientific literature for twelve years. Wakefield was also struck off the UK General Medical Council's register after the longest hearing in its history, banning him from practising as a doctor in the UK. This wasn't just for falsifying data, but also for subjecting a child to unnecessary medical procedures, including a colonoscopy, without proper approval.[136] The Council described him as acting with 'callous disregard', an apt description of the actions of all the fraudsters we've met in this chapter.[137] He has since found fame in the anti-vaccination movement in the US; for example, before it was pulled due to public outcry, his anti-vaccine film, *Vaxxed*, almost had a showing at New York City's 2016 Tribeca Film Festival, having been championed by the festival's founder, and vaccine sceptic, Robert De Niro.[138]

Wakefield's fraud ignited a vaccine scare that spread as fast as any virus. Many sections of the UK media began to run 'just-asking-questions'-style articles about the MMR, seeding doubts among parents. The *Daily Mail* was a ringleader, and even *Private Eye* magazine, which prides itself on not getting carried away on media bandwagons, produced a special issue on MMR in 2002, painting Wakefield as a Galileo-esque figure standing up against the combination-vaccine establishment.[139] The coverage had just the effect you'd expect.[140] In the late 1990s, the UK's MMR acceptance rate had been climbing towards the 95 per cent that's necessary for 'herd immunity' – the state where enough people are vaccinated that a disease becomes rare enough so as not to threaten those who, for reasons like allergies to the vaccine's ingredients, can't be inoculated against it. Post-1998, however, acceptance plunged as low as 80 per cent, and rates of measles correspondingly jumped upwards.[141] Outbreaks began across Europe and the wider world, and countries that had for a time been measles-free began to see new cases. The most recent estimate from the World Health Organisation is that over 140,000 people died from measles and its complications in 2018 – particularly tragic and frustrating since this is a disease we know how to prevent.[142] It's not too far a leap to say that the world is a substantially more dangerous place – especially for children and other vulnerable people, and for those in developing countries – as a consequence of Wakefield's scientific fraud.

The acceptance of Wakefield's study in as prominent an outlet as the *Lancet* will go down as one of the worst decisions in the history of scientific publication. There could hardly be a more concrete example of the importance of reliable science for society's wellbeing, and of the failure of the peer-review system to screen out bad research. It brings us back once again to the issue of trust – this time, the public's trust in science. Having your child vaccinated is an act of commission: you're actively having something done to them, trusting that the medics are right that it's safe.[143] If a study in a famous journal with the seal of approval of scientific peer review implies that it's not, it's only rational to take notice. For years after the publication, people genuinely didn't know who to trust about vaccines. Many still don't.[144]

This betrayal of public trust is perhaps the most pernicious thing about fraud. People have invested a great deal in science, up to and including the health of their children. Fraudsters make a mockery of that trust. Although the fraudsters themselves are obviously to blame, our scientific system also deserves condemnation. Not only do the most glamorous journals encourage people to send only the flashiest findings – more or less guaranteeing that some small fraction of scientists will turn to deception to achieve such flashiness – but journal editors often act with reluctance and recalcitrance when even quite solid evidence of wrongdoing comes to light.[145] Universities, too, shouldn't escape criticism: not only are their investigations often sluggish, but we've seen cases like that of Macchiarini, where they defended the indefensible and actively went after the whistleblowers.[146] Naturally, anyone accused of fraud should be considered innocent until proven guilty; there's no better way to ruin trust than to throw around allegations haphazardly. But the longer those in positions of responsibility drag their feet, and the longer each faked paper exists in the literature, the more our systems and institutions fail science – and by extension, society.

*

As if all the stories of the contamination of the scientific literature, the waste, the corrosive effect on trust and even the deaths aren't bad enough, here's that terrifying thought again: *these are just the ones we know about.* Could it be that the fraudsters we haven't caught are smarter, more cunning and more dangerous than any of those covered in this chapter? Many scientific fraudsters, after all, are exposed not by data sleuths or by scanning papers for faked images, but by whistleblowers who just happened to be in the right place at the right time to spot something suspect. It's overwhelmingly likely that some fraudsters exist who have evaded such detection, who have concealed their misdeeds more effectively and produced fake science that doesn't raise any questions. It's entirely possible that we'll never find them.[147]

Why begin our foray into science's many problems with fraud? Shouldn't these stories, which are often so dramatic and upsetting, be the culmination of our journey into the worst that science can offer? That might seem natural at first. But as scary as these stories of fraud are, what I'm about to describe in the next chapter is in some ways much worse. And it's worse precisely because it's *not* as flagrant or as eye-catching as faking images or making up data. Unlike fraud, which is a clear, unforgivable break from the business of science, what we're about to see is something that shades into good and honest scientific intentions. It's a problem that's far subtler, far more insidious and, worst of all, far more widespread.

4

Bias

An adopted hypothesis gives us lynx-eyes for everything that
confirms it and makes us blind to everything that contradicts it.

Arthur Schopenhauer, *The World as Will and
Presentation* (1818)

Science ... commits suicide when it adopts a creed.

T. H. Huxley, 'The Darwin Memorial' (1885)

Samuel Morton, the noted American physician and scientist,
published a series of lavishly illustrated books in the 1830s and
1840s that described his measurements of hundreds of human
skulls from all over the world.[1] His method involved filling
up the skull cavities with mustard seeds (and later, lead shot),
then inferring how large the brain inside the skull must have
been from the number of seeds or pellets he could cram inside.[2]
He concluded from his collection that the skulls of Europeans
were more capacious than those of Asian, Native American and
African people and argued that these differences showed the
varying 'mental and moral faculties' of the different groups.[3]
Morton's books, in which he also discussed his far-fetched
theories about the entirely separate origins of different human
races, were an international sensation and played a key role in
the rise of scientific racism, the movement that attempted to

split humans into a hierarchy of superior and inferior groups and helped fuel some of the worst horrors of the nineteenth and twentieth centuries.

Along with the average differences by group, Morton provided copious data on his measurements of most of the skulls. This degree of transparency was unusual for the time and it allowed future researchers to take another look at his data. In 1978, by which time Morton and his theories had largely been forgotten, the palaeontologist Stephen Jay Gould did just that.

Morton's analysis of the skulls, wrote Gould, suffered from a variety of inconsistencies. He split up the groups arbitrarily, for instance reporting results for some subgroups of the White skulls, all of which had high averages, but not doing the same for some subgroups of Native Americans who also had large skulls. In some groups he unfairly included more males, who we know have larger heads since they have larger bodies in general, illegitimately raising their average. He double-checked slip-ups in the calculations for some groups, but not others. There was also a discrepancy between the measurements made with seeds and those made with the more reliable lead shot – and this seed-shot discrepancy was larger for Black and Native American skulls than it was for Whites, implying that the seed mismeasurement occurred selectively. Gould later suggested a 'plausible scenario' for how this happened:

> Morton, measuring by seed, picks up a threateningly large black skull, fills it lightly and gives it a few desultory shakes. Next, he takes a distressingly small Caucasian skull, shakes hard, and pushes mightily at the foramen magnum [the hole at the base of the skull through which the spinal cord enters] with his thumb. It is easily done, without conscious motivation; expectation is a powerful guide to action.[4]

In so doing, Morton would have made the skulls from the White populations appear larger than those from non-Whites. Indeed, all his errors moved the results in that same direction. The

mistakes, as Gould put it, reflected 'the tyranny of prior preference': that is, Morton's assumptions about White superiority.[5] If
you analysed the data properly, there were only tiny differences
between the skulls of the ethnic groups – certainly nothing
upon which to build a racial hierarchy. This was no isolated
story. The same lessons about the effects of bias, Gould said,
likely applied right across science: 'I suspect that unconscious
or dimly perceived finagling, doctoring, and massaging are
rampant, endemic, and unavoidable in a profession that awards
status and power for clean and unambiguous discovery'.[6]

Gould was absolutely correct. Since he wrote those words
in the 1970s, it's become ever clearer that scientists regularly
run their studies in ways that stop short of conscious fraud,
but that weight the odds heavily in their favour. And although
we'll return to scientists' ideological biases below, political
views, including the sorts of racial prejudices Gould ascribed to
Morton, are not the main focus of this chapter. The biases that'll
chiefly concern us have to do with the process of science itself:
biases towards getting clear or exciting results, supporting a pet
theory, or defeating a rival's argument. Any one of these can
be enough to provoke unconscious data-massaging, or in some
cases, the out-and-out disappearance of unsatisfactory results.

The irony is palpable. Science, as we've discussed, is considered the closest we can come to objectivity: a process that
can overcome individual biases by subjecting everyone's work
to review and scrutiny. But by focusing too much on this ideal
of science as an infallible, impartial method, we forget that in
practice, biases appear at every stage of the process: reading
previous work, setting up a study, collecting the data, analysing the results and choosing whether or not to publish.[7] Our
tendency to overlook these biases turns the scientific literature,
which should be an accurate summary of all the knowledge
we've gained, into a very human amalgam of truth and wish-
ful thinking.[8]

We'll start this chapter by looking at a kind of bias that
affects the scientific literature as a whole. Then we'll zoom

in on how bias can affect the results of individual studies. To get there, we'll need to take a brief detour into statistics, to see how they're used, abused and misunderstood by scientists in the analysis of their data. Finally, we'll examine the many forces, both internal and external, that push scientists away from the truth.

There's an age-old philosophical question that goes: 'Why is there something rather than nothing?' We can pose a similar query about the scientific process: 'Why do studies always find something rather than nothing?' Reading the science pages in the newspaper, one could be forgiven for thinking that scientists are constantly having their predictions verified and their hypotheses supported by their research, while studies that don't find anything of interest are as rare as hens' teeth. That's understandable: the newspapers are supposed to be 'news', after all, not 'the record of absolutely everything that's happened'. The scientific literature, on the other hand, *is* supposed to be the record of absolutely everything that's happened in science – but it shows just the same bias towards new and exciting stories. If one looks through the journals, one finds endless positive results (where the scientists' predictions pan out or something new is found) but very few null results (where researchers come up empty-handed). In just a moment we'll dive into the technical, statistical definition of 'positive' versus 'null' results. For now, it's enough to know that scientists are usually looking for the former and are disappointed to end up with the latter.

Research has quantified just how positive the scientific literature is: the meta-scientist Daniele Fanelli, in a 2010 study, searched through almost 2,500 papers from across all scientific disciplines, totting up how many reported a positive result for the first hypothesis they tested. Different fields of science had different positivity levels. The lowest rate, though still high, at 70.2 per cent, was space science; you may not be surprised to discover that the highest was psychology/psychiatry, with positive studies making up 91.5 per cent of publications.[9] Reconciling this astounding success rate with psychology's replicability rate is tricky, to say the least.[10]

You might wonder why we shouldn't expect a fairly high success rate for scientific studies. After all, scientists have background knowledge of their field and hypotheses are usually educated guesses, rather than random stabs in the dark. But unless scientists are genuinely psychic, we'd hardly expect to see the levels of positivity reported by Fanelli. Where are all the blind alleys, the great ideas that didn't quite work out when put to the test? Where is all the trial and error? Where, for that matter, are all the false negatives, the studies that failed to find the expected result by mere bad luck, even though the scientists' hypothesis was correct? In other words, the proportion of positive results in the literature isn't just high, it's unrealistically high.[11]

There's a straightforward, but devastating, reason for this persistent positivity: scientists choose whether to publish studies *based on their results*. In a perfect world, the methodology of a study would be all that matters: if everyone agrees it's a good test of its hypothesis, from a well-designed piece of research, it gets published. This would be a true expression of the Mertonian norm of disinterestedness, where scientists are supposed to care not about their specific results (the very idea of having a 'pet theory' is an affront to this norm) but just the rigour with which they're investigated.

That's far from the reality, however. Results that support a theory are written up and submitted to journals with a flourish; disappointing 'failures' (which is how null results are often seen) are quietly dropped, the scientists moving on to the next study. And it isn't just the researchers themselves: journal editors and reviewers also make the decision to accept and publish papers according to how interesting the findings appear, not necessarily how meticulous the researchers have been in discovering them. This feeds back to the scientists themselves and the whole thing becomes a vicious circle: why bother submitting your null paper for publication if it has a negligible chance of being accepted?

This is *publication bias*. It's also known, now anachronistically, as the 'file-drawer problem' – since that's where scientists are said to be keeping all their null results, hidden from the eyes of the world.[12] Think of it as 'history is written by the victors', but for scientific results; or think of it as 'if you don't have anything positive to publish, don't publish anything at all'.

To understand how publication bias plays out in practice, we need to take a closer look at how scientists decide what's a 'positive' or a 'null' result. That is, how data are analysed and interpreted. We return to the idea we encountered in the previous chapter when discussing phony datasets: *numbers are noisy*. Every measurement and every sample comes with a degree of random statistical variation, of measurement and sampling error. This is not just hard for a human to fake – it's also hard to distinguish from the signal for which scientists are looking. The noisiness of numbers constantly throws up random outliers and exceptions, resulting in patterns that might in fact be meaningless and misleading. You might, for example, see an apparent difference in reported pain between the group taking your new drug and the control group taking a placebo, even though the difference is due entirely to chance. Or you might see what appears to be a correlation between two measurements that's merely a fluke in your dataset, which wouldn't appear again if you ran a replication study. Or you might think you see an energy signal in your particle accelerator that turns out to be due to random fluctuations. How do you tell the difference between the effect you're interested in and the vagaries of chance and error? The answer, for the vast majority of scientists, is: calculate a p-value.

Where does this p-value (short for 'probability value') come from? For example, imagine we want to test the hypothesis that Scottish men are taller than Scottish women. Of course, in reality we know it's true: on average, men are taller than women across the world. But we also know that not every man is taller than every woman; all of us can think of individual cases that are just the opposite.[13] Let's pretend, though, that we genuinely didn't know if there was a height difference between men and women in Scotland as a whole. Scotland only has a population of 5.5 million, but we still can't realistically measure every single one of those people, so for our study we'll draw a random sample of a more manageable size. Let's say we don't have much funding for this study, so we can only afford to sample ten men and ten women. Here's where the noise comes in. Because height varies quite a bit across individuals, we might, by chance – or more specifically, as we've learned, by *sampling error* – end up with a group of unusually tall women and a group of unusually short men. Not only that, but because we can never fully eliminate *measurement error*, we won't get the height of every single person exactly correct (recall our discussion in the last chapter of people slouching, measuring tapes slipping, and so on).

Let's say we found that the women in our sample were on average 10 cm shorter than the men.[14] How can we know if this result reflects a genuine difference in the population (meaning we picked up on a true signal), or if it's just noise (meaning all we're seeing is random chance)? We need to compare the two groups formally with a statistical test. There are endless types of statistical tests, such as Z-tests, *t*-tests, chi-squared tests and likelihood ratios; the choice depends, among other things, on the kind of data you're looking at. Essentially all of them are done these days by feeding your data into computer software. When you run one of these programs, its output will include, alongside many other useful numbers, the relevant *p*-value.[15]

Despite being one of the most commonly used statistics in science, the *p*-value has a notoriously tricky definition. A recent audit found that a stunning 89 per cent of a sample of introductory psychology textbooks got the definition wrong; I'll try to avoid making the same mistake here.[16] The *p*-value is the probability that your results would look the way they look, or would seem to show an even bigger effect, if the effect you're interested in weren't actually present.[17] Notably, the *p*-value doesn't tell you the probability that your result is true (whatever that might mean), nor how important it is. It just answers the question: 'in a world where your hypothesis *isn't* true, how likely is it that pure noise would give you results like the ones you have, or ones with an even larger effect?'[18]

Let's say that in the case of our height study, the *p*-value is 0.03. This means that, if there was in reality no height difference between men and women in the Scottish population, and we were to draw an infinite number of samples like ours, in only 3 per cent of cases would we see a height difference of 10 cm or more. In that 3 per cent of cases, we'd be making a mistake if we declared that Scottish men are taller on average than Scottish women. To put it another way, finding a height difference as large as, or larger than, the one in our sample would be quite unlikely (but not impossible) if men and women in Scotland were actually the same height.

And so, in most cases, the lower the *p*-value, the better. But how low does *p* have to be before we're confident our result isn't due to noise? Or, looking at it a different way, how high a probability of making a false-positive error (where we make a mistake by saying there's an effect when there isn't) should we tolerate?[19] To help scientists make a decision, the pioneering statistician Ronald Fisher suggested in the 1920s that a threshold be set, above which the associated result would be considered null (because it bears too much resemblance to what we'd see if there was actually nothing going on), and below which it would be regarded as 'statistically significant'.

That term has led to enormous confusion. To our modern ears, 'significant' sounds like it means an effect that's big or important in some way. But as we just saw, that's not what a *p*-value, however low, really means. The *size* of an effect (representing, for example, how much taller Scottish men are than Scottish women; in our case the effect size is 10 cm) is a different thing from the probability of seeing these results by chance if your hypothesis isn't true. It's perfectly possible, for example, that a drug has a very minor impact on an illness, but one that you're

quite sure isn't a false positive – a small yet statistically significant effect. Back when Fisher was writing, people understood
the word 'significant' somewhat differently: it implied that the
result 'signified' that something was happening in the data, not
that whatever was happening was necessarily noteworthy.[20]

In any case, Fisher originally suggested that the 'statistically
significant' threshold should be set at 0.05 – implying that we
should tolerate no more than a 5 per cent chance of a false-
positive error with each test (note that this would mean that
our height study, with its p-value of 0.03, would have had
a statistically significant result). In a highly influential 1926
paper, Fisher wrote that 'a scientific fact should be regarded as
experimentally established only if a properly designed experiment *rarely fails* to give this level of significance.'[21]

The 0.05 level is fairly arbitrary. It's a bit like the magnificent
Scottish website taps-aff.co.uk, which checks the weather for all
areas of the country and automatically declares any area where
the temperature is above 17 degrees Celsius (about 63 degrees
Fahrenheit) as 'taps aff' – meaning it's warm enough that
gentlemen can legitimately walk around topless outdoors.[22] 17
degrees is reasonable, but arbitrary: perhaps some would never
consider baring their torsos unless it was above 20 degrees;
perhaps hardier souls would do so at 15 degrees. Along the
same lines, Fisher later noted that some researchers might like
to set their significance criterion differently depending on what
they were testing.[23] The '5 sigma evidence' that CERN physicists
famously discussed after the discovery of the Higgs Boson in
2012, for example, was just a fancy-sounding way of talking
about the extremely stringent p-value threshold they used for
such a crucial result: '5 sigmas' corresponds to a p-value threshold of about 0.0000003.[24] Having poured vast resources into
constructing the Large Hadron Collider, the physicists *really*
didn't want to be hoodwinked by noise in their data, so they
set a very high standard for the evidence to pass.

Outside of exceptions such as the Higgs Boson, though,
the 0.05 threshold remains, through conformity, tradition and

inertia, the most widely used criterion today. It has scientists feverishly rifling through their statistical tables, checking for p-values lower than 0.05 so that they can report their results as being statistically significant. It's easy to forget the arbitrariness. Richard Dawkins has bemoaned the 'discontinuous mind': our human tendency to think in terms of distinct, sharply defined categories rather than the messy, blurry, ambiguous way the world really is.[25] One example is the debate around abortion, which often fixates on the question of when an embryo or foetus becomes 'a person', as if there could ever be a bright line by which we could make that decision. In Dawkins's own field of evolutionary biology, pinpointing the exact moment when one species evolves into another is likewise a fool's errand, no matter how satisfied we might feel if we could do so. It's the same for p-values: the 0.05 cut-off for statistical significance encourages researchers to think of results below it as somehow 'real', and those above it as hopelessly 'null'. But 0.05 is as much a convention as the 17-degree taps-aff rule – or, slightly more seriously, the societal decision that someone legally becomes an adult precisely on a particular birthday.

Before we embarked on that somewhat complicated (but necessary) statistical diversion, we learned about publication bias – the tendency of scientists to publish only positive results and hide away the nulls. Now we know how they usually make that decision: 'significant' results below the sacred 0.05 p-value threshold are excitedly submitted for publication, while those above it are file-drawered. Conflating Fisher's arbitrary statistical cut-off and the 'realness' or the importance of a result has had baleful consequences for the scientific record.

We sometimes see the characteristic traces of publication bias when we zoom out to look at a whole segment of scientific literature. This zooming-out often takes the form of a meta-analysis, which by combining results from multiple studies can calculate the overall effect (sometimes, perhaps tempting fate, called the 'true' effect) on a given topic. This could be, say, the overall impact a vaccine has on reducing mortality from a disease, or the overall link between climate change and crop yields.[26]

When gathering the relevant research, meta-analysts pay close attention to two numbers. The first is the *effect size*. For our two examples: does the vaccine reduce mortality by just a few cases per year (a small effect), or could it save thousands of lives (a large effect)? Does climate change have a small, manageable effect on crops, or a large, ruinous one? We know that because of sampling error and measurement error, different studies can produce wildly different estimates of effect size, so it's unwise to rely on the estimate from a single one. Because having more evidence on a question is usually better – and because the random fluctuations caused by error should cancel out across different samples – the overall effect size calculated in a meta-analysis is normally considered more reliable than the estimates from individual studies.

A meta-analysis doesn't calculate the overall effect simply by averaging the effect sizes reported by all the included studies. It also takes into account the second number that's of interest to meta-analysts: the *sample size*. All else being equal, large studies, because they contain more data, are expected to get closer to the 'true' effect (the average effect in the whole population). In other words, the large studies' best guess at the true effect will tend to be more accurate than that of small studies.[27] In our Scottish study with ten men and ten women, you can easily imagine how we might accidentally pick an unrepresentative sample of unusually short men or unusually tall women, and end up drawing the wrong conclusion. But imagine we sampled 1,000 men and 1,000 women: clearly, the risk of accidentally picking 1,000 people with an unusual characteristic is far lower than just picking ten. That general pattern holds in most situations: small studies, which are more limited snapshots and are more affected by sampling error, will have more variability, over- and under-estimating the true effect by wider margins.

A meta-analysis therefore gives more weight to the effect sizes from big studies, because they're likely to be more accurate.[28]

In the context of publication bias, what we're interested in is how the effect size and the sample size relate to one another. If you plot one versus the other, with one dot per study, you'd expect your graph to look something like Figure 2A below. (Note that this is an idealised version of a meta-analysis; real datasets almost never look this clear-cut.) Looking at this 'funnel plot' (so named for what are hopefully obvious reasons), you can see how all the smaller studies, towards the bottom of the y-axis, fluctuate widely; as you progress up the y-axis, to bigger studies, the dots begin to cluster around the average effect, illustrating what we've just discussed about larger studies being more precise. The variation on the x-axis is why it's a bad idea to take for granted the effect from any individual study: even though there is a real effect in this example, individual

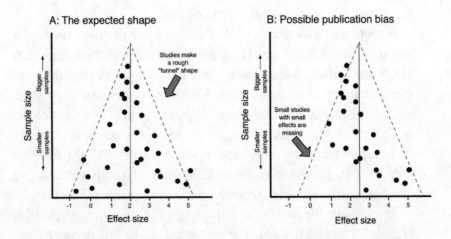

Figure 2. Funnel plots from an imaginary meta-analysis, in two different scenarios. In scenario A, the distribution of the thirty studies is about what you'd expect if every study ever done on the topic had been published. In scenario B, the six studies from the bottom-left section (studies with small samples and small effects) are missing – a pattern that might signal publication bias. The vertical line in the middle of each graph is the overall effect size calculated by each meta-analysis. In the case of scenario B, it's been shifted to the right, meaning that the meta-analysis is coming up with a bigger effect than it should.

studies have under- and over-shot its 'true' size by varying degrees (though the biggest studies do an admirable job). In any case, nothing seems to be missing here: the upside-down funnel shape is just what we'd expect if all the studies had converged upon a real effect.

Just as in an archaeological dig, where the *absence* of particular objects tells you interesting things about the historical people you're investigating – for instance, a lack of weapons might mean they were civilians rather than soldiers – we can learn a lot from what we don't see in a meta-analysis. What if our plot looks more like Figure 2B? Here, we've lost a chunk from the expected shape. The studies we'd expect to see in the lower left of our funnel, which had small sample sizes as well as small effects, are missing. Thinking like an archaeologist, a meta-analyst might infer that those studies were done, yet instead of being published, were file-drawered. Why? A likely explanation is that these small-sample, small-effect studies had *p*-values higher than 0.05 and were dismissed as unimportant nulls.

Perhaps the scientists who ran these studies thought something like: 'well, it was only a small study, and the small effect I found is probably just due to noisy data. Come to think of it, I was silly even to *expect* to find an effect here! There's no point in trying to publish this.' Crucially, though, this post hoc rationalisation wouldn't have occurred to them if the same small-sample study, with its potentially noisy data, happened to show a large effect: they'd have eagerly sent off their positive results to a journal. This double standard, based on the entrenched human tendency towards confirmation bias (interpreting evidence in the way that fits our pre-existing beliefs and desires), is what's at the root of publication bias.

If you consider the overall conclusions of a meta-analysis based on Figure 2B rather than 2A, you can see how publication bias deranges the scientific literature. If the studies with small effects have been removed from the funnel shape, the overall effect that shows up in the meta-analysis will by definition be larger than is justified. We get an exaggerated view of the importance of the effects and can be misled into believing something exists when it doesn't. By failing to publish null or ambiguous studies, researchers force blinkers onto anyone who reads the scientific literature.

One of the most striking recent funnel plots was featured in a meta-analysis by the psychologist David Shanks and his colleagues.[29] It examined yet another variation on the priming theme: 'romantic priming'. This is the idea that after being shown a picture of an attractive woman, men are willing to take more risks and to spend more on consumer goods (engaging in 'conspicuous consumption' to attract partners). Fifteen published papers, documenting forty-three separate experiments, seemed to support this hypothesis. However, when those studies were plotted for the meta-analysis, the funnel had a huge chunk missing: compelling evidence that many of the studies that hadn't found the effect had been nixed prior to publication. Indeed, when Shanks and his colleagues tried to replicate romantic priming in large-scale experiments of their own, their study found no effect from it whatsoever, with all their replication effect sizes converging around zero.

Publication bias appears no less prominent in medicine. An analysis in 2007, for instance, found that over 90 per cent of articles describing the effectiveness of prognostic tests for cancer reported positive results. In reality, we still aren't particularly good at predicting who'll get cancer, which should tell you that this literature was missing something.[30] Another study, which looked at forty-nine meta-analyses of potential markers of cardiovascular disease (for example, certain proteins in the blood that might be more common in those who are at risk of having a heart attack), found that fully thirty-six of them showed evidence of a bias towards positive results.[31] The published literature, in other words, appeared decidedly to inflate the apparent usefulness of these bio-markers.

It's the same for treatments. Doctors prescribing drugs to patients need to balance the benefits against the costs: for example, deciding whether antidepressants are worth taking despite their common side effects such as nausea and insomnia. If the medical literature gives doctors an inflated view of how much benefit a drug provides (as indeed appears to have been the case for antidepressants, which do seem to work, but not with as strong an effect as initially believed), their clinical reasoning will be knocked off track.[32]

If you hadn't heard of publication bias before now, it would be perfectly understandable: it is one of science's more embarrassing secrets. But a 2014 survey of reviews in top medical journals found that 31 per cent of meta-analyses didn't even check for it. (Once it *was* properly checked for, 19 per cent of those meta-analyses indicated that publication bias was indeed present.)[33] A later review of cancer-research reviews was even worse: 72 per cent didn't include publication bias checks.[34] It's often hard to know exactly what to do when you find hints of publication bias in your meta-analytic dataset – should you revise the estimate of the average effect downwards? If so, by how much?[35] – but it's doubtful that the proper answer is to ignore the issue entirely.

The trouble with the archaeological approach to publication bias is that it relies on conjecture to fill in the gaps in the funnel plot – those places where we would expect the small studies with small effects to appear. Funnel plots can have weird shapes for reasons other than publication bias, especially if there are a lot of differences between the assorted studies that go into the meta-analysis.[36] There are many cases where publication bias is more subtle, and thus harder to discern, than in those described above. Are there better ways to check for this kind of bias?

One alternative approach would be to take a set of studies you know for sure were completed, with a range of results from strongly positive to null, then check how many of each type made it through to get published. That's exactly what a group of Stanford researchers, led by political scientist Annie Franco, did in 2014, in a project they described as 'unlocking the file-drawer'.[37] They looked at studies whose authors had applied to a central government programme that facilitates survey research.[38] Using the successful applications between 2002 and 2012 as a log of which of these studies had gone ahead, Franco and her colleagues checked what happened with each one, contacting their authors directly if necessary. In all, it turned out, 41 per cent of the studies that were completed found strong evidence for their hypothesis, 37 per cent had mixed results, and 22 per cent were null. In a world where the methods, not the results, were what mattered, the percentages among the published articles would be similar. But they were nowhere near. Of the articles that were published, the percentages for strong, mixed, and null results were 53 per cent, 37 per cent and 9 per cent, respectively. There was, in other words, a 44-percentage-point chasm between the probability of publication for strong results versus null ones.[39]

From the scientists in question, Franco and her colleagues learned that 65 per cent of studies with null results had never even been written up in the first place, let alone sent off to a journal. Many of those scientists predicted they'd have no chance of publication. 'The unfortunate reality of the publishing world [is] that null effects do not tell a clear story,' said one. 'Given the discipline's strong preference for [positive results], I haven't moved forward with it,' said another. Many had also simply shrugged and moved onto the next project, with one admitting that 'I never got to [write up the study] due to a combination of time constraints, some loss of interest ... and lack of earth-shattering results.'[40] In other words, the file drawer is real.[41]

In business and politics, one of the great villains is the Yes Man. Countless books tell aspiring managers and leaders to be careful never to surround themselves with people who'll just nod along with all their decisions, even the bad ones. Winston Churchill proclaimed that 'the temptation to tell a chief in a great position the things he most likes to hear is one of the commonest explanations of mistaken policy. Thus the outlook

of the leader on whose decisions fateful events depend is usually far more sanguine than the brutal facts admit.'[42] In science,
publication bias makes Yes Men out of the articles that do get
published: we can see all the positives, but we don't get to
see the brutal null results. Making decisions on such partial
information is a recipe for disaster.

Last but not least, there's a moral case against publication
bias. If you've run a study that involved human participants,
particularly if it's one where they've taken a drug or under-
gone an experimental treatment, it could be argued that you
owe it to those participants to publish the results. Otherwise,
all the trouble they went to (including, in some cases, painful
procedures or side effects) was for nothing. A similar argument
applies to research you've done with someone else's money.

From every perspective – scientific, practical and ethical –
publication bias is thus a major problem. Unfortunately, it's
far from the only problem caused by science's persistent, deep-
rooted bias toward positive results.

For an aspirational, career-driven scientist, publication bias has
a major downside: hiding null results in the file drawer means
you won't get that all-important publication or that gratify-
ing extra line on your CV. To avoid that loss, and save your
results, you have another option: data manipulation. This isn't
the kind of outright fabrication and forgery that we saw in
the previous chapter. Rather, this is a sort of unconscious, or
semi-conscious, massaging of data – 'dimly perceived finagling',
to use Gould's words – into which scientists can fall entirely
innocently. Indeed, what's scary about data manipulation is not
only that it leads to false conclusions entering the literature, but
that so many scientists either do it completely unwittingly or, if
they're aware that they're doing it, are oblivious as to why it's
wrong.

I imagine you've heard of the idea that if you take a larger
plate to the buffet, you'll eat more than you otherwise would

have. Which means you've indirectly heard of Professor Brian Wansink. The head of Cornell University's Food and Brand Lab, Wansink was for a long time easily the world's most prominent voice on food psychology, with popular books, a two-year directorship of the US Department of Agriculture's Center for Nutrition Policy and Promotion under President George W. Bush, and hundreds of published papers, many of which were cited as part of the Obama-era 'Smarter Lunchrooms' movement in American schools.[43] He even won an Ig Nobel Prize – the fun version of the Nobels for studies that 'make people laugh, then think'. It was for a study in 2007 where he tricked people into consuming much more soup than they'd intended, by using a rigged soup bowl that continually refilled itself.[44] Just like the bottomless bowl, Wansink's ability to come up with quirky, eye-catching results on food psychology appeared to have no limit. As well as his portion-size research, he found that if you go food-shopping while hungry you'll buy more calories; that the eyes of characters on packets of sugary cereal tend to look downwards, so as to make eye contact with small children standing in front of the supermarket shelf; and that adding a sticker of Elmo from *Sesame Street* to an apple makes children more likely to choose it instead of a cookie.[45]

It wasn't until late 2016 that it all fell apart. Wansink wrote a blog post about how he'd encouraged one of his graduate students to analyse a dataset he'd collected in a New York pizza restaurant.[46] The original hypothesis, he said, had 'failed', but instead of publishing those null results or file-drawering the whole thing, he told the student that 'there's got to be something [in the dataset] we can salvage'. She complied, and 'every day … came back with puzzling new results … and every day … c[a]me up with another way to reanalyse the data'. By openly admitting to dredging through his datasets looking for anything that was 'significant', Wansink had unintentionally revealed a major flaw in the way he, and unfortunately many thousands of other scientists, conducted research.

That flaw has been dubbed 'p-hacking'.[47] Because the $p < 0.05$ criterion is so important for getting papers published – after all, it supposedly signals a real effect – scientists whose studies show ambiguous or disappointing results regularly use practices that ever so slightly nudge, or hack, their p-values below that crucial threshold. Such p-hacking comes in two main flavours. In one, scientists pursuing a particular hypothesis run and re-run and re-re-run their analysis of an experiment, each time in a marginally different way, until chance eventually grants them a p-value below 0.05. They can make these impromptu analysis changes in all sorts of ways: dropping particular datapoints, re-calculating the numbers within specific subgroups (for instance, checking for the effect just in males, then females), trying different kinds of statistical tests, or keeping on collecting new data with no plan to stop until *something* reaches significance.[48] The second option involves taking an existing dataset, running lots of ad hoc statistical tests on it with no specific hypothesis in mind, then simply reporting whichever effects happen to get p-values below 0.05. The scientist can then declare, often perhaps convincing even themselves, that they'd been searching for these results from the start.[49] This latter type of p-hacking is known as HARKing, or Hypothesising After the Results are Known. It's nicely summed up by the oft-repeated analogy of the 'Texas sharpshooter', who takes his revolver and randomly riddles the side of a barn with gunshots, then swaggers over to paint a bullseye around the few bullet holes that happen to be near to one another, claiming that's where he was aiming all along.[50]

Both kinds of p-hacking are instances of the same mistake and ironically, it's precisely the one that p-values were invented to avoid: capitalising on random chance. Although p-values are designed to help us sort signal from noise, this ability breaks down when you calculate a large number of them for the same research question. Each time you make some adjustment to your data or statistical test, you're giving yourself another roll of the dice, increasing your chance of picking up on a random fluctuation and declaring it 'real'. As we saw, the $p < 0.05$ threshold means that if our hypothesis is false (our new drug doesn't actually work, for instance), then 5 per cent of the time we'll get a false-positive result – we'll declare that our experiment succeeded when in fact it did not. But that 5 per cent value is for *a single test*. Some straightforward maths shows that in a world where our hypothesis is false, increasing the number of statistical tests snowballs our chances of obtaining a false-positive result.[51] If we run five (unrelated) tests, there is a 23 per cent chance of at least one false positive; for twenty tests, it's 64 per cent. Thus, with multiple tests, we go well beyond the 5 per cent tolerance level that Fisher proposed. Instead of helping us discern our signal, our p-values begin to sink back into the sea of noise.

This concept isn't exactly intuitive, so here's an example using the sort of analogy that statisticians love. Imagine I have a bag of coins, but I'm worried that they're all weighted to come up heads more often than tails. I take one coin out, flip it five times and get heads every time. That would be at least modestly compelling evidence that something strange was going on. But say on that first coin I got heads only three times and tails twice: not great evidence for my theory. Instead of abandoning that theory, though, I could respond by checking other coins – drawing more and more of them from the bag and flipping them until one finally does come up heads five times in a row. Hopefully you'll agree that this is far less convincing than having the very first coin I tested show me all heads. But what if I covered up my multiple testing with a story? 'In fact,' I could say, 'each time I drew a new coin from the bag I was testing a new variation of the hypothesis. The second time I was checking whether the coins are weighted to come up heads only when flipped left-handed. The third time I was checking whether it only happens when the room temperature is above 20 degrees. The fourth time ...' – you get the idea. I might even have convinced myself that I really was testing all these new, interesting hypotheses. But fundamentally, I was still giving myself multiple tries at the same test, thus increasing my chances of a false positive. And once I found one coin that produced the all-important five heads in a row, I might be tempted just to publish the data from that coin alone.

This is the same logic that makes seemingly amazing coincidences, such as 'I randomly thought about someone I hadn't contacted in months, and they sent me a text message at that exact moment!' less amazing, once you consider how many thousands, maybe millions, of people in the world have random thoughts about someone and *don't* receive such a message. A one-in-a million chance, after all, happens quite a few times if your population includes several million people. Increase the number of opportunities for a chance result to arise, and you can bet that eventually it will do so; cherry-picking specific instances from the vast multitudes is no proof that it wasn't just an accident. To be sure, even if you've definitely only run one test with the $p < 0.05$ criterion, you might still get misled by random chance. But this risk is substantially lower than in the case of p-hacking, where running multiple tests multiplies the risk that any one of them is misleading.

That's the fundamental insight that so many scientists don't seem to grasp: even when there's nothing going on, you'll still regularly get 'significant' p-values, especially if you run lots of statistical tests.[52] It's analogous to the 'psychics' who make thousands of predictions about what'll happen in the next year, then at the end of that year highlight only the ones they got correct, making it look as if they have magical forecasting skills.[53] Roll the statistical dice enough times and something will show up as significant, even if it's just a freak accident in your data.[54] Hide away all the times it wasn't significant and you have the perfect recipe to convince people your result is real, even if it's based on little more than noise.

Which brings us back to Brian Wansink, his grad student and the data from the pizza restaurant. Across all that student's daily attempts at analysing the data, although a great many p-values were being calculated, only a few of them were making it into the published literature. Given the unwritten rules of scientific publishing, these values were generally the ones that had $p <$ 0.05. We, as readers of the literature, have no idea how many tests have been done. Since so many results were being hidden from view, this was like publication bias happening within a single study. If we were able to see the whole process, null results and all, it'd look like a classic case of the Texas sharpshooter. In his fateful blog post, Wansink inadvertently made clear that his process involved drawing a target around the results that happened, by chance, to be statistically significant. To those who understood the statistics involved, though, Wansink's target wasn't on the side of a barn: it was on his own foot.

After Wansink's blog post appeared, some sceptical readers started digging into the numbers in his papers.[55] It turned out that the p-hacking was only one of a multitude of statistical screw-ups. Across the four papers he'd published using the pizza dataset, the team of sceptics found no fewer than 150 errors: a whole host of numbers that were inconsistent between the different studies (and sometimes within a single paper).[56] It got worse as researchers started looking at Wansink's other research. In one paper about cookbooks, a re-analysis of the data showed he'd reported almost every single number incorrectly.[57] In the paper about putting Elmo stickers on apples, he'd both misla-belled a graph and misdescribed the methodology of the study.[58] The retractions soon began: at the time of writing, eighteen of Wansink's articles have been pulled from the literature, and one suspects more will follow.[59] Just under two years after his notori-ous blog post, Wansink tendered his resignation to Cornell.[60]

As outrageous as they were, all of Wansink's errors and blunders were a distraction from the most widely applicable lesson of the affair: his p-hacking. Amid the media scandal that arose when the retractions started rolling in, a *BuzzFeed News* journalist published an incriminating and highly revealing email from Wansink to one of his co-authors, sent while they were writing the cursed Elmo paper. In the message, Wansink fret-ted that 'although the stickers increase apple selection by 71%, for some reason this is a p value [*sic*] of .06. It seems to me it should be lower. Do you want to take a look at it and see what you think[?] If you can get the data, and it needs some tweeking [sic], it would be good to get that one value below .05.'[61]

This is a rare instance of a scientist explicitly encouraging their colleagues to p-hack. But it was only remarkable because of its directness. When the Wansink story broke, I suspect many scientists were shifting uncomfortably in their seats, knowing that he was merely on the extreme end of a spectrum they occupied themselves. They might not have Wansink's sloppiness, and they might make their re-analysis requests more delicately in emails (or in person, where there isn't a record).[62] But when the desire to 'get that one value below .05' is strong – and, given journals' clear preference for exciting, flashy, positive results, that's definitely the case – p-hacking is the near-inevitable result.

Wansink's accidental confession ended badly for him. However, when another prominent scientist was forthright in admitting that they'd unwittingly p-hacked their data in the past, the scientific community reacted in heartening fashion. Recall from Chapter 2 the enormous success of the idea of 'power posing', which was based on a 2010 paper that didn't replicate. Although Amy Cuddy is the name that's become synonymous with the concept, she wasn't actually the lead author on the paper. That was Berkeley's Dana Carney, and in 2016 Carney released a statement about her changed views on power posing. Over the years, she had updated her beliefs to the point that, as she put it, 'I do not believe that "power pose" effects are real'. She went on to list some facts about the original experiment (which had a 'tiny' sample size of forty-two, and also an effect size that was 'barely there'), all of which add up to a clear-as-day story of p-hacking:

- They recruited participants 'in chunks and checked the effects along the way' (that is, they kept adding to the sample until they got a significant result).
- A few participants were excluded for arbitrary-sounding reasons.

- Some outlying datapoints were removed, but others were left in.
- Multiple measures were used and multiple statistical tests run, but only those with the lowest p-values were reported.
- Many different questions about self-rated power were asked but the only ones reported were the ones that showed an effect.[63]

'Back then,' said Carney, 'this did not seem like p-hacking' – even though it surely was. Did Carney receive a torrent of vile abuse, or lose her job, for admitting all this? No. In fact, the reaction was the precise opposite. A search of Twitter, often maligned as a haven for online bullies, reveals other researchers (rightly) calling Carney's statement 'brave', 'impressive', 'admirable', 'the way forward', an example of 'how to deal with failure to replicate one's work' and a 'remarkable demonstration of scientific integrity'. One neuroscientist called Carney an 'intellectual/academic hero'.[64] I couldn't find a single negative reaction – except the one from Amy Cuddy herself, who took the opportunity to distance herself from the p-hacking: 'I ... cannot contest the first author's [that is, Carney's] recollections of how the data were collected and analyzed, as she led both.'[65]

How common are the sorts of analytic biases that pervaded Wansink's work, and that undermined the study on power posing? In 2012, a poll of over 2,000 psychologists asked them if they had engaged in a range of p-hacking practices.[66] Had they ever collected data on several different outcomes but not reported all of them? Approximately 65 per cent said they had. Had they excluded particular datapoints from their analysis after peeking at the results? 40 per cent said yes. And around 57 per cent said they'd decided to collect further data after running their analyses – and presumably finding them unsatisfactory.

Similarly disheartening findings come from surveys in other fields. A survey of biomedical statisticians in 2018 found that 30 per cent had been asked by their scientist clients to 'interpret the statistical findings on the basis of expectations, not the actual results' while 55 per cent had been asked to 'stress only significant findings but underreport nonsignificant ones'.[67] 32 per cent of economists in another survey admitted to 'present[ing] empirical findings selectively so that they confirm[ed their] argument', while 37 per cent said they had 'stopped statistical analysis when they had [their] result' – that is, whenever p happened to drop below 0.05, even if it might've done so by chance.[68] If you collect together all the p-values from published papers and graph them by size, you see a strange, sudden bump *just under* 0.05 – there are a lot more 0.04s, 0.045s, 0.049s, and so on, than you'd expect by chance. This isn't knockdown proof, but it's a red flag for p-hacking: scientists appear to tweak studies just enough to limbo their results under the 0.05 line, then send them off for publication.[69]

The thing to remember is that p-hacking can feel to the scientist as if they're making their result, which is in their mind probably true, in some way clearer or more realistic. This is the phenomenon of confirmation bias again. That one participant? I'm *sure* I saw them staring out of the window instead of concentrating on the psychological test we gave them. That one petri dish? Right enough, it *did* have a speck of dirt and might be contaminated, so it's best to drop its results from the dataset. Yes, it *does* make more sense to run statistical test X instead of statistical test Y (and, oh look, it happens that statistical test X gives positive results!). You get the idea. As we discussed in the last chapter when we considered the motives of fraudulent scientists, if you already believe your hypothesis is true before testing it, it can seem eminently reasonable to give any uncertain results a push in the right direction. And whereas the true fraudster knows that they're acting unethically, everyday researchers who p-hack often don't.

There's never just one way to analyse a dataset. Do you delete those outlying datapoints because you reason that they make your sample less representative of the population? Or do you leave them in? Do you split the sample up into separate age groups, or by some other criterion? Do you merge observations from week one and week two and compare them to weeks three and four, or look at each week separately, or make some other grouping? Do you choose this particular statistical model, or that one? Precisely how many 'control' variables do you throw in? There aren't objective answers to these kinds of questions. They depend on the specifics and context of what you're researching, and on your perspective on statistics (which is, after all, a constantly evolving subject in itself): ask ten statisticians, and you might receive as many different answers. Meta-science experiments in which multiple research groups are tasked with analysing the same dataset or designing their own study from scratch to test the same hypothesis, have found a high degree of variation in method and results.[70]

Endless choices offer endless opportunities for scientists who begin their analysis without a clear idea of what they're looking for. But as should now be clear, more analyses mean more chances for false-positive results. As the data scientists Tal Yarkoni and Jake Westfall explain, 'The more flexible a[n] … investigator is willing to be – that is, the wider the range of patterns they are willing to 'see' in the data – the greater the risk of hallucinating a pattern that is not there at all.'[71]

It gets worse. So far, I've made it sound as though all p-hacking is done explicitly – running lots of analyses and publishing only those that give p-values lower than 0.05. This undoubtedly happens a lot, but the true problem is much trickier. It's this: even if you just run one analysis, you still need to consider all the analyses you *could* have run. The statisticians Andrew Gelman and Eric Loken compare the process of doing an unplanned statistical analysis to a 'garden of forking paths', from the Jorge Luis Borges short story of that name: at each point where an analytic decision is required, you might choose any one of the many options that present themselves. Each of those choices, as we've seen, would lead to slightly different results.[72] Unless you've set out very specific criteria for what a result favouring your hypothesis would look like, unless you say that you want 'a $p < 0.05$ with the variables treated *this* way under *these* precise conditions and with *these* controls', then you might end up accepting any one of the many possible results as evidence that you're right. But how do you know the one you ended up with, having followed your unique combination of forking paths, wasn't a statistical fluke? Even without the trial-and-error of classic p-hacking, then, scientists who don't come to their data with a proper plan can end up analysing themselves into an unreplicable corner.

Why unreplicable? Because when they reach each fork in the path, the scientist is being strung along by the data: making a choice that looks like it might lead to $p < 0.05$ *in that dataset*, but that won't necessarily do the same in others. This is the trouble with all kinds of p-hacking, whether explicit or otherwise: they cause the analysis to – using the technical term – *overfit* the data.[73] In other words, the analysis might describe the patterns in that specific dataset well, but those patterns could just be noisy quirks and idiosyncrasies that won't generalise to other data, or to the real world. This is useless. Most of the time we're not interested in the workings of one particular dataset (we don't want to know 'what is the link between taking antipsychotic drugs and schizophrenia symptoms measured in this specific sample of 203 people between April and May 2019 in Denver, Colorado?') – we're looking for generalisable facts about the world ('what is the link between taking antipsychotic drugs and schizophrenia symptoms in humans in general?').

Figure 3, below, illustrates overfitting. As you can see, we have a set of data: one measurement of rainfall is made each month across the space of a year. We want to draw a line through that data that describes what happens to rainfall over time: the line will be our statistical model of the data. And we want to use that line to predict how much rain will fall in each month next year. The laziest possible solution is just to try a straight line, as in graph 3A – but it looks almost nothing like the data: if we tried to use that line to predict the next year's measurements, forecasting the exact same amount of rain for every month, we'd do a terribly inaccurate job. Next, we might try a curved line that goes through the data like in graph 3B, and this turns out to be a decent approximation. This curve would be a useful model for making our subsequent-year prediction. The danger, though, is when we don't stop there. We could go on to graph 3C, where we draw a line that touches every single datapoint, twisting and turning as it goes. This model fits *our* dataset beautifully: it's a perfect description of the points. But what are the chances that the data next year would have those exact same ups and downs? Not high. By fitting our line so closely to the points, we're just modelling the random noise that exists in our dataset. The model overfits the data.

This is what scientists are unwittingly doing when they *p*-hack: they're making a big deal of what is often just random noise, counting it as part of the model instead of as nuisance variation that should be disregarded in favour of the real signal (if a signal exists in the first place). Woe betide anyone who takes a *p*-hacked, overfitted model and tries to replicate it in

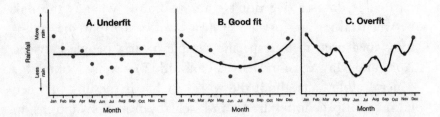

Figure 3. The overfitting problem, illustrated by models of yearly rainfall. Graph A is a poor model – it 'underfits', because it doesn't do a good job of describing the data. Graph B is a far better model, because it describes the phenomenon in a way that probably generalises across several different years. Graph C shows 'overfit': whereas it's a great model of the data from this particular year, there's little chance that other years will show the same exact pattern of ups and downs. Note that these data are just made up for illustration.

a different sample: its results were contingent on the specific forking paths its creators followed through their noisy data, so it'll probably tell us very little about the world beyond that single dataset.

You can see why scientists are tempted by overfitting. If you focus only on your own data and forget that your job is to make general statements about the world, a perfectly fitting model like graph 3C is very alluring: there are no uncertainties, no messy datapoints that evade the line you've drawn. The neatness alone is not what makes it so compelling, though: simply connecting the dots in the graph doesn't require any scientific knowledge. But a paper that sounds as if you had come up with the specific shape of the line (your theory) *before* collecting the data? Now you've got the scientific world's attention – and as we know, a major goal in science is to convince other scientists that your model, theory or study is worth taking seriously.

The same motivation exists for p-hacking more generally: studies that aren't marred by the occasional non-significant result amongst all their sub-0.05 p-values seem far more compelling. Recall that Stephen Jay Gould referred to science as 'a profession that awards status and power for *clean and unambiguous discovery*' [my italics]. The social psychologist Roger Giner-Sorolla concurs: 'In a head-to-head competition between papers ... the paper with results that are all significant and consistent will be preferred over the equally well-conducted paper that reports the outcome, warts and all, to reach a more qualified conclusion.'[74]

Here we see how publication bias and p-hacking are two manifestations of the same phenomenon: a desire to erase results that don't fit well with a preconceived theory. This phenomenon was illustrated by a clever meta-scientific study carried out by a group of business-studies researchers. They took advantage of the fact that some results are included in students' dissertations before they're formally written up and sent for publication in scientific journals. The researchers called what happened between dissertation and journal publication 'the chrysalis effect'. By the time they reached their final publication, initially ugly sets of findings had often metamorphosed into handsome butterflies, with messy-looking, non-significant results dropped or altered in favour of a clear, positive narrative.[75] In most cases, the students probably thought that by nixing such results they were letting their data more clearly 'tell a story' – and they were likely being taught by senior researchers that this was the right thing to do to persuade peer reviewers that a study should be published.[76] In reality, they were leaving future scientists with a hopelessly biased picture of what went on in the research.

The desire for attractive-looking results affects even the 'hardest' sciences. In her book *Lost in Math*, the physicist Sabine Hossenfelder argues that physicists have gotten high on their own supply, focusing on the elegance and beauty of models such as string theory at the expense of being able to test, in practice, whether they're actually true.[77] Although the lofty, mathematical work of these string theorists feels like it could hardly be further from the (almost literally) kitchen-sink science of Brian Wansink, both kinds of research can become saturated with the same kinds of all-too-human biases.

Nor are fields where lives are at stake safe from such biases. Generations of medical students have been taught, quite rightly, that the double-blind, randomised, placebo-controlled trial is the gold standard of evidence for new treatments. If done properly, such a trial rules out placebo effects, biases on the part of the doctors administering the intervention, spurious results due to factors other than the treatment (so-called 'confounding'), and many other problems that bedevil clinical research. But even the most tightly controlled clinical trial can't rule out bias that occurs after the results are in: bias when the data from the trial are being analysed.

Texas sharpshooter-style behaviour in medical trials is often called 'outcome-switching' (another name for what amounts to p-hacking). Let's go back to our example of running a study of height differences between men and women. Maybe you also

happened to measure some other facts about your subjects –
for example, recording their weights, the number of hours
they spent watching TV per week, their levels of self-reported
stress. These were secondary datapoints – interesting, to be
sure, but not central to the study. What happens if you don't
find your desired statistically significant result for height, but
you *do* find a significant difference between men and women
in, say, their amount of TV watching? Outcome-switching is
when you decide to present the study as if it had always been
about TV time. However, this has the now-familiar flaws: it
buries potentially useful knowledge about the lack of a height
difference and the fact that additional statistical tests have been
run means that we also need to take extra care in our interpretation. If you hide the full extent of your testing from the
reader, they won't be on their guard for the upwardly creeping
risk of false positives.

From 2005, the International Committee of Medical Journal
Editors, recognising the massive problem of publication bias,
ruled that all human clinical trials should be publicly registered
before they take place – otherwise they wouldn't be allowed to
be published in most top medical journals.[78] The idea was to
create an incentive against studies being file-drawered, because
everyone would know the trial had been done. The edict produced a useful side effect: a published list of the plans for
each study, including which outcomes they intended to test
for.[79] By searching back through these registries, observers are
able to detect inconsistencies between the proposed plans and
what was actually written up. Ben Goldacre's 'COMPare Trials'
project took all the clinical trials published in the five highest-
impact medical journals for a period of four months, attempting
to match them to their registrations.[80] Of sixty-seven trials,
only nine reported everything they said they would. Across all
the papers, there were 354 outcomes that simply disappeared
between registration and publication (it's safe to assume that
most of these had *p*-values over 0.05), while 357 unregistered
outcomes appeared in the journal papers *ex nihilo*.[81] A similar

audit of registrations in anaesthesia research found that 92 per cent of trials had switched at least one outcome – and, perhaps predictably, that the switching tended to be towards outcomes with statistically significant results.[82]

It's impossible to know for sure how many patients have been given useless medical treatments – and false hope – because of p-hacked clinical trials, but the number is surely enormous.[83] Think back to the meta-analyses we covered earlier. Even leaving aside the fact that some research is often missing due to publication bias, if the studies included in the meta-analysis are all themselves exaggerated by p-hacking, the overall combined effect – in what's supposed to be an authoritative overview of all the knowledge on the topic – will end up far from reality.[84] You might wonder how doctors and their patients are supposed to trust a medical literature that's permeated with bias in this way, outside of a minority of clear, well-replicated findings. My response is that I have no idea.

The drive to publish attention-grabbing, unequivocal, statistically significant results is one of the most universal sources of bias in science. Yet other distorting forces also exert their pull. The first one that comes to mind is money. In the US, where numbers are easily available, just over a third of registered medical trials in recent years were funded by the pharmaceutical industry.[85] To what extent does this funding, from companies who plan to market the drug if it works, influence the results? The consensus of meta-scientific research on clinical trials is that industry-funded drug trials are indeed more likely to report positive results. In a recent review, for every positive trial funded by a government or non-profit source, there were 1.27 positive trials by drug companies.[86] It's possible that this bias comes in while the study is being designed: there's some evidence that industry-funded trials are more likely to compare their new drug to a useless placebo rather than to the next-best-alternative drug, artificially making their new product look better.[87] But much of the reason is probably due to the factors we've seen earlier in this chapter: industry-sponsored trials have, for example, been shown to be file-drawered more often than those funded by other sources.[88]

It's currently a requirement at most journals that you disclose at the end of any published paper, in a conflict-of-interest section, any money you've received for, say, consulting for a pharmaceutical company.[89] But other kinds of financial conflicts aren't treated the same way. Many scientists, for example, forge lucrative careers based on their scientific results, producing bestselling books and regularly being paid five- or six-figure sums for lectures, business consulting and university commencement addresses.[90] People should of course be allowed to pay whatever they like for book deals, speakers and consultants. But when a lucrative career rests on the truth of a certain theory, a scientist gains a new motivation in their day job: to publish only studies that support that theory (or p-hack them until they do). This is a financial conflict of interest like any other and one aggravated by the extra reputational concerns. One could argue that scientists in this situation should, for the sake of transparency, include a conflict-of-interest statement at the end of every relevant future study.[91]

Outside of financial or even reputational interests, though, there's another bias that's rarely explicitly discussed. It's the bias of a scientist who really *wants* their study to provide strong results, because it would mean progress in fighting a disease, or a social or environmental ill, or some other important problem. It's not even a matter of wanting to find significant results for the sake of a publication (though that's an important pressure, as we'll see in later chapters). It's simply that the scientist means well and wants to feel like their research is beneficial. We could even call this the 'meaning well bias'. It's crushing when the trial you've designed to test your new treatment provides null results, meaning we're no closer to helping sick people. It's disheartening when you hypothesise about the link of some biological factor to a disease, and it turns out you've been barking up the wrong tree. Or, at least, it might feel that way if you have the wrong attitude to science. The currency of positive, statistically significant results in science is so strong that many researchers forget that *null results matter too*. To know that a treatment doesn't work, or that a disease isn't related to some bio-marker, is useful information: it means we might want to spend our time and money elsewhere in future. If it's properly designed, a study should be of interest whether it produces positive *or* null results.

What we've seen so far are biases that affect individual scientists. But remember: science is a social thing. Although sharing results among a community of researchers can, at least in part, compensate for the biases of individual scientists, when those biases become shared among a whole community, they can develop into a dangerous groupthink. In 2019 the science writer Sharon Begley wrote an eye-opening report on the controversy surrounding the 'amyloid cascade hypothesis' of Alzheimer's Disease. It's the idea that build-up of the amyloid-beta protein, which can be observed as 'plaques' in the brain (and which was first noted over a century ago by Aloysius Alzheimer, for whom the dementia is named) is the ultimate neurological cause of the devastating memory loss and other impairments that come with the disease.[92] Begley points out that, while important leaps have been made in treatments for other ageing-related illnesses like cancer and cardiovascular disease, Alzheimer's disease has remained stubbornly untreatable, with only a series of disappointments from clinical trials of drugs that attempt to relieve the symptoms by breaking down the amyloid.[93] Why? According to several researchers interviewed by Begley, it's simply because the amyloid hypothesis is wrong. Although amyloid plaques are *associated* with the symptoms, they're not the cause: targeting the amyloid won't help treat the disease.[94]

The dissenting researchers described how proponents of the amyloid hypothesis, many of whom are powerful, well-established professors, act as a 'cabal', shooting down papers that question the hypothesis with nasty peer reviews and torpedo-ing the attempts of heterodox researchers to get funding and tenure. Begley argues that this isn't necessarily because of any conscious decision: the researchers are simply so attached to the amyloid hypothesis – in which they genuinely believe, and which they think is the best route towards breakthroughs in Alzheimer's treatment – that they've developed a strong bias in its favour.

It's probably going too far to imply, as some of Begley's interviewees do, that if scientists had discarded the amyloid hypothesis years ago we'd be well on our way to a treatment (or cure) for Alzheimer's. It could simply be the case that the disease, which after all affects the brain, the most complex organ of all, is a particularly vexing one – or that the trials of amyloid-busting drugs that have been published so far haven't been asking the right question (they might have targeted people too late in the lifespan, for whom the amyloid has already had its detrimental effects).[95] But the stories of bullying and intimidation that result when researchers challenge the amyloid hypothesis hint at a field where bias has become collective, where new ideas aren't given the hearing they deserve, and where scientists routinely fail to apply the norm of organised scepticism to their own favoured theories.

What about researchers who have a more ideological or political stake in the truth or falsity of a result? One of the more remarkable conflict-of-interest sections I've come across is in a public health paper on the so-called 'Glasgow Effect'. This is the phenomenon whereby people from Glasgow, and Scotland more generally, die younger on average than those in other similar cities or countries, even after accounting for levels of poverty and deprivation. After reviewing the evidence on this effect, the paper concluded that the root of the unique problem was the 'political attack' on Scotland from Prime Minister Margaret Thatcher's Conservative government in the 1980s, with its policies of deindustrialisation and pushback against organised labour. There are no financial conflicts listed in the conflict-of-interest section – instead, it's noted that the lead author, Gerry McCartney, 'is a member of the Scottish Socialist Party.'[96] Good on him, I say, for this rare degree of honesty.[97]

My own field, psychology, is no stranger to scientists who identify as left-wing. The skew this way in psychology is very large indeed: surveys in the United States have found the ratio of liberals to conservatives to be around 10:1. (This isn't found in other fields, such as engineering, business studies and computer science; in all of those, as in America as a whole, the ratio is far more equal.) Psychology, being about humans and their behaviour, is normally much more intertwined with political concerns than, say, theoretical physics or organic chemistry. A 2015 piece by José Duarte and several other prominent psychologists argued that, partly for this reason, political bias could be particularly damaging for psychology.[98] They posited that, similarly to the example of groupthink we just encountered, if the vast majority of a community shares a political perspective, the important function of peer review – to hold claims to the harshest scrutiny possible – is substantially weakened. Not only that, but priorities for what to research in the first place might become skewed: scientists might pay disproportionate attention to some politically acceptable topics, even if they're backed by relatively weak evidence, and avoid those that go against a particular narrative, even if they're based on solid data.[99]

Critics of the liberal bias in psychology have turned their fire, for instance, on the idea of *stereotype threat*.[100] It's the idea that girls' mathematics test performance suffers when they're reminded of the stereotype that 'boys are better at maths'. This idea would seem much more intuitively plausible to someone who has socially liberal political views, in which stereotypes and sexist prejudice are considered powerful forces that affect individuals and shape society. The evidence for the phenomenon is quite weak, and possibly subject to publication bias, for a 2015 meta-analysis reviewing all the relevant stereotype threat studies found a clear gap where the small, null studies on the subject – those that showed girls were equally good at maths with and without the stereotypes being mentioned – should have been (meaning the funnel plot looked rather like our Figure 2B above).[101] It's certainly possible that these small, null studies were done and then discarded since they didn't fit the scientists' overwhelmingly liberal preconceptions, skewing our picture of the evidence on this important educational question.[102] However, just as a misshapen funnel plot isn't necessarily evidence of bias, nor should publication bias itself be interpreted as direct evidence of political bias. We know, after all, that scientists in general will favour positives over nulls, regardless of political perspective. Nevertheless, the discussion around stereotype threat alerts us once more to the possibility that shared assumptions, political or otherwise, might sometimes impede the self-criticism necessary for science to progress.

The discussion of stereotype threat leads us to the issue of sexism, which, alongside political bias, is one of the most discussed of all biases among scientists. One important debate about sexism in science is about the representation of women within different scientific fields, and at higher versus lower levels of seniority in those fields. But there's also discussion of how sexist biases might affect the practice of the science itself.[103] For example, neuroscientists working with animals such as mice have a tendency to study only the male of the species on which they're experimenting. This is because females are often thought to be more affected by hormonal fluctuations – a hard-to-control source of variability in the animals' brains and emotions, which might compromise the results. This ignores the fact that in male animals, the presence of hormones such as testosterone varies massively too, usually in accordance with their status in a dominance hierarchy. This potentially produces just as much behavioural variability overall.[104] If this bias leads researchers to study only males, their results might not apply universally, since females do differ from males in many aspects of their brain and their behaviour.[105] As the neuroscientist Rebecca Shansky observes, 'illnesses such as major depressive disorder and post-traumatic stress disorder are twice as prevalent in women, but common behavioural tests designed to model their symptoms in rodents were developed and validated in males.'[106]

The psychologist and philosopher Cordelia Fine – a perspicacious critic of shoddy, often *p*-hacked studies that purport to explain behavioural differences between the sexes by straightforwardly linking them to testosterone levels, in the process forgetting about social explanations – has also addressed the issue of males being treated as the 'default' in medical research, with females being considered a secondary concern or even an aberration.[107] In a 2018 opinion piece in the *Lancet*, Fine suggested that 'feminist science' could redress the balance by highlighting these kinds of omissions. She acknowledges that some will be sceptical: 'common thinking is that while feminism is all very well for Gender Studies, it should be kept away from science, lest political preferences of how women, men, and the world should be lead to distortion of scientific evidence as to how they actually are.'[108] But, she argues, aren't we all biased anyway? Indeed, she says, 'everyone has a standpoint; no one enjoys a "view from nowhere". Opening the door to feminist science doesn't open the door to political bias. It just means we're not all looking through the same blurred window.'[109]

In a sense, this is a political mirror-image of the concern expressed by José Duarte and his colleagues about the paucity of conservatives in psychology research. Though their emphasis differs, both Duarte and Fine are arguing for a greater presence in science of an underrepresented perspective. And that's valuable: one of the great things about the process of debating scientific findings is that people come at questions from different angles. To my mind, though, calling for scientists, or science in general, to take on any set of socio-political commitments is unwise, even if it might help solve specific problems in the short term. We should instead do our best to limit the effects of our own prejudices on our scientific decisions and analyses. (Whereas some, like Fine, might claim this is impractical or impossible, we'll see some ideas for how to do this later in the book.)

Trying to correct for bias in science by injecting an equal and opposite dose of bias only compounds the problem, and potentially invites a vicious cycle of ever-increasing division between different ideological camps. Not only that but the suggestion that scientists should feel proud to let their ideological views impinge on their research seems to offend against both the Mertonian norms of disinterestedness (since it involves allowing non-scientific concerns to encroach on research) and universalism (since it might involve holding scientific arguments to a different standard depending on the political affiliation of their source). The reason that Gerry McCartney decided to declare his Socialist Party membership on his paper about Margaret Thatcher is because it would have been embarrassing – given the paper's scientific conclusions, which were very much in line with Socialist Party ideology – if it had come out after publication. Even where they're not as directly relevant to the study as in that case, scientists should regard merging their political beliefs with their science with the same sense of embarrassment.

The idea that every scientist has an ideological perspective that affects their research brings us back full circle to the case of Samuel Morton and his skull measurements, and the criticism of his biases by Stephen Jay Gould. In 2011, the anthropologist Jason Lewis and his colleagues went back not just to Morton's numbers, as Gould had, but to the actual skulls from his collection at the University of Pennsylvania, remeasuring about half of them with modern techniques.[110] Lewis and his team agreed that Morton's ranking of the different groups of people was obviously racist, and confirmed that he did indeed make measurement mistakes. However, they contended, the errors weren't systematic in the way Gould had argued: instead, the mismeasurements were present across many of the skulls and didn't seem to favour one racial group over another. They could also have been due to an assistant who Morton mentions as having made errors, rather than to Gould's 'plausible scenario' about Morton stuffing more seeds into the White peoples' skulls.

Furthermore, Lewis and his team argued that Morton simply hadn't manipulated the sample groupings (omitting to mention groups from non-White races with high average skull sizes) in the way Gould had charged. In fact, Lewis and colleagues alleged that Gould made his *own* mistakes, splitting up Morton's sample in ways that suited his preferred beliefs about the equality of the skull sizes. In the foreword to his book, *The Mismeasure of Man*, Gould had freely admitted to having a strong commitment to social justice and liberal politics.[111] The Lewis paper concluded that 'ironically, Gould's own analysis of Morton is likely the stronger example of a bias influencing results'.[112]

Those were fighting words. Could it really be true that a legendary analysis by as well-respected a figure as Gould could be so wrong? Not everyone thought that Lewis and colleagues' case was a slam-dunk. The philosopher Michael Weisberg, while accepting that the new skull measurements were correct and agreeing that Gould had fumbled some of his analyses, argued that the main thrust of Gould's argument was still valid.[113] The idea that an assistant might have innocently made some errors was just speculation, after all; the evidence was still consistent with Morton (or perhaps the assistant) being biased against giving non-Whites larger skull sizes. And after the mistakes were corrected, there was still very little difference in the skull sizes by race, which was the main thrust of Gould's critique. A final twist (for now) came in 2018, when some additional skull measurements made by Morton himself were rediscovered. When these new data were taken into account, the idea that the discrepancy between Morton's seed-based and shot-based measurements was larger for disfavoured racial groups, which formed a large part of Gould's case for Morton's bias, no longer held water.[114]

Of course, in terms of answering any substantive scientific question, this whole debate over dusty skulls is moot. Even if we were to grant Morton his dubious association between skull size and 'mental and moral faculties' between the groups, his collection isn't a representative sample of skulls from across the world; thus, very little – perhaps nothing – can be concluded from it about any general differences between groups.[115] The back-and-forth does, however, have a clear moral for addressing bias in science: the watchers must also be watched and the debunkers debunked. And even then, it's worth checking whether those who debunked the debunkers got their facts straight. As we now recognise, everyone is subject to their own biases; in the Morton saga alone we've seen examples of what we might call 'racism bias', 'egalitarianism bias' and 'wanting to prove a famous historical scientist wrong bias', all of which might have contributed to distortions of the truth.[116]

Biases are an unavoidable part of human nature and it's naïve to think they could ever be eradicated from anything that we do. We do, however, have tools that are supposed to bring a little more objectivity to the table. Statistical analysis itself is aimed at taking decisions out of our biased human hands – yet we've seen how easy it is for the numbers to be nudged in our favour. Peer review is supposed to act as a check on our prejudices – yet we've seen how the attempt to persuade reviewers and editors to publish our work leads to inconvenient results being hidden away entirely or else forced to fit our preconceptions. Each of these biases – whether caused by scientific dogma, political slant, financial pressures, or just the desire to see statistically significant results – can be utterly unconscious. Indeed, the fact that they're unconscious might be what gives them their power: if you want to convince reviewers and the world of your case, it helps first to have convinced yourself. This fact is also what makes these biases so unnerving.

Impartiality, like honesty, is one of the qualities that science exists to ensure. The lamentable truth is that the way we do science often encourages its direct opposite. In the next chapter we'll add another entry to our list of scientific ironies: despite science being about getting the facts right, our studies are often replete with the most basic kinds of error.

5

Negligence

Ignorance is a blank sheet, on which we may write, but error
is a scribbled one, on which we must first erase.
                    Charles Caleb Cotton, *Lacon, or Many
                              Things in Few Words* (1820)

Physics has laws, mathematics has proofs, and social science has
'stylised facts': statements such as 'people with more education
tend to get a higher income during their lifetime' and 'democracies tend not to go to war with one another'.[1] These may not be
as ironclad as laws or proofs, but they're supposed to express
broad, important and replicable findings in simple language.
Just as physicists would love to discover a new law (or a way to
break the ones we already know), and just as mathematicians
work endlessly to prove their theorems, many social scientists,
particularly economists, long to discover a stylised fact that can
be associated with their name – and that the people who make
important decisions can easily keep in mind. When they published a major paper in 2010, the economists Carmen Reinhart
and Kenneth Rogoff thought they'd hit the stylised-fact jackpot.

For two years, politicians had been frantically trying to
address the fallout of the 2008 financial crisis and the ensuing
Great Recession. Amid all the conflicting advice, Reinhart and
Rogoff's paper, entitled 'Growth in a Time of Debt', was a god-
send, providing strong evidence to recommend one particular

course of economic action: austerity.[2] Reinhart and Rogoff had studied the debt-to-GDP ratio – the relationship between what a country owes to its creditors (its public debt, which, perhaps confusingly, is also known as its government debt or its sovereign debt) and what new goods and services it can produce (its Gross Domestic Product). They investigated the relation between this ratio and the rate of growth in the country's economy. Using historical data from several dozen countries, their paper showed that when the debt-to-GDP ratio was low – say, when debt was between 30 and 60 per cent of GDP – there wasn't much of a link to be drawn with growth, but countries with ratios above a very specific threshold – 90 per cent – had shrinking economies.

Perhaps because its main result lent itself so easily to becoming a stylised fact ('debt-to-GDP ratios above 90 per cent are bad for growth'), the study became enormously influential. Not only was it covered extensively in the media, but it helped shape many states' policies of austerity in response to the recession – the idea being that governments should try to pay off their debts (by cutting spending, increasing taxes, or both) to get that ratio below the crucial number of 90 per cent. The study was explicitly mentioned in a major speech by the UK's Chancellor of the Exchequer of the time, George Osborne, and in a statement by Republican members of the US Senate and House Budget Committees.[3] According to the economist Paul Krugman, a critic of Reinhart and Rogoff's message, so many pro-austerity politicians referenced the study that it 'may have had more immediate influence on public debate than any previous paper in the history of economics'.[4]

Which makes what happened next all the more concerning. In 2013, critics of the paper discovered a mistake in the Microsoft Excel spreadsheet Reinhart and Rogoff had used in their analysis: the debt of several countries had been left out of the equations.[5] Specifically, the spreadsheet omitted the debt of Australia, Austria, Belgium, Canada and Denmark from its calculation. The horribly banal reason? A typo. When this was corrected, along with amendments to another couple of debatable analytic choices Reinhart and Rogoff had made, the relation between the debt ratio and growth changed dramatically.[6] The paper had said average growth above the 90 per cent ratio was -0.1 per cent; after the corrections it was +2.2 per cent. There was nothing magic about that 90 per cent number after all; growth didn't suddenly turn negative after that threshold. In reality, there was 'a wide range of GDP growth performances at every level of public debt'.[7] If it had always had a more circumspect claim like this, one much more complicated than the original stylised fact, it's hard to imagine the paper would have received so much attention.

So, did a typo change the world economy? Not exactly. Although the paper, and its stylised 'fact', had an unusually wide influence, the case for keeping the debt-to-GDP ratio low doesn't hinge on a single study.[8] The discovery of the typo just weakened, rather than completely invalidated, Reinhart and Rogoff's conclusions. And, as noted above, the critics didn't *only* focus on the typo. We can't re-run world politics to find out whether those with a preference for austerity would have found other reasons to implement it had the Reinhart-Rogoff study never existed; doubtless many would have. Still, the fact that such a basic error got all the way to the desks of powerful politicians should be deeply worrying, not least because it inevitably makes us wonder how many *other* such errors must be out there, compromising the scientific literature and maybe even influencing real-world decisions.

The answer is: all too many. This chapter is concerned with two kinds of scientific negligence. The first is what we just encountered: unforced errors that are introduced to scientific analyses by inattention, oversight or carelessness. The second kind is when scientists, who should know better, bake errors into the very way their studies are designed. This latter kind of mistake could be due to poor training, apathy, forgetfulness or, much as it seems cruel to say, sheer incompetence. These negligent errors serve as further, painful evidence of our scientific system failing to deliver on the core purpose for which it was conceived.

How common are numerical errors in scientific papers? In 2016, a group of Dutch researchers led by the psychologist Michèle Nuijten took a stab at working it out. They unveiled an algorithm called *statcheck*, a kind of 'spellchecker for statistics'.[9] If you feed a scientific paper into statcheck, it'll trawl through the numbers in the paper and flag up mistakes in its p-values. It can do this because many of the numbers in statistical tests are dependent on one another, so knowing just some of them always allows you to reproduce the others (in the same way that, thanks to Pythagoras's theorem, you can always calculate the hypotenuse of a triangle if you know the lengths of the other two sides). If the p-value and the other related numbers are inconsistent with one another, something probably isn't right. Nuijten and her colleagues fed statcheck over thirty thousand papers: a gigantic sample of studies published in eight major psychology research journals between 1985 and 2013.[10] What they found makes for awkward reading.

Nearly half of the papers that included relevant statistics had at least one numerical inconsistency. To be fair, most of the mistakes were minor and made little difference to the overall results. But some of the inconsistencies had a big effect on the study conclusions: 13 per cent had a Reinhart-Rogoff-style serious mistake that might have completely changed the interpretation of their results (for example, flipping a statistically significant p-value into a non-significant one, or vice versa). Of course, these inconsistencies could have come about for all sorts of reasons, from simple typos and copy-and-paste slips all the way to conscious fraud. Statcheck is just a way of highlighting, not getting to the bottom of, errors in a scientific text.

One of the most interesting results from Nuijten's analysis shows how negligence connects with bias. The inconsistencies that statcheck flagged tended to be in the authors' favour – that is, mistaken numbers tended to make the results more, rather than less, likely to fit with the study's hypothesis. If these were just entirely random typos, we wouldn't expect them to go in any particular direction on average. But as we might have predicted from what we know about bias, it seems as though the scientists were more likely to take a second look when the results didn't go their way. Mistaken results that *supported* their theory were, on the other hand, simply too good to check.

Another particularly elegant method for testing whether the numbers reported in a paper check out has the decidedly inelegant name of 'Granularity-Related Inconsistency of Means' – or 'the GRIM test' for short.[11] Devised by the data sleuths Nick Brown and James Heathers, the GRIM test can be used to check whether the average (specifically, the arithmetic mean) of a set of numbers makes sense, given how many numbers the set contains. Imagine you're asking people to rate how happy they are with their job, on a scale of 0 to 10 (and you've only given them the option to respond in whole numbers – say, '4' or '5', but not '3.7'). In the simplest case, let's say you gave this survey to just two people, and you report the average of their scores: that is, you add up their ratings and divide the total by two. If you take that result and look at the digits after the decimal point, there are only so many ways they can look: with two people, the average of their answers can only end in .00 or .50. If you said that the average was, say, 4.40, something must have gone wrong: there's no way to divide a whole number in half that would produce that fraction.

The GRIM test applies this same logic to bigger samples. For example, if twenty participants rate something on a 0-to-10 whole-number scale, there's no way you can arrive at an average of 3.08. Dividing by twenty means that the decimal values can only be increments of .05; it's plausible to get an average of 3.00, or 3.10, or 3.15, but 3.08 is an impossibility.[12] Brown and Heathers used the GRIM test to check a selection of seventy-one published psychology papers and found that half of them reported at least one impossible number, while

20 per cent contained several. As with statcheck, GRIM errors can have benign causes, but they are red flags that signal the need for further investigation.

The number 3.08 in my example was a deliberate choice, because it's a notable one from the history of the GRIM test – and psychology research in general. In 2016, the psychologist Matti Heino applied the GRIM test to one of the most famous psychology papers of all time: Leon Festinger and James Carlsmith's 1959 paper on 'cognitive dissonance'[13]. This is the now widely known idea that forcing someone to say or do something inconsistent with their true beliefs will make them psychologically uncomfortable and they'll do their best to alter those beliefs to make them fit with what they've been made to say or do. In the 1959 study, participants were made to complete some tedious, pointless tasks, such as endlessly twisting pegs around on a pegboard. When they were finished, some were paid $1 to tell the next waiting participant that they'd found the tasks really interesting and enjoyable. In an interview afterwards, the participants who'd been paid $1 to lie about the tasks reported thinking the task was much more enjoyable than those who were paid nothing. They'd reduced their dissonance, in other words, by making themselves believe they'd had fun.[14] Alas, Heino's use of the GRIM test showed that it wasn't just the participants' beliefs that were inconsistent – it was Festinger and Carlsmith's numbers.[15] They reported an average score of 3.08 for a sample of twenty people filling in a scale of 0-to-10, which as we just saw isn't possible, alongside several other averages that failed the GRIM test.

Cognitive dissonance is a remarkably useful concept that makes intuitive sense, and the experiment was clever and memorable. But would the thousands of researchers who've cited Festinger and Carlsmith's study over the years have done so if they'd known it was riddled with impossible numbers?[16] The story reminds us once more that even 'classic' findings from the scientific literature – the ones that you would hope had been examined most rigorously – can be wholly unreliable, with what should be the most important part, the numbers and the data, acting as mere window-dressing in service of an attention-grabbing story.

Numerical errors are also disturbingly common in scientific fields with much higher stakes. You may recall that the world's most prolific known scientific fraudster, at the time of writing anyway, is the anaesthesiologist Yoshitaka Fujii. The analysis that ended his long spree of data forging was run by the anaesthetist John Carlisle, who devised a statistical technique to check whether randomised trials really are randomised.[17] Randomisation is done by essentially flipping a coin for each participant in order to assign them to a group (say, the active drug group versus the placebo group) at random, rather than in some pre-planned way that might be prone to bias. This process is crucial: the point is to ensure before the trial begins (in the jargon, 'at baseline') that there are no substantial differences between the groups. If one group is healthier, better educated, older or markedly different in any other way that might affect the results, the trial won't be a fair test.[18] And so, if there *are* big differences between the groups at the beginning of a randomised controlled trial, there's an issue: the randomisation process must've failed. On the flipside, if the groups are *perfectly* matched, inexplicably avoiding the iron rule of the noisiness of all numbers, that's problematic too: even after randomisation you'd still expect there to be tiny differences between the groups, just by chance. This is what Carlisle's method relies upon. When he checked Fujii's papers, he found data of completely implausible consistency: the reported age, height, and weight distributions for Fujii's patients, for instance, were almost perfectly synchronised. The odds of that happening in reality were less than one in ten to the thirty-third power (that is, one in a billion trillion trillions).[19] Sure enough, it turned out that Fujii was a fraud.

In 2017, Carlisle applied his error-spotting technique to 5,087 medical trials from eight journals, again checking for randomisation that was either faulty or suspiciously perfect.[20]

Of course, it remains the case that *some* trials might look suspicious just by bad luck. But even after taking this into account, Carlisle found that 5 per cent of trials had suspicious data: his results thus pointed to hundreds of studies that might have been completely corrupted – their results rendered meaningless – by a failure to randomise their groups properly. Fujii-like fraud was responsible only for a small proportion of these broken trials; in all likelihood, Carlisle had mainly uncovered innocent mistakes. Considering what's at stake in a medical trial, though, with doctors using the results to choose treatments for their patients, those innocent mistakes could end up being extraordinarily serious.[21]

The great thing about statcheck, the GRIM test, and Carlisle's method is that they can all be performed using just the summary data that are routinely provided in papers: things like *p*-values, means, sample sizes and standard deviations. No access to the full, original data spreadsheets is required. That's just as well, because scientists are notorious for their reluctance to share their data, even when other *bona fide* researchers ask them nicely. A study in 2006 found that a paltry 26 per cent of psychologists were willing to send their data to other researchers upon an email request, and similarly dismal figures come from other fields. You're also much less likely to be able to access the data the older a study gets.[22] This reluctance to share data is a block on the vital processes of self-scrutiny – those Mertonian norms of communalism and organised scepticism again – that lie at the heart of science. And as clever as the above three methods are, they pale in comparison to the full audit one would be able to conduct with the whole, detailed dataset at one's disposal. At the moment, though, the incentives for keeping data private (including, perhaps, the fear that someone will find errors in your published work) clearly outweigh the Mertonian reasons for sharing it.

*

Essentially all scientific fields suffer from numerical mistakes. Some fields, though, have their very own brands of errors. Consider cell lines, for example – the effectively immortal biological cultures that multiply indefinitely and are used as models for studying various kinds of cells, both healthy and cancerous. In 1958, a few years after the first immortalised cell line was created, it was noted that cells from different lines – indeed, different species – could sometimes become mixed up if the scientists weren't paying close enough attention.[23] It could happen easily: someone mislabels a sample, several researchers are working in too-close proximity or in a cluttered lab, or perhaps equipment isn't being properly cleaned and sterilised. Contamination is particularly bad news because some cell lines grow faster and more robustly than others, so one can completely take over another before you know it. Needless to say, conducting an experiment on the wrong type of cell is likely to undermine its results completely – you thought you were doing research on bone cancer but it turns out the cells you're using are from colon cancer; you thought you were doing research on human cells but it turns out the ones you're using are from pigs or rats.[24] It would hardly be a surprise if results from those experiments failed to replicate. An alarming editorial in the prestigious *Nature Reviews Cancer* put it bluntly: 'The scientific community has failed to tackle this problem and consequently thousands of misleading and potentially erroneous papers have been published using cell lines that are incorrectly identified.'[25]

How many thousands? A 2017 analysis which scoured the literature for studies using known misidentified cell lines found an astonishing 32,755 papers that used so-called impostor cells, and over 500,000 papers that cited those contaminated studies.[26] (Because so many cell lines studied by scientists are of cancer cells, the field of research with the biggest number of contaminated papers, unsurprisingly, was oncology.) A more specific survey of Chinese labs found that up to 46 per cent of the cell lines there were misidentified.[27] Another study noted that up to 85 per cent of cell lines thought to be newly established in China were contaminated with cells from the US; labs share cell lines promiscuously, so this kind of error is rapidly compounded.[28] It's far from just a Chinese problem, though – it was also found that among published papers using contaminated cells, 36 per cent were from the US. And it isn't just the cell lines that are multiplying: so is the number of scientific studies that make the mistake, with the toll growing over time, in spite of how long scientists have known about the problem.[29]

That's the most frustrating thing about cell line misidentification: how thoroughly relaxed the scientific community has been, for over half a century, about such an urgent issue that sets back medical research and wastes vast sums of money on tainted experiments. More than two decades after the first identification of impostor cells, *Nature* published an editorial – in response to an incident where owl monkey cells had been confused with human ones – that both downplayed the issue and dismissed whistleblowers who had pointed out the contamination as 'self-appointed vigilantes'.[30] Certainly, not all misidentifications have a grievous impact on the study's results; you could mix up, say, different kinds of human prostate cancer cells and still make some worthwhile discoveries. Many of them do render the findings of the study completely useless, though, as is borne out by a glance at the growing list of misidentification-related retractions.[31]

It's extraordinary that the issue of cell line mix-ups continues to haunt cell biology. In the past decade alone, I found editorials and calls to action about cell line misidentification in prominent journals from 2010, 2012, 2015, 2017 and 2018, continuing the calls that go back to the 1950s.[32] There are now better, cheaper ways, using new DNA technology, to authenticate each cell line, so it's possible that more of these mistakes will be prevented in future.[33] But cell biologists' lackadaisical response in the past means that it might be unwise to hold your breath. This isn't just a case of an unfortunate error, but a decades-long denial of a serious failing. If you compare this to what happens in other fields – such as aviation, for example, where enormous efforts are poured into investigating the cause of every plane crash so that the same problem never happens again – then scientists come out looking very inadequate indeed. And since this obstinacy is slowing down cancer research, the whole affair takes on a very unhappy moral dimension.

That moral case – that making errors in science is much more than just an academic matter, because of the harm it can cause – applies similarly to fields of research that directly sacrifice lives. I'm referring, of course, to research on non-human animals, where the subjects are often 'euthanised' – that is, killed – as part of the experiment (for example, to examine their brains after a new drug has been administered). This kind of research is usually strictly regulated by government agencies, since virtually everyone agrees it would be immoral to kill lab animals, or even just to cause them to suffer, for no good scientific reason.[34] So animal studies don't just carry the usual scientific burden of trying to produce accurate, replicable results without wasting resources. They also have an additional responsibility: ensuring that errors in their design and analysis don't render pointless the pain and death that they inevitably cause. Unfortunately, a considerable proportion – by some measures, a majority – of animal research studies fail this test.

A team of researchers led by the neurologist and neuroscientist Malcolm Macleod published in 2015 a survey of animal research studies that checked whether they'd reported following some basic principles of study design – techniques that are essential to make sound inferences from the results of these types of experiments.[35] The first principle they examined was one that we discussed earlier: randomisation. Everyone familiar with study design knows the importance of randomisation, yet Macleod's research found that only 25 per cent of the relevant studies reported randomising their groups.[36] Might this alarmingly low figure be due to studies using randomisation but simply failing to report it? This seems unlikely: given what a crucial difference randomisation makes to the quality of one's data and therefore the validity of one's findings (and, more pragmatically, to the way the paper will be received by peer reviewers), it's unlikely that a researcher who went to the trouble of randomising their groups would omit to mention it.

Another technique Macleod and his colleagues examined is something known as 'blinding'. In a blinded study, the scientist who collects the data is unaware either of the hypothesis being tested or of which group is the treatment group and which the control.[37] The scientist has all the information they need to run the study, and no more; only once all the data are in is it revealed to them which one was which. (In some cases, blinding is also applied to the process of data analysis: the researcher running the statistical tests doesn't know which group was the experimental one until all the tests are completed.) Blinding should be done wherever possible in a study: it acts as a firewall, stopping the researchers' conscious and unconscious biases from infecting the conduct or analysis of an experiment. It is a well-known concept, taught in every class on experimental design (at least with reference to data collection, if not analysis), yet only 30 per cent of the animal-research papers reported using it.[38] A clue as to why might be the following: in some of his prior meta-research, Macleod found that when scientists did their randomisation and blinding properly, their experiments tended to find much smaller effects.[39]

Without randomisation and blinding, even studies with enormous sample sizes can be misleading.[40] But that's not to say that the size of a study doesn't matter. In fact, it's one of the most vital considerations in the design of an experiment and it's another aspect that Macleod and his team investigated. Did the papers tell the reader how they decided how many animals to include in their study? Only 0.7 per cent did. There are two reasons why this is disappointing. The first is something we've seen before: p-hacking. Not setting the sample size beforehand allows the researchers to continue collecting data and testing it, collecting data and testing it, again and again in an open-ended way until they get their desired $p < 0.05$. The second reason is a related concept we haven't yet discussed: *statistical power*. To put it simply, far too many scientific studies are far too small.

Picture a perfect headache pill that instantly clears up all pain, under any circumstances. We wouldn't need p-values or statistics to pick up on this super-strong effect: we would notice it every time we compared even one headache sufferer who took the pill to a control patient who took a placebo or a less effective brand of headache pill. Thinking back to the study in the previous chapter where we compared men and women's heights, this would be akin to every man in the world being taller than any woman in the world. In reality, of course, it doesn't work that way: real statistical effects are almost always smaller and harder to spot. An actual pill might reduce headache pain by an average of, say, half a point on a pain-rating scale from 1 to 5. It would be impossible to disentangle this small effect from random noise in a two-person comparison; running such a study would be useless. Even if we compared two groups of ten people, there would be numerous ways for the small effect to be overwhelmed by noise. Maybe a couple of participants aren't paying attention and circle the wrong number on the pain questionnaire; maybe one gets hit on the head before the survey, worsening their headache; maybe one gives up drinking, easing theirs.

If we looked at many more people, however – say, 500 getting the pill and 500 the placebo – the modest effects of the pill would be far easier to distinguish from chance fluctuations. This is because the effects of the drug – our signal – would be *systematic*: they'd go in the same direction in a large enough number of people taking the pill. The noise, on the other hand, would be *random*: circumstances unrelated to whether they took the pill or the placebo would sometimes make the pain worse, sometimes better, for people in either group. With many people in the study, these random variations would tend to cancel out, so the average in the large sample will be closer to the 'true' effect. A statistician would say that the test in the bigger study had higher *statistical power* than that in the small one: a higher chance to detect a difference between the groups if the new drug really works better than the placebo.

As we saw in the last chapter, a *p*-value describes the chance that we'd find results that looked like ours (or even more impressive results) if in fact there was nothing going on, so we usually want it to be as low as possible (at least, lower than the standard threshold, normally set at 0.05). On the other hand, statistical power describes how likely we are to see a statistically significant signal when it really *is* there, so we want it to be as *high* as possible. Smaller effects – weaker signals – are far trickier to detect when you don't have much data, so usually the more nuanced the effect you're looking for, the bigger the sample that's going to be required.

Here's a more concrete way to think about it. In 2013 the psychologist Joseph Simmons and colleagues asked an online sample of participants to answer a set of questions about their preferences in areas such as food and politics, and also collected their basic demographics (gender, age, height, and so on).[41] Simmons then split the sample into various groups (such as male versus female, or liberal versus conservative), and noted how much the groups differed on a selection of variables. From there, he worked out how many participants you would need to be confident that you could detect a given difference, if you didn't already know it existed.[42] For instance, it turns out you could reliably establish our now-familiar link between height and sex – that men are taller than women on average – with just six men and six women from the survey; this effect, as we know, is large and therefore obvious (our twenty-person study from the previous chapter, therefore, had high statistical power). Another straightforward one: 'Do the older people in the sample tend to say that they're closer to retirement age?' They do, and Simmons found that you would only need nine older and nine younger people to detect this. But here are some effects that would require larger numbers of participants to be detected:

- People who like spicy food are more likely to like Indian food (twenty-six spice-likers and twenty-six spice-dislikers needed).

- Liberals tend to think social justice is more important than do conservatives (thirty-four of each political persuasion needed).
- Men weigh more than women on average (forty-six of each sex needed).

The point of the exercise was to get scientists to be realistic about the size of the effect they're looking for in any given study and, therefore, the sample size they would need for their results to be meaningful. If your sample size wouldn't be enough for a reliable test of 'do men weigh more than women?', it probably doesn't have enough statistical power to detect the specific esoteric effect that's implied by your theory.

Running a study with low statistical power is like setting out to look for distant galaxies with a pair of binoculars: even if what you're looking for is definitely out there, you have essentially no chance of seeing it. Sadly, this point seems to have passed many scientists by, not least in Macleod's chosen field of animal research. A 2013 review looked across a variety of neuroscientific studies, including, for example, research on sex differences in the ability of mice to navigate mazes.[43] To have enough statistical power to detect the typically expected sex effect in maze-navigating performance, a study would require 134 mice; it's a much subtler effect, in other words, than 'men weigh more than women'. But the average study the researchers looked at included only twenty-two mice. This isn't specific to mice in mazes: it seems to be a problem across most types of neuroscience.[44] Large-scale reviews have also found that under-powered research is rife in medical trials, biomedical research more generally, economics, brain imaging, nursing research, behavioural ecology and – *quelle surprise* – psychology.[45]

If studies in these fields are so underpowered, how come so many of them still find effects? The first reason is that they might be *p*-hacked: the researchers didn't find an effect in their original analysis, so they got creative with their numbers.[46] But even without *p*-hacking, underpowered studies will still occasionally find effects and it's for a troubling, though somewhat convoluted, reason. Think back to our discussion of *sampling error*. Imagine the average effect of our headache pill, in the population, really was half a point on the 1-to-5 scale. We might sometimes take samples where the effect, by chance, is below average: maybe it looks as if there's no effect at all. We might also sometimes take samples where the effect is above average because we happen to have only included people for whom the drug had big benefits. In a low-powered study, we can *only* find a positive result – a significant *p*-value – if the sample shows an unusually and spuriously large effect.

At the risk of sounding tautological: since underpowered studies only have the power to detect large effects, those are the only effects they see. This is where the logic leads. If you find an effect in an underpowered study, *that effect is probably exaggerated*.[47] Then comes publication bias: since large effects are exciting effects, they're much more likely to go on to get published. That's why, when reading the scientific literature, so many tiny studies seem to be reporting big effects: as we saw with the funnel plots in the last chapter, journals are often missing all the small studies that were discarded for failing to detect anything 'interesting'.

The situation poses problems for follow-up studies. Researchers look at the past literature to see what kind of effect size they can expect in their own experiments. If an initial, small study exaggerates the size of an effect, future researchers will use small samples to follow up on it, thinking they'll have enough statistical power. However, the reported effect, if it exists at all, is probably far smaller in reality and so too subtle for a small-sample study to find.[48] These are the long-term domino effects of underpowered research: study after study wastes time, effort and resources chasing an effect that's like the giant shadow projected by a moth sitting on a lightbulb.

The use of small samples wouldn't be so bad if we lived in a world where there really were lots of large-sized effects. But large effects usually come from very obvious factors – such as our example of the height difference between men and women, for instance. Most effects are far less obvious. One study that looked at clinical trials found that the average medical effect was small-to-moderate. This roughly means that if you had a trial with 100 people taking the treatment and 100 people taking the placebo, and twenty got better with the placebo, about six more (so, around twenty-six) would get better with the treatment.[49] Even for well-supported treatments, like antipsychotics for schizophrenia, benzodiazepines for insomnia, and corticosteroids for asthma, the effects are still only moderate in size: for those three, an additional eighteen (so, around thirty-eight) of the patients in the treatment group would get better.[50] In psychology research, the average effect is also pretty modest and the situation is probably similar in many other fields.[51]

When it comes to studies of such extraordinarily complex systems as the body or the brain, or an ecosystem, the economy or society, it's rare for scientists to find one factor that has a massive effect on another. Instead, most of the psychological, social and even medical phenomena we're interested in are made up of lots of small effects, each of them playing a small role. For example, if economists want to explain why different people in their sample have different incomes, they'd need to take into account where the participants live, their family backgrounds, their abilities, personalities and education, their country's tax system and its changes over time, and a whole host of other factors and experiences that might have nudged the participants' fortunes in one direction or another throughout their lives. The fact that small effects are so much more common and, collectively, so much more influential than big ones makes underpowered studies, which portray a world full of large effects, all the more misleading.

One of the most embarrassing examples of how low-powered research can lead scientists astray is the excitement over 'candidate genes'. Over the last decade, geneticists have learned a punishing lesson about the danger of low-powered studies. We've known for a long time, mainly due to studies of twins, that people's height and weight, their cognitive (IQ) test scores, their likelihood of developing various diseases and psychiatric disorders, and many other characteristics are related to genetic differences.[52] It was only about twenty years ago, however, that technology became widely available for geneticists to attempt to pinpoint precisely *which* sections of DNA related to which traits. The early attempts to do so were studies in which a specific gene – the 'candidate' – was isolated and measured, in the hope of establishing whether variation in that gene might be linked to variation in the trait.

Initially, it looked as if these attempts were successful. Positive candidate gene studies kept popping up, with variation in the *COMT* gene being linked to cognitive test scores, variation in the *5-HTTLPR* gene being linked to depression, and variation in the *BDNF* gene to schizophrenia, to name just a few well-studied examples. Many hundreds of studies began to accumulate.[53] The effects they found were often impressive: for example, a study in the prestigious journal *Nature Neuroscience* in 2003 claimed that memory performance was 21 per cent poorer in people who had one particular variant of the *5-HT2a* gene.[54] With effects as big as this, we were surely well on our way to understanding the genetic basis of important traits. Geneticists also started elucidating the biological 'pathways' between the genes and the traits: for example, the *5-HTTLPR* gene was found to have its effect on depression by making the amygdala (a region of the brain that's related to emotion) more reactive when its owner is threatened.[55]

When I was an undergraduate student, between 2005 and 2009, candidate gene studies were the subject of intense and excited discussion. By the time I got my PhD in early 2014, they were almost entirely discredited. What happened? The main factor was that technology improved, making it possible to measure people's genotypes much more cheaply.[56] Consequently, we could use much larger samples in genetic studies, with sample sizes of many thousands, or tens of thousands, now in reach. Geneticists also started taking a different approach: instead of looking at just one or a handful of candidate genes, they looked simultaneously across many thousands of points on the DNA that vary between people, checking which of them were most strongly related to the traits in question. This approach is called a genome-wide association study (or GWAS), and the analyses in these studies had far better statistical power – so they could find genetic variants that had much smaller effects on the traits, in addition to the large effects of the well-established candidate genes.[57]

Except the GWASs didn't find the large effects of those candidate genes.[58] They were nowhere to be seen, whereas if they were real, they'd have stuck out like the sorest of thumbs. Instead, the conclusion of the new, high-power GWASs was that, with a few very rare exceptions, complex human traits are generally related to many thousands of genetic variants, each of which appears to contribute only a minuscule effect.[59] There was no space for large effects of single genes, which was completely at odds with the results of all those previously lauded candidate gene studies. Since then, efforts that specifically tried to replicate the candidate gene studies with high statistical power have produced flat-as-a-pancake null results for IQ test scores, depression and schizophrenia.[60]

Reading through the candidate gene literature is, in hindsight, a surreal experience: they were building a massive edifice of detailed studies on foundations that we now know to be completely false. As Scott Alexander of the blog Slate Star Codex put it: 'This isn't just an explorer coming back from the Orient and claiming there are unicorns there. It's the explorer describing the life cycle of unicorns, what unicorns eat, all the different subspecies of unicorn, which cuts of unicorn meat are tastiest, and a blow-by-blow account of a wrestling match between unicorns and Bigfoot.'[61]

The whole sorry tale is a textbook example of the perils of low statistical power. The initial candidate gene studies, being small-scale, could only see large effects – therefore, large effects were what they reported. In hindsight, these large effects were extreme outliers, freak accidents of sampling error. The follow-up studies expected to see large effects too, so the sample sizes stayed relatively small. In this way, the studies capitalised on chance results, and built a chain of misleading findings that became the mainstream, gold-standard science in the field. To be sure, there were some null findings, and some meta-analysts sounded alarm bells about low power.[62] But most candidate-gene researchers ploughed on regardless. Had these geneticists known their history, they would've been highly suspicious of the large-effect genes: Ronald Fisher, the statistician who popularised the p-value and the idea of 'statistical significance', worked out that complex traits must be massively *polygenic* – that is, must be related to many thousands of small-effect genes – as far back as 1918.[63]

Luckily for geneticists, the technological improvements that brought down the price of genotyping meant that their candidate-gene ideas were eventually tested in GWASs, with proper statistical power, leaving them in no doubt as to whether they were on the right track (and they weren't). They've since moved on to the routine use of large samples and while there are still a few redoubts of candidate-gene belief, that type of study has very nearly gone extinct.[64] Think how many other scientific fields haven't yet faced this kind of ultimate test, though. Large chunks of scientific literature – ones based on small studies with implausibly large effects – could well be just as mistaken and mirage-filled as the research on candidate genes.

Is it fair of me to label researchers who run underpowered studies as negligent, or even incompetent? Neuroscientists, for example, might respond that their research is very expensive, given that they often must pay for lab animals and their upkeep, or use exorbitantly expensive equipment like MRI brain scanners. With that kind of outlay, small-scale studies are all they can afford. Worse, many studies are run by PhD students and postdoctoral researchers, who don't have much, if any, grant money. When I've made comments about low statistical power in scientific seminars, I've often heard replies along the lines of 'my students need to publish papers to be competitive on the job market, and they can't afford to do large-scale studies. They need to make do with what they have.' This is a prime example of well-intentioned individual scientists being systematically encouraged – some would argue, forced – to accept compromises that ultimately render their work unscientific.

Be that as it may, explaining *why* low-powered research is done doesn't justify it. We'll return to the question of who (or what) is to blame for negligence later in the book. But for now, we should acknowledge that when the scientific community gave its collective approval to these low-powered studies, it neglected, or even reneged on, one of its defining responsibilities. Their propensity to mislead means that low-powered studies actively *subtract* from our knowledge: it would often have been better never to have done them in the first place. Scientists who knowingly run low-powered research, and the reviewers and editors who wave through tiny studies for publication, are introducing a subtle poison into the scientific literature, weakening the evidence that it needs to progress. And in the case of research on animals, it's extremely difficult to justify 'euthanising' all those lives if the studies we do are so low powered that they never had any hope of answering the relevant scientific questions in the first place.[65]

In surveys of the public, scientists are generally described as being very competent.[66] The most frustrating thing about the errors documented in this chapter is that the overwhelming majority of scientists do indeed know better. They know to check thoroughly for typos and other slip-ups, they know about the importance of randomisation and blinding, and they know (and have known since the 1950s) that contamination of cell lines is a grave problem for fields like oncology. They know, even from their undergraduate statistics lectures, that statistical power is a vital consideration, especially in a world of small effects. Tiny samples just can't provide a meaningful answer to most scientific questions.

And still, negligence – on the part of scientists, reviewers and editors – means that papers with these kinds of glaring errors appear in the literature with depressing regularity. Just as with bias, we'll never be able to avoid *all* mistakes: the occasional typo or other screw-up is inevitable in any complex human activity. But given science's unique societal position as a clarifier of mysteries, an arbiter of disputes and a fingerpost towards objectivity, the stakes are uncomfortably high – society takes science remarkably seriously. Scientists need to reciprocate by holding themselves to far higher standards.

What explains the paradox of scientists who know better, and who work in a community whose *raison d'être* is organised scepticism, still ending up surrounded by avoidable mistakes? How have scientists, so often stereotyped as ultra-careful and conscientious about numbers and data, been found so many times to be asleep at the wheel? We'll see some potential explanations in Part III of the book. Before that, though, there is one more of science's afflictions that we need to diagnose. Part of being careful and conscientious is to resist leaps beyond one's data, to remain restrained about what can be concluded from any given study and about its implications for the world. You might think, then, that scientific culture would be one of modesty and humility above all else. Unfortunately, you'd be wrong.

6

Hype

The following tale of alien encounters is true. And by true, I mean false. It's all lies. But they're entertaining lies. And in the end, isn't that the real truth? The answer is no.

Leonard Nimoy, *The Simpsons*

In 2010, *Science* published a paper by researchers at NASA describing a strain of mysterious bacteria living in California's Mono Lake.[1] The lake is an odd place: not only is the water highly alkaline and around three times as salty as the ocean, but rising above its surface are irregular, stalagmite-like towers of limestone that look to be straight from the set of a science-fiction movie.[2] It was an apt location for scientists seeking to investigate some of the world's most exotic species and see what lessons they might teach us about the chemistry of life on other planets.

According to the study, led by the microbiologist Felisa Wolfe-Simon, the bacteria's adaptation to the lake's extreme environment had caused two unique things to happen. First, they had grown and multiplied in an environment that contained virtually no phosphorus, one of the elements thought to be essential for all forms of life. Second, and even more sensationally, the reason for the growth was that the bacteria, which Wolfe-Simon named GFAJ-1, had altered the very backbone of their own DNA, replacing the phosphorus with an element abundant in Mono Lake: arsenic. Arsenic is, of course, well known to be a poison, and this made the result doubly surprising: not only was it unheard of for another element to replace phosphorus in the DNA molecule but arsenic, particularly in such high levels, is usually toxic. Not so, apparently, for GFAJ-1, where arsenic played the opposite role, sustaining its life.[3] If the results held up, they would do a lot more than 'Give Felisa a Job' – which, by the way, is what the bacteria's name stands for.[4] They would change the way we thought about life itself. Wolfe-Simon certainly seemed aware of the implications: at a news conference announcing the study, she described how the discovery had 'cracked open the door to what's possible for life elsewhere in the universe.'[5]

Time to tear up the textbooks? Not so fast. From the get-go, other researchers were sceptical about 'arsenic life'.[6] The University of British Columbia microbiologist Rosemary Redfield noted flaws in the study and described them in a series of detailed blog posts.[7] Wolfe-Simon's response was to disregard the criticism. 'We're not going to engage in this sort of discussion,' she told a journalist. 'Any discourse will have to be peer-reviewed in the same manner as our paper.'[8] This seemed a little rich, given that NASA had clearly sought to pique broad public interest when announcing the paper's publication. A few days beforehand, they'd told the world in a tantalising press release that a new finding 'will impact the search for evidence of extraterrestrial life.'[9] It led to a frenzy of speculation: one much-read blog immediately suggested that signs of life might have been found on Saturn's largest moon, Titan.[10] And although the findings, when revealed, weren't quite that exciting, NASA still managed to frame them in portentous terms. 'The definition of life has just expanded,' enthused a NASA administrator immediately after the paper appeared. 'As we pursue our efforts to seek signs of life in the solar system, we have to ... consider life as we do not know it.'[11]

Still, it wasn't long before Wolfe-Simon got her wish and the discussion moved to the journals. In a relatively rare example of a journal publishing strong critiques of one of its previous articles – an example, one might suggest, of the scientific process working just as it should – *Science* printed no fewer than eight 'Technical Comments' on the study, including one by Redfield, along with a defiant response from Wolfe-Simon and her colleagues.[12] Redfield and her team then put the arsenic-bacteria claims properly to the test.[13] One of Wolfe-Simon's main observations had been that whereas the bacteria wouldn't multiply in an environment that had neither phosphorus nor arsenic, they would when arsenic was added to the mix. Redfield and colleagues attempted but failed to replicate this in their lab: they found that GFAJ-1 didn't grow at all unless phosphorus was provided. And as far as GFAJ-1's DNA was concerned, they found only minimal arsenic there after they'd washed the samples in water. The best explanation was a banal one: simple contamination. The arsenic with which Wolfe-Simon fed the bacteria might have been contaminated with enough phosphorus to allow some growth; Wolfe-Simon's DNA samples, conversely, might have been contaminated with arsenic from the Mono Lake-like environment she had created in her lab. A parallel replication attempt at the Swiss Federal Institute of Technology in Zürich gave closely similar results to Redfield's, providing the final empirical nail in arsenic-life's coffin.[14] Life was still life, just as we knew it.

There's a lot in the arsenic-life episode that should be celebrated: a surprising claim was immediately subjected to harsh testing by the scientific community. Making us revise our opinions is precisely what science is there to do, and this was in many ways a textbook case of the process going right. And yet the fallout was harsh: Wolfe-Simon published only one more paper after the arsenic-life affair and has since moved into teaching rather than research. NASA, meanwhile – probably the world's best-known scientific institution, inspiring awe and admiration in equal measure – seriously damaged the credibility of any of its future press releases. The problem, of course, was that they had overhyped the finding, backing their researchers into a defensive corner and ultimately devaluing the currency of their scientific publications.

We shouldn't discount the possibility, concerning as it might be, that NASA genuinely thought it was a good idea to hype up such preliminary results. A likelier culprit, though, is financial pressure. Scientific institutions work hard to convince their funders – in NASA's case, the US government, who could cut their grant in any given federal budget – that they're doing something worthwhile. As one post-mortem of the arsenic-life study put it, NASA must 'project an ongoing sense of relevance'.[15] It's easy to see how this urge, taken to an extreme, could lead to such an overcooked press release.

Scientists, of course, are subject to the same kinds of demands. They're dependent on grants to support their research, and work in an atmosphere that favours showy and ostentatious findings over workhorse studies that only add small pieces to our knowledge. In the next chapter, we'll see how these kinds of explanations might help us understand, though certainly not justify, the scientific problems we've encountered in the book. First, though, we'll look at the various ways that scientists misleadingly promote their research. It's the contention of this chapter that, far from their work just being misrepresented by the popular press, it's scientists themselves who have become a major source of hype. Every time they hype or misrepresent one of their results, they risk making a dent in our collective trust in science. And sometimes, when the hype reaches fever pitch, they risk corrupting the credibility of entire fields of research.

At the centre of the arsenic-life story was NASA's press release. Many people don't realise that scientific press releases aren't written only by press officers and PR agents: the scientists themselves are heavily involved. Indeed, sometimes they draft the entire release. The scenario where an innocent researcher is minding their own business when the media suddenly seizes on one of their findings and blows it out of proportion is not at all the norm.[16] The main problem with press releases isn't that they routinely report earth-shaking findings that turn out to be erroneous. Instead, it's that they puff up the results of often perfectly serviceable scientific papers, making them seem more important, ground-breaking, or relevant to people's lives than they really are. A 2014 study led by researchers at Cardiff University looked through hundreds of press releases for health-related scientific studies, matching them to the studies they were describing and to the eventual news stories they produced.[17] It found that the press releases commonly engaged in three kinds of hype.

The first was *unwarranted advice*: press releases gave recommendations for ways readers should change their behaviour – for example, telling them to engage in a specific kind of exercise – that was more simplistic or direct than the results of the study could support. This was found in 40 per cent of the audited press releases. Another variety of hype was the *cross-species leap*. As we've seen previously, lots of preclinical medical research is done using non-human animals like rats and mice – a practice known as translational research, or animal modelling.[18] The idea is that the basic principles of how, say, the brain or the gut or the heart work can be studied in the animal 'model' and then, with lots of work, the findings will eventually translate to humans, helping us design better treatments. Yet there are a lot of steps between making a discovery in mice (or in cells in a dish, or in computer simulations) and it being relevant to humans. There's a whole cycle of development, validation and trials that must occur first, a painstaking process that can take decades.[19] The vast majority of results from mice, somewhere around 90 per cent, don't end up translating to human beings.[20]

Animal researchers, of course, are well aware of this. Nevertheless, the Cardiff team found that it didn't stop them from hyping up their press releases to imply, or even claim explicitly, that their initial-stage animal-based results had important human implications: 36 per cent of press releases did this. News stories about health research, in turn, frequently bury the admission that the study they're describing wasn't done in humans somewhere in the eighth or ninth paragraph. The psychophysiologist James Heathers has set up a novelty Twitter account that exists solely to retweet misleading news headlines from translational studies, such as 'Scientists Develop Jab that Stops Craving for Junk Food' or 'Compounds in Carrots Reverse Alzheimer's-Like Symptoms' with a simple but accurate addition: '… IN MICE'.[21]

The third kind of hype found by the Cardiff team was possibly the most embarrassing. Everyone, especially scientists, is supposed to know that *correlation is not causation*.[22] This basic insight is taught in every elementary statistics course and is a perennial feature of public debates about science, education, economics and more. When scientists look at an *observational* dataset, where data have been gathered without any randomised experimental intervention – say, a study charting the growth in children's vocabulary as they get older – they're generally just looking at correlations. That's nothing to be ashamed of: there's a lot we can learn about how things relate to each other in the world and building up an accurate picture of patterns of correlation is an essential foundation for understanding systems like the brain or society. We need to be awfully careful about how we interpret those correlations, however. If we find that drinking more coffee is correlated with having a higher IQ (which, by the way, it is), we can't conclude that 'coffee raises your IQ'.[23] The causal arrow could just as easily point in the opposite direction, with being smarter making you drink more coffee. Alternatively, a third factor – such as being from a more affluent socioeconomic class, which might make you healthier and thus give you a higher IQ, as well as leading you to drink more coffee because it's fashionable in your social circles – could be causing both of the others.[24] These points are straightforward and boringly well-rehearsed, yet 33 per cent of the press releases in the Cardiff study threw causal caution to the wind and made it sound as if their observational, correlational results came from a randomised experiment that could reveal what caused what.[25]

Hype in press releases was linked to hype in the news. The Cardiff researchers found that if the press release exaggerated the claim first, similar exaggeration in the media was 6.5 times more likely for advice claims, 20 times more likely for causal claims, and a whopping 56 times more likely for translational claims. (When the press release was more circumspect, journalists only exaggerated a small amount.) And although this itself was merely a correlational study, the Cardiff team followed up in 2019 with an impressive randomised trial.[26] They worked with university press offices to modify randomly selected press releases by adding unwarranted causal statements into them, and compared their effects to releases that were more aligned with the evidence. As went the press releases, so went the headlines: exaggeration caused exaggeration. Another trial from 2019 filled in the next part of the story: hyped health news stories really did make readers more likely to believe a treatment was beneficial.[27]

In an age of 'churnalism', where time-pressed journalists often simply repeat the content of press releases in their articles (science news reports are often worded virtually identically to a press release), scientists have a great deal of power – and a great deal of responsibility.[28] The constraints of peer review, lax as they might be, aren't present at all when engaging with the media, and scientists' biases about the importance of their results can emerge unchecked. Frustratingly, once the hype bubble has been inflated by a press release, it's difficult to burst. A study in 2017 found that only around 50 per cent of health studies covered in the media are eventually confirmed by meta-analyses (that is, 50 per cent are found to be broadly replicable). This finding would be scandalous enough on its own, but what makes it even worse is the fact that those meta-analyses are rarely, if ever, covered in the press.[29] By that time the damage might already have been done. With apologies to Jonathan Swift: hyped science flies, and the refutations come limping after it.

*

Arguably, the kind of ephemeral hype that's spread by news articles isn't the most concerning sort. The hype that makes its way into books, however, is of a different order. When popular books by scientists hit the zeitgeist, they can create ideas that stick. At their best, books can translate complex scientific results for the general reader in ways that neither exaggerate nor distort, providing new means to think about ourselves and the world. At their worst, they become part of a Wild West of hype, far outside the jurisdiction of the peer-review sheriffs.[30] Once again, my own field of psychology, which easily lends itself to the hugely successful genres of self-help and life advice, is one of the worst offenders.

A particularly influential example is the idea of the 'growth mindset'. Having a growth mindset means believing that it's possible, if you work hard, for your brainpower to improve rather than remain static throughout your life. What you want to avoid is a 'fixed mindset', where you have no faith that you can develop your abilities. The originator of the mindset concept, the Stanford psychologist Carol Dweck, has published hundreds of scientific papers on it, but her book *Mindset: The New Psychology of Success* is where the idea really hit the big time. She makes the notion of mindsets sound potentially life-altering. 'The view you adopt for yourself profoundly affects the way you lead your life,' writes Dweck. And: 'When you enter a mindset, you enter a new world.'[31] Indeed, when you learn about mindsets, 'You'll suddenly understand the greats – in the sciences and arts, in sports, and in business – and the would-have-beens. You'll understand your mate, your boss, your friends, your kids. You'll see how to unleash your potential – and your children's.'[32]

Although it mainly consists of illustrative anecdotes, the success and influence of Dweck's book (and of the idea of mindsets in general) is based on Dweck being a scientist. And not just any scientist: a world-leading professor at a top university, who is, as she writes at the very beginning of her book, sharing her scientific research. Dweck's ideas have become a veritable craze in education: a 2016 survey of American teachers found that 57 per cent had received training on growth mindset principles, and that 98 per cent agreed that using ideas about growth mindset in the classroom would improve their pupils' learning. Thousands of schools in the UK mention their growth mindset policy on their websites.[33]

What does the best research on growth mindset tell us? Over 300 studies on mindset were meta-analysed in 2018.[34] The meta-analysts looked at studies on the correlation between people's growth mindset (measured using questionnaires) and their school or university achievement, and also at experiments that tried to induce more of a growth mindset in students to try to improve their grades. In both cases, the effects were real – but weak. The correlational part found that mindset accounted for about 1 per cent of the variation in grades. The effort to induce a growth mindset in students, which compared an experimental group that received mindset training with a control group that did not, didn't fare much better. If there was no mindset effect at all, we'd expect 100 per cent overlap between the distributions of school grades in these two groups (the distributions would be identical). And while the intervention shifted the grade distributions apart, it was only slightly: after mindset training, there was a 96.8 per cent overlap.[35] These aren't big impacts.

Even with such a small benefit, if you could roll it out across thousands or millions of students, then on aggregate you might do a decent amount of good.[36] But that's not how Dweck chose to frame growth mindset, nor would that kind of framing have made parents and teachers flock to buy her book. Instead, she hyped up its individual effects, making it sound almost revelatory.[37] The risk of such overhyping is that teachers and politicians begin to view ideas like mindset as a kind of panacea for education, focusing time and resources on them that might be better spent on dealing with the complex web of social, economic and other reasons that some children fail at school. Reality can't help but pale in comparison to the bombastic claims in Dweck's book – claims, incidentally, that go against the intellectual humility that science demands. As we saw in the previous chapter, complex phenomena are made up of lots of small effects: scientists should know better than to promote the idea that there's a single 'quick fix' for anything as complex as a child's education.[38]

To be charitable to Dweck, the meta-analysis came over a decade after the publication of *Mindset* in 2006. Perhaps it was unclear then how things would turn out for the findings (though this might be a good argument for that humility I've just mentioned). Other scientists-turned-authors have no such excuse. The Yale psychologist John Bargh was the lead author of the 'priming the idea of elderly people makes you walk more slowly' study – a study, as we saw in Chapter 2, that failed to replicate in 2012 when tested with a bigger sample and a more rigorous experimental setup.[39] But in 2017, years after that failed replication effort and the unfolding of the replication crisis in psychology more generally, Bargh published a bestseller called *Before You Know It: The Unconscious Reasons We Do What We Do*.[40] The book makes the case for strong subliminal influences on human behaviour, yet not only fails to mention any of the big replicability problems in the field, it blithely continues to cite social-psychological studies – often ones with tiny sample sizes and borderline results – to make head-turning claims about human behaviour.[41]

In the book's introduction, for instance, Bargh writes that unconscious influences 'can even affect your future employment and the salary you're able to negotiate – all depending on what kind of drink your prospective employer is holding in his or her hand, or the type of chair they're sitting in.'[42] The chair claim comes from a study of fifty-four participants, which found that certain types of people tended to admit more racist attitudes when sitting in Bargh's office chair, which apparently 'primed' them to feel more powerful, as opposed to a smaller chair on the other side of the desk, which made them feel less so.[43] The cup claim – specifically, that after holding a hot drink, people judge others as more likeable, thus being metaphorically 'warm' towards them – came from a study with forty-one participants that has failed to replicate in much larger studies.[44] Even setting aside concerns about replicability, though, note that neither study had anything to do with 'prospective employers'. Bargh breezily transplanted the findings from the context of his small-scale studies on undergraduates to a setting where they simply haven't been tested – a classic case of overclaiming based on a limited set of results.

The books we've talked about so far are examples of authors hyping up a thin evidence base.[45] But here's an example of a hugely popular book whose claims have been criticised as blatant misinterpretations of the underlying science. In 2017, the neuroscientist Matthew Walker, a professor at the University of California, Berkeley, published *Why We Sleep*, a book making the case that we should all get eight hours per night or suffer terrible health (and other) consequences.[46] Like the other books we've discussed, *Why We Sleep* was a bestseller across the world. Walker also gave a successful TED talk entitled 'Sleep is Your Superpower', garnering around 10 million views.[47] Richard Smith, ex-editor of the *BMJ*, called *Why We Sleep* 'one of those rare books that changes your worldview and should change society and medicine'.[48]

Walker's claims are anything but soporific. Just in the first chapter of the book, he writes that 'the shorter you sleep, the shorter your lifespan' and that 'routinely sleeping less than six or seven hours a night demolishes your immune system, more than doubling your risk of cancer'.[49] Both statements go against the evidence. In an article in 2019, the writer and researcher Alexey Guzey attempted to track down the sources for many of Walker's claims.[50] First, he found that the studies actually show a U-shaped relation between sleep length and risk of mortality: people who sleep for longer than eight hours each night have shorter lifespans, just like those who sleep for five hours or less.[51] Second, the claim about shorter sleep increasing cancer risk by 'demolishing' the immune system (which, by the way, is an example of causality being drawn erroneously from correlational data) is inconsistent with the evidence: the increase in cancer risk for those with less sleep is weak at most, and probably non-existent.[52] Guzey also criticised a host of other claims made throughout the book and noted that in one case, Walker showed only part of a graph about the relation between sleep and risk of injury – leaving out the inconvenient portion of it showing that people with five hours of sleep per night were less likely to get injured than those with six.[53]

Naturally, this isn't to say that sleep, or getting more of it, doesn't matter. It's not to say that other parts of *Why We Sleep* aren't more accurate. But it serves to illustrate how scientific hype, far from being restricted to self-help narratives, can sometimes take the form of catastrophising. Walker *could* have written a far more cautious book that limited itself to just what the data showed, but perhaps such a book wouldn't have sold so many copies or been hailed as an intervention that 'should change science and medicine'. As it is, the book could mislead people into needless anxiety about how much sleep they're getting, or into wasting their time sleeping more than is necessary. As a matter of factual accuracy, the sheer density of misleading statements in a book that sold in such numbers should keep us all awake at night.

Maybe I'm missing the point here. Maybe popular-science books, which are commercial enterprises, don't need to be 100 per cent rigorously accurate, resistant to every nit-picking criticism. Maybe writing easy-to-digest treatments of scientific findings, even if they're a little on the simplified side, is beneficial overall, since it promotes science and makes it relevant to people's lives. And wouldn't we rather that such books were being written by people who at least *nod* at the evidence? There's some merit in this kind of argument, but it's bad news in the long run. Letting the facts slide in favour of a good story risks a race to the bottom, with science books getting published that are ever more inaccurate and ever more divorced from the data. When these books are inevitably debunked, or the lifestyle changes they recommend fail to live up to the hype, damage is done to the reputation of science more generally. The books we've just discussed were by professors at Stanford, Yale and Berkeley, respectively. If even top-ranking scientists don't mind exaggerating the evidence, why should anyone else?

The simplicity of pop-science books also runs counter to one of the facts that should, by this point in the book, be deafeningly clear: science is complicated. Even the best writers struggle to get across in an engaging way the sawtooth of scientific progress, where findings are often contradictory and confusing, and where our best evidence is sometimes suddenly undercut by new data. By smoothing out these intricacies, and by implying that complex phenomena have simple, singular causes and fixes, such books contribute to an image of science as something it isn't.[54] Unfortunately, as we'll see next, the hyped expectations nurtured by pop-science might even have begun to affect the practice of science itself.

Good news! Scientific innovation is increasing! Or, at least, that's what you'd conclude if you took seriously the language used in scientific journals. An analysis in 2015 charted how frequently particular positive-sounding words appeared in the abstracts of scientific papers each year.[55] Abstracts are those prefatory summaries in scientific papers, where scientists are trying to grab the readers' attention; in the ever more crowded scientific marketplace, they need to work harder and harder to do so. The 2015 analysis plotted the proportion of abstracts that contained particular terms, starting in 1974. The use of the words 'innovative', 'promising' and 'robust' increased exponentially; 'unique' and 'unprecedented' (perhaps paradoxically) became vastly more common; 'favourable' rose steadily.[56] 'Groundbreaking' trundled along near zero until around 1999, when for some reason it suddenly shot upward. On average, the use of positive words in abstracts increased by nearly nine-fold over the analysed period: just 2 per cent of abstracts in 1974 had contained such self-praise, compared to 17.5 per cent in 2014. The authors wryly concluded that by 'extrapolating the upward trend of positive words over the past forty years to the future, we predict that the word 'novel' will appear in every [abstract] by the year 2123'.[57]

It seems doubtful that scientific innovation has genuinely accelerated alongside the dramatic upsurge in hyperbolic language.[58] A more likely explanation is that scientists are using this kind of language more frequently because it's a great way to make their results appeal to readers and, perhaps more importantly, to the reviewers and editors of big-name journals. The most glamorous journals state on their websites that they want papers that have 'great potential impact' (*Nature*); that are 'most influential in their fields' and 'present novel and broadly important data' (*Science*); and that are of 'unusual significance' (*Cell*) or 'exceptional importance' (*Proceedings of the National Academy of Sciences*).[59] Conspicuous by their absence from this list are any words about rigour or replicability – though hats off to the *New England Journal of Medicine*, the world's top medical journal, for stating that it's looking for 'scientific accuracy, novelty, and importance', in that order.[60]

The steep rise in positive-sounding phrases in scientific journals tells us that hype isn't just restricted to press releases and popular-science books: it has seeped into the way scientists write their papers. In the scientific community, this species of hype is often referred to with a term borrowed from politics: *spin*. One analysis from 2010 took a representative sample of randomised medical trials with null results (in other words, trials that found no difference between the treatment being tested and a placebo), and examined how much spin – defined as the presence of language whose purpose was to distract the reader from the absence of positive findings – they contained.[61] 68 per cent of the papers' abstracts and 61 per cent of their main texts contained attempts to highlight the benefits of the treatment, even though it had failed the trial. 20 per cent of the articles contained at least one instance of spin in every section

(Introduction, Method, Results and Discussion). 18 per cent had spin even in their titles.

A common form this spin takes is the weasel-wording that scientists use about non-significant p-values. Recall from Chapter 4 that you normally need to have $p < 0.05$ for your effect in order to declare it 'statistically significant'. The statistician Matthew Hankins has amassed a collection of genuine quotes from published papers where p-values remained stubbornly above that threshold, but whose authors clearly had a strong desire for significant results:

- 'a trend that approached significance' (for a result reported as '$p < 0.06$')
- 'fairly significant' ($p = 0.09$)
- 'significantly significant' ($p = 0.065$)
- 'narrowly eluded statistical significance' ($p = 0.0789$)
- 'hovered around significance' ($p = 0.061$)
- 'very closely brushed the limit of statistical significance' ($p = 0.051$)
- 'not absolutely significant but very probably so' ($p > 0.05$)[62]

There's a whole literature of studies by scientific spin-watchers, each of them highlighting spin in their own fields. 15 per cent of trials in obstetrics and gynaecology spun their non-significant results as if they showed benefits of the treatment.[63] 35 per cent of studies of prognostic tests for cancer used spin to obfuscate the non-significant nature of their findings.[64] 47 per cent of trials of obesity treatments published in top journals were spun in some way.[65] 83 per cent of papers reporting trials of antidepressant and anxiety medication failed to discuss important limitations of their study design.[66] A review of brain-imaging studies concluded that hyping up correlation into causation was 'rampant'.[67] Some forms of spin shade into fraud, or at least gross incompetence: a 2009 review showed that, of a sample of studies published in Chinese medical journals that claimed to be randomised controlled trials, only 7 per cent actually used randomisation.[68]

Even meta-analyses aren't safe, as we've seen before. A 2017 review of meta-analyses on diagnostic tests (for example, blood tests for Alzheimer's disease) found that 50 per cent of them drew a positive conclusion about how well the test worked despite finding trivial, statistically non-significant effects in their analyses. The spin, the review concluded, 'could lead to unjustified optimism about the performance of tests in clinical practice.'[69] This seems to be another example of how scientists' urge to hype their results misleads the people who most rely on them.[70]

All this spin serves the same ultimate purpose as exaggerations in press releases and books: scientists want to emphasise the impressiveness and 'impact' of their work because impressive and impactful work is what attracts grants, publications and plaudits. The problem is that this can create a feedback loop: the hype nurtures an expectation of straightforward, simple stories on the part of funders, publishers and the public, meaning that scientists must dumb down and prettify their work even more in order to maintain interest and continue to receive funding. The science itself comes out of this feedback loop looking very unhealthy indeed.

Now that we've seen how hype in the media and in scientific papers can link together, let's examine one scientific field where that feedback loop is particularly strong.

At any given time, there's usually an 'emerging' field that's subject to the worst hype. Typically, a few publications with easy-to-grasp results in big-name journals get picked up by the media, public interest intensifies, and scientists in the field develop a kind of recklessness, feeding the hype cycle with careless and overblown statements. Then big claims fail to replicate in later experiments, the furore dies away and normal science resumes. Ultra-hyped fields include stem cells, genetics, epigenetics, machine learning and brain imaging; for the past few years, a strong contender for the 'most hyped' award has been research on *the microbiome* – the countless millions of microbes that inhabit our bodies.[71]

Thanks to the hype, the microbiome has been targeted by a plethora of products and treatments. So-called 'probiotics', drinks or pills that top up the 'good bacteria' in your intestines, have become a multi-billion-dollar industry.[72] There's also increasing interest in a therapy known as a 'faecal transplant'.[73] This is where stool samples from a healthy donor, replete with their various microbes, are transferred to a patient – usually via a colonoscopy but sometimes via swallowable capsules.[74] Though at first the idea might sound as implausible as it does unpleasant, there *is* solid evidence for the effectiveness of at least one type of faecal transplant, in the case of recurrent gut infections by the *Clostridium difficile* (or *C. diff*) bacterium. In severe cases where antibiotics have failed to defeat the *C. diff* but have cleared out all the good bacteria in the process, the faecal transplant allows the patient to have a dose of microbes from someone whose gut is in better shape, helping them win the battle against the bad bacteria.[75]

Where we should be wary, though, is when the microbiome is invoked as a contributing factor to diseases and conditions that have no obvious link to the gut. This is where the claims and the reality truly begin to part ways. Reading the scientific literature, one could get the impression that the microbiome is the cause of, and solution to, a truly remarkable array of mental and physical problems. For example, studies have appeared claiming links between the microbiome and depression, anxiety and schizophrenia, while faecal transplants have been proposed for the treatment of, among other conditions, heart disease, obesity, cancer, Alzheimer's disease, Parkinson's disease and autism.[76] The idea is that the various activities and fermentations of microbes in the gut can produce harmful chemicals that travel throughout the body, causing problems well beyond the intestines.[77]

The evidence underlying these claims is usually less than impressive. In the case of autism, the gap between the data and the hype is especially wide.[78] In 2019, a paper in the world-leading journal *Cell* reported an experiment where the authors had taken faecal samples from sixteen children either with or without an autism diagnosis and transplanted them into mice.[79] They then bred those mice in germ-free conditions, meaning that the offspring had known only the humans' microbes for their whole lives (autism is a developmental disorder, so any influence of the microbiome would need to be in effect from the very beginning of life). Depending on whether their guts had been colonised by microbes from autistic or non-autistic donors, the offspring mice were reported to behave differently in tasks that were designed to elicit the rodent equivalent of the symptoms of autism. When placed in a cage with another mouse, for instance, those mice that had received the microbes of autistic children were less likely to approach it, thus mimicking autism's social impairments; they also spent more time burying marbles in a sawdust-filled cage, which apparently relates to the repetitive behaviours often seen in those with the condition.[80]

You might think that the link between these behaviours in mice and autism in humans is, to say the least, somewhat tenuous. You might also wonder whether such a tiny number of donors could possibly be representative of people with autism in general.[81] Nevertheless, the authors were happy to draw an impressive conclusion: 'microbiota-based interventions such as probiotics [or] fecal microbiota transplantation ... may offer a timely and tractable approach to addressing the lifelong challenges of [autism spectrum disorder].'[82] They put out a press release discussing the 'profound' effects of the faecal transfer, claiming again that their research implied that probiotics might one day be used to treat the symptoms of autism.[83] This was all the purest hype. Even setting aside both the study's trifling sample size and its rather questionable assumptions about the equivalence of human and mouse behaviour, the experiment had at no point tested the power of probiotics or transplants to *reduce* any of the supposed rodent-autism symptoms – let alone the 'equivalent' in humans.

The authors also appeared to attempt some spin by omission. Their study had included a second test of sociability, in which the mouse could choose to spend time with either a fellow mouse or 'a small object'. The hypothesis was that the mice with microbiomes from autistic people would choose the object over the mouse companion – but they showed no difference. As the science writer Jon Brock noted in a detailed critique of the study, the authors quickly skipped over this inconvenient result in a single sentence, whereas all the results that turned out to be significant were featured in a full-colour graph in the paper.[84]

Hyping and spinning such a tiny, preliminary study would have been bad enough, but it gets worse. When the biostatistician Thomas Lumley obtained the authors' data and attempted to reproduce their analysis, he found that they'd fouled up their statistical tests. They'd run the analysis as if every mouse had its own human faecal donor, but there were actually a very small number of donors shared between 100 mice.[85] When the proper tests were applied, only the marble-burying result survived, and with the somewhat borderline p-value of 0.03. Despite those strong criticisms, to my knowledge the authors haven't responded.

Not all studies of the microbiome are fundamentally statistically flawed like the mouse-autism paper, though many are just as shaky in terms of the over-the-top conclusions they draw. A 2019 study that followed a similar methodology to the autism paper argued that transferring the microbiomes of schizophrenia patients to mice can cause the rodents to display symptoms of psychosis. It concluded by saying that the results 'may lead to new diagnostic and treatment strategies' for schizophrenia, which seems more than a little premature.[86] Nevertheless, it could still turn out that differences in the microbiome do play some role in the complex causes of autism or schizophrenia symptoms, or those of some of the other conditions listed above, in mice or in people.[87] However, microbiome researchers need to accumulate solid research over time instead of thrusting into the media every small, possibly *p*-hacked study that finds an effect, claiming it's a huge scientific breakthrough. It might even be the case that the number of press releases is related to the immaturity of a scientific field, with more media attention given to fields with lots of 'promising' results, but fewer well-replicated ones.

There have been recent calls from within the scientific community to cool down the gee-whiz hype surrounding the microbiome and its associated treatments, and to improve the quality of the research.[88] In the meantime, the grossly exaggerated claims of these papers and press releases provide the semblance of scientific backing to a host of useless, harmful, or just plain daft microbiome-related remedies: a probiotic drink made using microbes found in the guts of elite athletes that can supposedly boost your performance; the craze for 'colonic irrigation', which involves flushing out your bowels with water and comes with ghastly-sounding risks like 'rectal perforation'; and a direct-to-consumer microbiome testing company that allows you to discover 'the nationality of your microbiome'.[89]

Fads like microbiome mania wax and wane, but there's one field of research that consistently generates more hype, inspires more media interest and suffers more from the deficiencies outlined in this book than any other. It is, of course, nutrition. The media has a ravenous appetite for its supposed findings: 'The Scary New Science That Shows Milk is Bad For You'; 'Killer Full English: Bacon Ups Cancer Risk'; 'New Study Finds Eggs Will Break Your Heart'.[90] Given the sheer volume of coverage, and the number of conflicting assertions about how we should change our diets, little wonder the public are confused about what they should be eating. After years of exaggerated findings, the public now lacks confidence and is sceptical of the field's research.[91]

Nutritional science, like psychology, has been going through its own replication crisis. Some of this may have to do with fraud: for instance, the cardiologist Dipak Das, who published dozens of much-cited papers on the heart-health benefits of resveratrol, a substance abundant in red grape skins (and thus in red wine), was fired by the University of Connecticut in 2012 after he was found to have faked data in nineteen studies.[92] Some is likely due to biases: many studies are funded by the food industry.[93] Moreover, many researchers adhere to the diets they are themselves investigating, giving them a personal incentive to find evidence of their benefits.[94]

Finally, some of the problem is due to the many other kinds of bias and error with which we're now familiar. Take, for example, the idea that we should eat fewer saturated fats and more unsaturated ones. This is a cornerstone of nutritional advice, repeated in countless dietary guidelines.[95] But it didn't hold up in a 2017 meta-analysis that compared saturated fatty acids to polyunsaturated fatty acids for their effects on heart disease and death.[96] This was probably for three reasons. First, there were the tell-tale signs of publication bias, with lopsided funnel plots suggesting small-sample, small-effect papers were being file-drawered.[97] Second, one of the trials that claimed to be randomised was subject to an error which meant that it probably wasn't.[98] Third, many trials were designed incompetently, with changes in factors other than the diets that could have affected their results.[99] The conclusion of the meta-analysis was that there was little compelling evidence for the benefits of replacing saturated with unsaturated fat, and that previous meta-analyses – the ones on which several governments had based their nutritional advice – simply hadn't noticed all the problems.

It's safe to say that a substantial portion of media-hyped nutritional studies are also affected by *p*-hacking. Because so many large datasets exist with so many relevant variables – it's typical in nutrition research for participants to fill in a so-called Food Frequency Questionnaire that indexes everything they've eaten in, say, the preceding week – there is plenty of opportunity to dredge through the data for anything that happens to be statistically significant. This is part of the reason for the muddled, self-contradictory smorgasbord of correlational studies that exists in nutrition research. In a now-classic paper entitled 'Is everything we eat associated with cancer?', researchers Jonathan Schoenfeld and John Ioannidis randomly selected fifty ingredients from a cookbook, then checked the scientific literature to see whether they had been said to affect the risk of cancer.[100] Forty of them had, including bacon, pork, eggs, tomatoes, bread, butter and tea (essentially all the aspects of that Killer Full English). Some foods apparently raised the risk, some reduced it, and some had different effects in different studies. We know that numbers are noisy, so we'd expect the literature to look untidy. We should, however, ask ourselves what's more likely: that 80 per cent of a sample of commonly consumed foods genuinely have an impact, often a large one, on our risk of cancer; or that the low standards of the field led to the publication of poor-quality, *p*-hacked studies that capitalised on chance, misleading us into thinking many staple foods are either dangerous or guarantors of a long, healthy life?[101]

Nutritional research often fails to live up to its hype because it relies so heavily on observational studies rather than experiments. In other words, many studies simply collect data on what people eat, without any randomised controlled trials. There's at best a shaky correlation between the foods identified as healthy and unhealthy in observational research and those identified as such in randomised trials – so one of the two types of research must be getting it wrong.[102] There's also the issue that differences in people's health may be driven not by the foods they eat but rather by cultural or socioeconomic factors that affect both their health *and* their diets. Indeed, any given aspect of someone's diet is probably correlated with some other aspect, thereby complicating the analysis of effects: people who eat more eggs probably also eat more bacon and sausages, and probably many other foods and nutrients that you haven't thought to ask about. There are statistical methods to 'adjust for' these kinds of confounds, but they're tricky to get right and they rely on you having measured every kind of food and nutrient intake that might be important.[103] The accuracy of such measurement is itself a big source of contention and there's a surprisingly vituperative debate about how obser- vational studies record food consumption. Some researchers have called Food Frequency Questionnaires 'fatally flawed', in part because they depend on people's often inaccurate memory of the foods they've eaten.[104] They're also distorted by social desirability bias, where participants are reluctant to disclose that they ate, say, five double cheeseburgers in the last seven days.[105]

One suggestion that's been made for improving nutritional epidemiology is for researchers to take the resources they've been pouring into the curation of observational datasets and instead run a series of very large, straightforward 'megatrials' that'll nail down, to everyone's satisfaction, a set of facts about the optimal diet.[106] The problem is that large-scale nutritional trials are anything but straightforward.[107] One of the largest ever randomised trials in nutritional epidemiology, with over 7,000 participants, was published in the *New England Journal of Medicine* in 2013, and focused on the Mediterranean Diet.[108] Compared to controls who were advised to follow a low-fat diet, the participants who followed the Mediterranean Diet (more white meat and seafood; more nuts and legumes; more olive oil) had a substantially lower score on a measure that reflected the number of strokes, heart attacks and deaths due to cardiovascular events over the next five years. The Spanish researchers running the study announced that they themselves had converted to the Mediterranean Diet after they'd seen the results.

The study, called 'Prevención con Dieta Mediterránea', or PREDIMED for short, got all the media attention you might expect: 'Mediterranean Diet Shown to Ward Off Heart Attack and Stroke;'[109] 'Mediterranean Diet Reduces Cardiovascular Risk.'[110] Even the California Walnut Commission put out an excited press release.[111] And who could blame them? This was, we were told by PREDIMED's authors, a 'major study' that had 'proved' with 'strong evidence,'[112] the benefits of the Mediterranean Diet.

Then along came John Carlisle. You may remember Carlisle as the intrepid data-detective who audited thousands of 'randomised' controlled trials and found that many weren't randomised at all. PREDIMED was one of the trials caught by Carlisle's method: the baseline numbers just weren't consistent with the trial having been properly randomised.[113] The authors went back to their data and sure enough, they found some serious mistakes. Among other things, instead of each participant in the trial being randomly assigned one of the two diets, it turned out that whenever several participants came from the same household, they were all assigned the same diet. At one particular study site, meanwhile, the researchers had accidentally randomised by clinic rather than by participant, so that everyone who attended the same clinic got the same diet. Because people from the same household or the same clinic necessarily share all sorts of other influences, it becomes impossible to ascribe differences in outcome between them and others specifically to the diet itself.[114] These and other mistakes applied to 1,588 of the participants – fully 21 per cent of the sample.

The paper, which by this point had been cited over 3,000 times, was retracted, and replaced by a corrected version in 2018.[115] As it happens, the authors argue that the corrected results provide even *stronger* evidence for the benefits of the Mediterranean Diet. But there are reasons to remain cautious. For instance, when the three kinds of adverse effects tracked by the study are disentangled, the diet curiously only seems to affect the risk of stroke; there are no effects on heart attacks or mortality.[116] The study was also stopped earlier than originally planned due to the effects of the diet being so impressive, which is a controversial practice in the clinical trial literature.[117] And worryingly, 250 further papers have now been published using the PREDIMED data to look at other effects of the Mediterranean Diet, some of which appear to have unexplained numerical inconsistencies.[118] Whether they too have been affected by the randomisation problems remains unclear.

I include PREDIMED here not because it was a particularly bad example of hype, but because it's an example of some of the best research in an extraordinarily hyped field – and because it shows us how even a poster child for rigour might be subject to hidden flaws. Rather like psychology, nutritional epidemiology is *hard*. An incredibly complex physiological and mental machinery is involved in the way we process food and decide what to eat; observational data are subject to enormous noise and the vagaries of human memory; randomised trials can be tripped up by the complexities of their own administration. Given that context, the sheer amount of media interest in nutritional research is particularly unfortunate. Perhaps the very scientific questions that the public wants to have answered the most – what to eat, how to educate children, how to talk to potential employers, and so on – are the ones where the science is the murkiest, most difficult, and most self-contradictory. All the more reason that scientists in those fields need to take more seriously the task of sensibly communicating their findings to the public.

It's all well and good to advise against hype and to point out how complicated reality is, but scientists are still under pressure to get their results 'out there' into the world. The public deserve to be kept up to date with the latest scientific results, to which they've often contributed their tax money. Is there a way of communicating findings while avoiding the excesses we've encountered? Here's an example of how it should be done.

According to scientific consensus, nothing can move faster than the speed of light. This is the basis of Einstein's theory of special relativity and every result we've had from physics so far has supported it. This was what made a 2011 observation from the OPERA experiment so bizarre.[119] OPERA was a collaborative particle physics project that studied subatomic particles as they passed through the Earth's crust between the CERN laboratory in Geneva, Switzerland and a detector in Gran Sasso in Italy. The team working on it, consisting of around 150 scientists from many universities, had found that neutrinos (particles similar to electrons but without a charge) were getting to their destination far too quickly – they were reaching Italy 60.7 nanoseconds (that's 60.7 billionths of a second) faster than it would've taken light to travel the same distance.[120]

After a period of frantically checking their calculations and equipment, the OPERA physicists couldn't pin down any mistakes. The faster-than-light neutrino result appeared to be real. Since rumours were beginning to swirl, the scientists voted to inform the world. They published a working paper that described the finding, then put out a press release.[121] At this point, they could've followed NASA's 'arsenic-life' press strategy, and written about how our fundamental understanding of the universe had been overturned by these jaw-dropping new findings. Instead, they were cautious. There was no hype or spin. Rather, the announcement explicitly emphasised uncertainty: 'Given the potential far-reaching consequences of such a result, independent measurements are needed before the effect can either be refuted or firmly established. This is why the OPERA collaboration has decided to open the result to broader scrutiny.'[122]

The accompanying statements from OPERA researchers expressed bafflement with the results, calling them 'a complete surprise' and 'apparently unbelievable'. They then waited to see how the media would treat the news. Although there were some unfortunate headlines ('CERN Scientists "Break the Speed of Light"', said the *Daily Telegraph*; 'Could Time Travel be a Reality?' asked *Good Morning America* on ABC News), the scepticism still managed to sing through.[123] Most news articles carried quotations from the researchers saying that the results were really weird, special relativity had never before been proved wrong, verification from other experiments was required, and so on.

And then, of course, they found the mistake. It was a faulty connection: a loose fibre-optic cable was causing an underestimate of the time taken by the neutrinos. Once it was properly plugged in, the timings clicked into conformity with Einstein's theory.[124] A second experiment using different measurements confirmed it: neutrinos don't exceed the speed of light. Alas, this wasn't the end of the story: the whole affair had become something of an embarrassment to many of the physicists on the team, some of whom argued that they should never have written up the findings in the first place.[125] In spite of the caution with which they'd handled the media attention (and in spite of the collaboration voting in favour of publicising the results), the OPERA chairman and another coordinator marginally lost a no-confidence vote, and subsequently resigned.[126]

The resignations are a pity, because the way OPERA handled its unexpected result was close to exemplary. The physicists drew attention to a strange finding that needed replication while avoiding hype and expressing all the necessary reservations, giving the world a valuable lesson in scientific uncertainty. The initial flurry of reports was followed by further coverage of the story's resolution.[127] Is it too far-fetched to suggest that, if the OPERA scientists had been psychologists instead of physicists, they'd have skipped over the double-checking of their results and rushed to sign book deals for titles like *Breaking Barriers: What the Speedy Neutrino Means for Your Self-Confidence* and *Superluminary: The New Science of Subatomic Success*?

It might be unfair to hold up the OPERA story as an example of press releases done well, since it involved an exceptionally unusual event – the apparent breaking of the laws of physics – that would have attracted ample attention regardless of any hyping. But it still serves as an example of how scientists can inject caution into the news cycle right from the very beginning, going directly against the all-too-prevalent instinct to overplay findings. Imagine if all press releases – and all scientific papers, for that matter – came with built-in hype-busting statements reminding readers that the findings are tentative, and warning against taking them too far.

That, of course, would be to go against the way the scientific system is organised. Even though caution, restraint and scepticism are basic virtues of science, we have a system that incentivises the precise opposite. Scientists are pushed into publishing as many papers as possible, and hyping them up to the high heavens, by an academic system that's become an impediment to getting science right. That system, and how we might go about fixing it, is what we'll turn to in the final part of this book.

PART III

CAUSES AND CURES

7

Perverse Incentives

> If the rule you followed led you to this of what use was the rule?
> Cormac McCarthy, *No Country for Old Men* (2005)

The California wildfires of 2017 spread across more than a million acres, destroying thousands of buildings and killing forty-seven people. There followed the most expensive disaster clean-up effort in the state's history, costing $1.3 billion.[1] The U.S. Army Corps of Engineers was drafted in to organise it, and they in turn hired local contractors to remove the vast piles of debris the blaze had left behind. But the Corps of Engineers made a critical mistake: they paid by the ton. The heavier the load, the more money the contractors earned – and some of them exploited this to an absurd degree. A witness reported seeing workers 'inflate their load weights with wet mud'. Other contractors began to 'over-excavate' – instead of just picking up debris, they dug huge new holes, in some cases even ripping up the foundations of people's houses, packing their trucks with the resulting soil and concrete. In the end, the Californian government had to pay an additional $3.5 million *after* the clean-up to have more workers come in and refill all the holes the previous contractors had dug.

It was a classic case of a *perverse incentive*. The Corps of Engineers had incentivised not the clean-up, but the mere fact of having a heavier truck, inadvertently creating new problems. It isn't hard to think of similar examples from other fields: incentives for journalists that reward revenue rather than original reporting, leading to flimsy clickbait articles; incentives for teachers that reward school rankings rather than learning, leading to questionable marking; incentives for politicians that reward short-term vote gains rather than long-term solutions, leading to the subsidy of fossil-fuel industries.[2] In this chapter, we'll delve into the incentives that are built into today's scientific practice and ask whether they reward objectivity – or something else entirely.

So far in the book, we've seen scientists fabricating data, file-drawering and *p*-hacking their studies, failing to check for errors, and exaggerating their results. Taken together, we have a picture of scientific practice that's fundamentally at odds with the scientific ideal, and we've seen in depth how it happens. The piece of the puzzle that's only been hinted at so far, though, is *why*. Most scientists say that they chose their career because of a lifelong interest in nature, because of an inspiring science teacher or mentor, or because they want to make a difference to society.[3] And when they're explicitly asked, the overwhelming majority endorse all four of the Mertonian norms of universalism, communality, disinterestedness and organised scepticism.[4] So why do people who became scientists for the love of science and its principles end up behaving so badly?

Part of the answer was apparent back in the Preface, with the story of my psychic replication study and its instant rejection by the journal: null results and replication studies are of little interest to scientific journals, despite their crucial importance for a full picture of what the evidence shows. Because studies reporting positive, flashy, novel, newsworthy results are rewarded so much more than others, scientists are incentivised to generate them to the detriment of everything else. To convince the reviewers and editors that their papers really do have all those qualities, too many of them end up bending, or breaking, the rules.

This chapter goes even further. What we'll find is that the scientific incentive system engenders an obsession not just with certain kinds of papers, but with *publication itself*. The system incentivises scientists not to practise science, but simply to meet its own perverse demands. These incentives are at the root of so many of the dubious practices that undermine our research.

It's sometimes said, mostly seriously, that Charles Darwin was the last true scientific expert. He knew everything that there was to know at the time about his field of natural history, in no small part due to his global fact-finding voyages and his network of scientific correspondents whom he would – to use his own words – 'pester with letters'.[5] Nowadays, an all-knowing expert like Darwin couldn't exist in any scientific subject. That's because we're now drowning in scientific papers. A modern Darwin would have to keep up with 400,000 or so new studies published annually in the biological and biomedical sciences; across all scientific fields, a snapshot in 2013 found 2.4 million new papers published that year.[6] Another analysis, which zoomed out to encompass the entire history of science, showed that the upward trend is accelerating: there was a 0.5% annual growth rate between 1650 and 1750, a 2.4% growth rate between 1750 and 1940, and since then an 8 per cent growth rate. That latter rate means that the entire scientific literature will double in size every nine years.[7] In some ways this is a positive development: collectively, humans know vastly more about the world than we did in centuries past. But one has to wonder: is this massive proliferation of papers *only* a reflection of our growing knowledge?

There are reasons to think otherwise. Perhaps the most notorious example of a worrisome incentive in science is the cash-for-publications scheme. Since the early 1990s, Chinese universities have had a policy of paying scientists (at least those in the natural sciences) a cash bonus for every paper they publish in mainstream, international scientific journals.

The full details are unclear – one of the more comprehensive studies notes that many of the payment arrangements are kept secret – but the basic idea is that the cash reward increases as a function of the prestige of the journal where the paper is published, increasing substantially for the very high-end outlets.[8] If a scientist gets a paper into *Nature* or *Science*, they can, at least at some Chinese universities, look forward to a reward many times their annual salary.

This policy in China appears to be the most widespread and potentially lucrative, but versions of the same direct cash-for-publication scheme have been reported as government policy in Turkey and South Korea, and at certain universities in other countries, including Qatar, Saudi Arabia, Taiwan, Malaysia, Australia, Italy and the UK.[9] Paying by the paper sits very uneasily with the Mertonian norm of disinterestedness: scientists are not supposed to be in this game for their own pecuniary interests.[10]

The direct cash-for-publication scheme is among the more gauche policies that universities use to encourage scientists to publish as often as possible, but researchers are also under many subtler – yet still keenly felt – financial pressures. In the academic job market, hiring and promotion decisions are based in no small part on how many publications you have on your CV, and in which journals they're published. Publish too few papers, and publish them in too-obscure outlets, and you've a far lower chance of getting or retaining a job. In the American system, tenure – what happens when an assistant professor, the lowest rank in the academic faculty system, becomes an associate professor, and is at that point essentially guaranteed a job for life – is decided in large part using the same kinds of productivity measures.

Why, you might wonder, do universities prioritise this publication-based measure over others that might have more to do with the quality of research – for example, whether a scientist's work meet standards like randomisation or blinding, or even replicability? The answer is that they're subject to financial pressures as well. In many countries, including the UK, universities themselves are ranked by the government on the prestige of the papers their academics produce, with taxpayers' money divvied up accordingly.[11] All of this is what gives rise to the clichéd phrase 'publish or perish': keep cranking out the papers, and in the most impressive journals you can, or you'll never survive in the massively competitive world of modern academia.[12]

It's not just papers, either. We've previously seen that the first, necessary step on the road to doing a scientific study is usually getting a grant to pay for equipment, materials, data access, participant rewards and staff salaries. This means that scientists constantly need to be applying for grants to keep their research alive. And again, universities experience the same pressure. They take a slice of whatever grant money their scientists manage to win and use it to subsidise teaching, hiring, building maintenance, and so on. For that reason, they lean heavily on their academics to bring in the cash. In one study in the US, it was estimated that scientists spent on average around 8 per cent of their total working time and 19 per cent of their research time writing grant applications – and those both sound to me like rather low estimates.[13]

Having perpetually to seek funding is not just an expenditure of time: it leads to an enormous amount of failure and disappointment. The problem is compounded by what's known as the Matthew Effect: when scientific grants are allocated, the already rich get richer. (The name is a reference to Matthew 25:29: 'For unto every one that hath shall be given, and he shall have abundance: but from him that hath not shall be taken away even that which he hath.')[14] There's good evidence that this occurs: in one large study, scientists whose early-career grant applications were judged to be *just above* the arbitrary threshold for being funded subsequently got more than twice as much money in other grants across the following eight years as scientists whose scores had been *just below* that threshold, even though the quality of their initial applications can't have been all that different.[15] In this climate, many scientists quit the profession out of frustration, while those who stay are forced to hype up their grant applications to compete for money with the well-funded old guard. It's a toxic atmosphere, and it's not difficult to imagine how the accuracy of the science itself becomes an afterthought.[16]

Financial incentives aside, let's not forget the role of human nature: people have a natural tendency to compete intensely for status and credit, to collect reputation-burnishing achievements, and to work towards even objectively meaningless targets – in this case, a large number of publications and grants. Thus, for the more ambitious and competitive amongst us, a long CV can become its own reward. For some, simply getting one's name on a scientific paper, any scientific paper, feels like a meaningful accomplishment.

At any rate, incentivising scientists to publish more would appear to be having an effect. Not only has there been a huge increase in the rate of publication, there's evidence that the selection for productivity among scientists is getting stronger. A French study found that young evolutionary biologists hired in 2013 had nearly twice as many publications as those hired in 2005, implying that the hiring criteria had crept upwards year-on-year.[17] It's apt that the evidence was from evolutionary biology, since the competition for jobs is exactly the kind of selection process that gave the peacock its tail and the moose its antlers. In a competition for a scarce resource (for the peacock and the moose, mating opportunities; for scientists, grants and jobs), those with the more ostentatious traits kept winning, until ridiculously showy features – or ridiculously long CVs – had evolved. And those scarce resources are becoming ever-scarcer: as the number of PhDs awarded has increased (another consequence, we should note, of universities looking to their bottom line, since PhD and other students also bring in vast amounts of money), the increase in university jobs for those newly minted PhD scientists to fill hasn't kept pace.[18]

Maybe you're wondering why academics shouldn't perish if they don't publish. Isn't it a good thing that universities want their researchers to produce more findings, and more exciting ones at that, and share them with the world by publishing them in high-profile journals? Isn't this an appropriate and effective way to measure their success? And why *shouldn't* academics be paying their own way by competing for grants, with only the best ideas winning money? Otherwise, wouldn't we just get freeloaders loafing around at universities, never contributing anything to our knowledge?

In a perfect world, these kinds of productivity incentives would make sense. Journals would have such a degree of quality control and scientists would have such an innate integrity that despite the expanding number of papers published, quality would never be compromised. In reality, though, something has to give. In cognitive psychology experiments where participants must press a button as quickly as possible when, and only when, a light flashes, researchers talk about the 'speed-accuracy trade-off'. When the subjects focus on haste, their accuracy suffers; when they focus on getting it right, they have to slow down. (And by the way, this is a nice straightforward psychological effect that really does replicate, in almost every relevant experiment.)[19] It's no different in scientific publishing.[20] Time is finite. Pushing scientists to publish more and more papers and bring in more and more money – alongside all their other responsibilities, like teaching, mentoring, and administrative tasks – almost necessarily implies that they'll spend less time on each study. Pushing peer reviewers (who are, of course, busy scientists themselves) to review more and more submissions means that more research that's mistaken, overhyped or even fraudulent will get past the filter. It can't be surprising, in both cases, if standards slip.[21]

Evidence that quantity overrides quality for at least some scientists can be found in the cunning ways that many have found to exploit the system. The study cited earlier that analysed the Chinese cash-for-papers scheme gives one example:

> Professor Gao from Heilongjiang University published 279 papers in a single journal, *Acta Crystallographica Section E*, and received more than half of the total cash rewards given by Heilongjiang University between 2004 and 2009 ... Prof. Gao's only research focus in these five years was to find new crystal structures in his lab and always report the results of this to the same journal, because he could accomplish the goal of winning the cash bonus in a short term as contrasted with receiving fewer awards by conducting long-term research projects.[22]

In this, Professor Gao is the academic equivalent of the band Status Quo, who found success by churning out endless minor variations on the same couple of basic rock songs from the 1970s onwards. It's very far indeed from Charles Darwin. Gao was engaging in a process that's become known as 'salami-slicing': taking a set of scientific results, often from a single study, that could have been published together as one paper, and splitting them up into smaller sub-papers, each of which can be published separately.[23] It's the closest analogue to the disaster-clean-up drivers who loaded their trucks with wet mud just to make them heavier: it artificially expands your CV, making it look like you've done much more research and, in some academic systems at least, maximises your profits. Basically, imagine I published each chapter of this book, as well as its preface and epilogue, separately and announced I'd written ten books, then was paid separately for each of them.

One of the more ridiculous salami-slicing incidents I've recently encountered is that of some geneticists studying the genes linked to psychiatric disorders, using the genome-wide association study (GWAS) method that we discussed in Chapter 6. Humans have twenty-three pairs of chromosomes that carry their genetic material, and a standard GWAS involves scanning every pair at once (hence the 'genome-wide' part) to find any links to the trait in question. These geneticists, however, instead of doing this standard, large-scale analysis, wrote up an analysis of *every chromosome pair separately*. That is, they turned the single paper you'd usually get from such a study into a potential twenty-three individual publications. At the time of writing, they've successfully published six of them.[24] Although the sheer cheek of this is amusing, and although it's surely a boon for the authors' CVs and perhaps their bank balances, it's a drag on science. Interested readers will eventually have to read nearly two dozen papers to find the information that should've been included in one. It's a needless annoyance, wasting of the time of the editors and peer reviewers who have to vet each individual paper. Moreover, the more scrupulous scientists who aggregate their own results into larger, fuller papers are at a disadvantage in a world where the sheer volume of publications can be such a help on the job market.

Salami-slicing doesn't in itself mean that the science contained in each of the slices is necessarily of poor quality (though the fact the researchers are willing to take advantage of the publication system so blatantly doesn't exactly speak to their trustworthiness). However, some salami-slicing may have a more sinister purpose than mere CV-building. In clinical trials, it's been alleged that pharmaceutical companies and other drug researchers use tactical salami-slicing to take advantage of the fact that readers aren't paying full attention to every publication. Split up your study into several publications and you'll give the impression that there's stronger support for the efficacy of your drug than if there were just one or two papers published on it. It's a devious, but probably effective, tactic: busy doctors who see that there are six papers claiming support for one drug and only one paper for another might be more likely to prescribe the former, without necessarily noticing that the six papers on the latter are all sliced-up reports of the same study. Moreover, since not every doctor reads the same journals, splitting up the publications might help you reach a wider audience.

One investigation found extensive salami-slicing of studies on the antidepressant duloxetine. To take just one of many examples, the duloxetine researchers published one paper examining Black-White ethnic differences in the effect of the drug, and another examining Hispanic-White differences, even though the data all came from the same trials.[25] There seems to be no reason why those two analyses couldn't have been published together in one paper: no reason, that is, except that a larger number of papers is part of the 'publication strategy' that pharmaceutical companies set up for their drugs. This is marketing, not science, and patients, whose doctors are deliberately being manipulated into prescribing drugs that may be much less effective than they think, pay the price.[26]

Another kind of evidence that quantity is compromising quality is the emergence of so-called 'predatory journals'. Over the last fifteen years, there's been an upsurge in websites that appear to the untrained eye as if they're real scientific outlets, but that apply none of the usual peer-review or editorial standards.[27] These Potemkin publications are run by unscrupulous businesses that attempt to exploit scientists' well-known desire for more publications. They pepper inboxes with spam emails, often in broken English, that implore scientists to submit their research, boasting about how rapidly their journal accepts articles. Alas, many inexperienced, careless or desperate researchers are ensnared into publishing in them (and, of course, paying the charges the journals demand for 'processing' the article), tainting their own reputations in the process: seeing that a scientist has published in a fake journal implies that they're either gullible or unprincipled.[28]

Usually, it's easy to tell a predatory journal from a legitimate one: the websites are poorly designed, the articles are badly typeset, and the editorial board come from universities either that virtually nobody has heard of or that, in some cases, simply don't exist. Often, though, the line between real and predatory becomes blurred. Attempts to create lists of predatory journals are made much more difficult by the inability to define a 'predatory' journal to everyone's satisfaction, by legal threats from the publishers in question, and by the rate at which new fake journals keep appearing.[29]

The worst predatory journals will publish literally anything, even the most obvious hoax. In 2014, the computer scientist Peter Vamplew became so irritated by the constant stream of junk emails from the predatory journal *International Journal of Advanced Computer Technology* that he submitted a joke article entitled 'Get Me Off Your Fucking Mailing List'. The paper consisted entirely of the sentence 'Get me off your fucking mailing list' repeated over eight hundred times (including a helpful flowchart figure with boxes and arrows portraying the message Get → me → off → Your → Fucking → Mail → ing → List). The journal rated it as 'excellent' and accepted it for publication.[30]

Neither salami-slicing nor publishing in predatory journals is strictly against any rules and it's as hard to define what counts as salami-slicing as it is to categorise all journals into 'predatory' and legitimate. But that isn't to say that no genuine rule-breaking – that is, fraud – is going on in some scientists' quest for ever-more publications. As we have seen, fraud is ever present in science and it affects the publication process just as much as it does the process of collecting data and writing papers. For example, fraudsters take advantage of the perhaps surprising fact that scientists can often suggest peer reviewers for their own paper when they submit it to a journal. The editor is free to send the paper to these suggested reviewers or to contact their own choices, but often goes with the former: the whole idea of suggesting reviewers is to take the pressure off time-pressed editors in finding the required relevant experts. This system is, of course, ripe for abuse: authors can suggest their friends or colleagues as reviewers, giving themselves an easy ride to publication. That's bad enough, but as always, fraud can take it to a new level. One editor described the case of the biologist Hyung-In Moon (number 13 on the Retraction Watch Leaderboard with thirty-five retractions), who suggested preferred reviewers ... [who] were him or colleagues under bogus identities and accounts. In some cases the names of real people were provided (so if Googling them, you would see that they did exist) but he created email accounts for them which he or associates had access to and which were then used to provide peer review comments. In other cases he just made up names and email addresses. The review comments submitted by these reviewers were almost always favourable but still provided suggestions for paper improvement.[31]

Editors began to suspect something was fishy because the reviews often appeared within twenty-four hours. This was a schoolboy error on Moon's part: real scientists, who are notoriously busy and sometimes weeks or even months late with peer reviews, would never be so persistently punctual. Moon is far from alone: fake peer review is a staple of the Retraction Watch Database.[32] In 2016, the major scientific publisher Springer despaired so deeply of the rampant peer-review fraud occurring at one of their journals, *Tumor Biology*, that, after retracting 107 tainted articles from just four years' worth of issues, they gave up publishing the journal and sold it off to another company.[33]

A somewhat depressing fact about scientific papers might be an unlikely saviour here, helping to reduce some of the damage done by such shady publication practices. It's this: huge numbers of these papers receive barely any attention from other scientists. One analysis showed that in the five years following publication, approximately 12 per cent of medical research papers and around 30 per cent of natural- and social-science papers had zero citations.[34] It's possible that these lonesome papers will get cited eventually, or that maybe the analysis missed some citations.[35] But whereas it's probably a good thing that these low-quality products of the quantity-maximising system don't have much influence, it should be a signal that something is amiss. Is our time, and our research money, being well spent on these studies that are making so little contribution to the literature? A low citation count doesn't necessarily say anything about the quality of a paper, of course. It could, for example, be an underappreciated work. However, if scientists are publishing useless papers just to secure jobs or grants rather than advance science, it's no wonder that so many are of no interest to their peers.

The trifecta of salami-slicing, dodgy journals and peer-review fraud makes it clear that we shouldn't be rating scientists on their total number of publications: quantity is far too easily gamed. One response to this has been instead to rate scientists according to the number of *citations* their papers receive. As we've just seen, this measure should give us a better indication of their actual contribution to science or to the community. In an extreme case, however, a scientist could have a single highly successful paper with thousands of citations, then follow it up with dozens of worthless papers that nobody ever reads. In that situation, their total citation count wouldn't be a good representation of their broader contribution to science.

In 2005, the physicist Jorge Hirsch came up with a way around this problem, which he called the *h*-index.[36] A scientist with an index of *h* has *h* papers that have each been cited at least *h* times. That is, if you have an *h*-index of 33, as I do at the time of writing, thirty-three of your papers have been cited at least thirty-three times each. The clever thing about the *h*-index is that it becomes harder and harder to progress higher and higher. To increase my *h* to 34, I'd have to have an additional paper cited thirty-four times, but also raise the minimum number of citations for all my other top-cited papers to thirty-four as well. It thus takes a lot of work – and a lot of attention from other researchers – to end up, as some very prominent scientists have, with *h*-indices in the hundreds. Google's specialist academic search engine, Google Scholar, automatically calculates your *h*-index, and many scientists – and in this, alas, I must include myself – are slightly ashamed of the regularity with which they check their Scholar page for updates as new citations come in. (In my experience, if a scientist, even one who scoffs at the very idea of metrics for rating researchers, tells you they're not at all interested in checking their own h-index, they're probably either lying or haven't heard of Google Scholar.)[37]

As you might expect, a scientist's h-index is often explicitly taken into account for hiring and promotion decisions. Scientists, then, have a strong incentive to procure citations, as well as to publish more and more papers that might be cited. Once again, well intentioned as the h-index is, the incentives it produces can lead to behaviours that cater to the system itself rather than to the goals of science.

Obviously, the best way to procure citations is to have important, ground-breaking results; and as we've seen, some researchers spend an inordinate amount of time trying to convince journals (and the world) that their results are exactly that. Spin, of the kind that we saw in the previous chapter, helps bring in the citations: one study found that papers with significant results were cited 1.6 times more often than those reporting nulls, but papers where the authors *explicitly* concluded that the results supported their hypothesis were cited 2.7 times as often.[38] The lesson is clear: if you want citations, write your paper more positively – even if that means having verbally to sand down all the rough, but realistic, edges of your results.

A far more effective way to increase your citations, though, is simply to cite yourself. Self-citation, which one analysis found makes up around a third of all citations in the first three years after a paper's publication, is a grey area.[39] Science is incremental, and researchers work for many years on specific topics. It would be senseless to prevent them citing their own previous work when they're taking the next step in their research programme. But some take it too far. The line between acceptable and problematic self-citation is often blurry, but some cases are clear cut.[40] The psychologist Robert Sternberg stepped down as the editor of the prestigious journal *Perspectives in Psychological Science* in 2018 after receiving strong criticism for, among other issues, his self-citation practices.[41] Here was the problem: it's common for journal editors to write editorials offering their own comments on the papers being published in that issue. When Sternberg wrote such editorials, he frequently stuffed them with references to his own articles – across seven editorials, 46 per cent of the citations were to his own papers, with one reaching as high as 65 per cent.[42] Since as an editor you're in charge of what gets published in your journal, it takes a degree of self-control to avoid abusing the position for the benefit of your own *h*-index. Some editors appear to have less of this self-control than others.

If gaming your own *h*-index in this way feels a bit too obvious, you could instead pressure others to do it for you. Almost any academic will be able to tell you about a time that an anonymous peer reviewer just happened to recommend they cite papers X, Y and Z, which oddly enough just happened all to have an author in common – an author who, surely, couldn't *possibly* be that self-same anonymous reviewer. Beyond anecdotes, we have some evidence: a study of peer reviews that included citation 'suggestions' found that 29 per cent were to the reviewer's own work, and that self-citation suggestions were more common in positive than negative reviews (that is, reviewers were more likely to throw in the suggestion to cite their own work in reviews of papers that they were endorsing for publication).[43]

Robert Sternberg also engaged in a kind of hybrid of salami-slicing and self-citation: self-plagiarism. In new papers, he re-used chunks of text that he'd previously published elsewhere. You may wonder how one can self-plagiarise: isn't the whole point of plagiarism that you steal ideas and phrasing from other people? Recycling text might be lazy, but at least it's not increasing the number of bad or wrong ideas in the world. However, self-plagiarism breaks the author's contract – sometimes a literal one in the case of copyright forms, but more importantly the metaphorical one with the reader – stipulating that the work is original. It can make you look more productive when in fact you're just regurgitating the same ideas; and like salami-slicing, it produces an uneven playing field for comparing researchers' CVs.

In recent years many scientists have been caught red-handed republishing sizeable paragraphs of text, sometimes even entire papers, in multiple journals without any indication that they were doing so. In one of Sternberg's cases, he took an article he'd published in the *Journal of Cognitive Education and Psychology*, combined it with some text from an older book chapter, changed 'Cognitive Education' in the paper's title to 'School Psychology', and got it published in *School Psychology International*.[44] The latter journal's editor eventually had the paper retracted 'for reasons of redundant publication'.[45] One relatively small-scale analysis of papers from a sample of Australian academics found that if self-plagiarism is defined as reusing 10 per cent or more of the text from a previous paper in a subsequent one without attribution, six out of ten of the authors examined were guilty of it.[46]

Many of the methods we've seen used by individual scientists to take advantage of the system of publications and citations can also be used by journals. This is particularly disconcerting because journals are supposed to be the guarantors of scientific standards – further evidence that the problems science faces are systemic, indicating an entire culture gone awry.

The journal-level equivalent of the *h*-index is called the *impact factor*. It was originally conceived as a tool to help university librarians, who have a limited budget, choose to which journals to subscribe.[47] Over time, though, it mutated into the officially recognised quantification of the importance and prestige of any given journal. Broadly, the impact factor, calculated annually, is the average number of citations that the journal's recent articles garner in a year.[48] As I write this, the impact factors for the super-high-status *Nature* and *Science* are 43.070 and 41.063, respectively; journals nearer the bottom of the publishing hierarchy might have citation factors in the single digits.[49]

So the impact factor is an average – but because different papers published by the same journal can have wildly different fates, it can be the average of a very wide-ranging set of numbers. Citation distributions look a lot like income distributions: a few papers at the top get the lion's share, and the majority get relatively few if any at all.[50] Those high-end high-earners drag the average upwards – meaning that despite the journal's impact factor, you can't necessarily expect your newly minted *Nature* article to pick up forty-three citations in the near future, just as you can't expect most people you meet in a country to earn as much as that country's average income.

Nevertheless, in the current system, the bigger a journal's impact factor, the better for its brand. The majority of journals are run by for-profit companies like Elsevier or Springer: it's no surprise, then, that editors might come under intense pressure from the publishers to improve the impact numbers, even when this is at odds with scientific integrity. Some editors use Sternberg's trick of packing their editorials with citations, which 'coincidentally' refer only to articles published in that same journal – and only those published within the past two years, because the annual recalculation of the impact factor doesn't count any articles older than that.[51] Some act like the reviewers we saw above, engaging in what's been called 'coercive citation': demanding during peer review that authors cite a list of previous papers published in that journal, whether or not they're strictly relevant to the work at hand. In one survey, around a fifth of scientists said this had happened to them.[52]

In some extreme cases, editors have set up 'citation cartels', where backroom agreements are made to cite articles across several different journals. One egregious example was noted by the publishing consultant Phil Davies in 2012, and his following-the-citation-trail down the tortuous rabbit hole is worth quoting at length:

> In 2010, a review article was published in the *Medical Science Monitor* citing 490 articles, 445 of which were to papers published in *Cell Transplantation*. All 445 citations pointed to papers published in 2008 or 2009 ... Of the remaining 45 citations, 44 cited the *Medical Science Monitor*, again, to papers published in 2008 and 2009 ... Three of the four authors of this paper sit on the editorial board of *Cell Transplantation*.
>
> In the same year, 2010, two of these editors also published a review article in *The Scientific World Journal* citing 124 papers, 96 of which were published in *Cell Transplantation* in 2008 and 2009. Of the 28 remaining citations, 26 were to papers published in *The Scientific World Journal* in 2008 and 2009. We are beginning to see a pattern.[53]

In the face of a growing number of citation cartels, Thomson Reuters, the company that calculates impact factors, has begun to exclude certain journals from its rankings because of their 'anomalous citation' practices.[54]

So just like publication counts and *h*-indices, impact factors can be deliberately gamed. And as soon as scientists start to artificially inflate these numbers by self-citation, coercive citation and other suspect practices, they lose their meaning as measures of scientific quality. They begin to say less about which scientists and which journals are the best, and more about which have the most single-minded focus on boosting their metrics. It's a clear example of Goodhart's Law: 'when a measure becomes the target, it ceases to be a good measure'.[55] As we've seen, these measures have very much become the explicit targets in our modern scientific culture, creating unforeseen consequences: a perverse incentive structure that favours meaningless metrics and superficial stats over replicability, rigour and genuine scientific progress.

What's particularly disconcerting is that the people entangled in this thicket of worthless numbers are scientists: they're supposed to be the very people who are most *au fait* with statistics, and most critical of their misuse. And yet somehow, they find themselves working in a system where these hollow and misleading metrics are prized above all else. At first, having numbers that can quantify a scientist's, or a journal's, level of contribution might seem scientifically appealing: objective quantification, after all, is one of the unique strengths of science. But as Goodhart's Law states, once you begin to chase the numbers themselves rather than the principles that they stand for – in this case, the principle of finding research that makes a big contribution to our knowledge – you've completely lost your way. The fact that these metrics aren't just the preserve of individual scientists jockeying for status but are woven into the fabric of both the university and publication systems, is yet another example of how badly the scientific system is failing in its cardinal purpose.

Throughout the book, we've seen an abundance of factors that lead to bad research. There are scientists so deluded by their own theories, or who desire so strongly to feel they've made a difference, that they use fraud or *p*-hacking to vanish any bothersome ambiguities. There are scientists driven primarily by a desire for money, prestige, power, or fame, and who care about the truth as little as any charlatan does; scientists who are too busy or stressed to check for errors in their work; scientists who won't question the way they've been trained, and who carry on with the same old erroneous practices. Is it right to cast the scientific publication system as some kind of *ur*-problem that underlies all of the above? Can we truly say that the perverse incentives created by prioritising publications, citations, and grant money have led *directly* to acts of fraud, bias, negligence, and hype?

We can never know for sure what's going through the mind of a scientist when they commit one of the problematic practices we've seen in the preceding chapters. But we can try to make an inference to the best explanation. Scientists are human, and humans respond to incentives. The problems in science that we've observed are so widespread, across the world and across scientific disciplines, that they must have an explanation at the level of science's wider culture: we're not just talking about a few bad apples ruining science for everyone else. When we look at the overall trends in scientific practice in recent decades – the exponential proliferation of papers; the strong academic selection on publications, citations, h-indices and grants; the obsession with impact factors and with new, exciting results; and the appearance of phenomena like predatory journals, which are of course just catering to a demand – wouldn't it be strange if we *didn't* see such bad behaviour on the part of scientists? Though we shouldn't stop looking for other explanations – perhaps, for example, the problem lies more with the policing of the system, rather than the incentives themselves – the theory that the publishing-related incentives are what's degrading science does a pretty good job of explaining the state in which we find ourselves. At the very least, we can say that the motive fits all the crimes.

It would be inordinately difficult to run a meta-scientific experiment on this whole system, given that it would have to cover entire careers and thousands of universities and journals, across different countries and diverse fields of research. Yet we can do better than just speculating and some ingenious scientists have designed computer models to simulate the publication system, examining how its incentives might affect research.

Some of these models think of the scientific system in terms of evolution. Above, I compared the process of the lengthening of CVs required to get academic jobs to sexual selection, where increasingly flamboyant displays evolve to attract mates. But there's another evolutionary analogy to be drawn. As we've seen, the system of science is now set up to reward those who engage in underhand methods. If the more trustworthy researchers – those who are in it for the science, rather than status, money, or other non-scientific goals – can't compete in this system, they'll be more likely to drop out of the world of academia and get another job elsewhere. At the very least, they'll be less competitive for the top jobs. Meaning that as well as pushing everyone towards unreliable research practices, the system selects against researchers who have strong convictions about getting it right, filling their places instead with those who are happier to bend the rules.

The model built by the cognitive scientist Paul Smaldino and the ecologist Richard McElreath is probably the most vivid illustration of how this process might look over time.[56] It resembles a game with multiple turns. At the beginning, there are several labs, each one testing new hypotheses with varying degrees of effort to prevent false-positive results, and then trying to get them published. If a lab finds positive results, it gets rewarded with a publication; if its experiment comes up null, it gets no such reward. At each turn in the model, the labs with more publications are more likely to 'reproduce' – that is, to send off their successfully trained PhD students to form their own labs, thereby spreading their methodological techniques (and their level of meticulousness) through the scientific community. As the model ticks along, the incentives work their pernicious magic: rewarding the virtual labs with 'reproduction' if they publish more papers means that more and more labs put in less and less effort to ensure the quality of their science. This is because, perversely, false-positive results are just as publishable as true positives, but easier to come by. Eventually, the rate of published false discoveries skyrockets. Smaldino and McElreath call it 'the natural selection of bad science'.

Other computer models of the publication system draw similar conclusions. One found that, given the strong preference for novelty in the scientific literature, the optimal strategy for an aggressively ambitious scientist is to 'carry out lots of underpowered small studies to maximise their number of publications, even though this means around half will be false positives'.[57] Another model showed how scientific journals' fixation on positive results led to 'perverse rewarding of false positives and fraudulent results at the expense of diligent science'.[58] Computer models aren't reality, of course, which has vastly more moving parts. These simplified simulations can, though, put some mathematical flesh on the bones of the inferences I made above, showing how weaknesses in the incentive structure could, over time, cause the quality of science to deteriorate.

In the Getty Center in Los Angeles there's a painting from the Dutch Golden Age, *The Alchemist*, by the artist Cornelis Bega.[59] The titular alchemist sits in his shambolic laboratory, surrounded by the cracked pots, chipped bowls and smashed bottles that are the remnants of his failed attempts to turn base metals into gold. Contrary to popular belief, alchemy wasn't all as worthless as this and the line between some alchemical activities and the earliest appearance of what we now call chemistry is, at the very least, hazy.[60] But Bega was commenting on the futility of the gold obsession. It's a neat analogy for the modern scientific incentive system. The pursuit of academic treasure like publications and citations has left us with a detritus of broken and useless scientific research.[61]

Perverse incentives work like an ill-tempered genie, giving you exactly what you asked for but not necessarily what you wanted. Incentivise publication quantity, and you'll get it – but be prepared for scientists to have less time to check for mistakes, and for salami-slicing to become a norm. Incentivise publication in high-impact journals, and you'll get it – but be prepared for scientists to use *p*-hacking, publication bias and even fraud in their attempts to get there. Incentivise competing for grant money, and you'll get it – but be prepared for scientists to hype and spin their findings out of all proportion in an attempt to catch the eye of the funders. On the surface, our current system of science funding and publication might seem like it promotes productivity and innovation, but instead

Cornelis Bega, *The Alchemist*, 1663. Getty Museum.

it often rewards those who are following only the letter, rather than the spirit, of the endeavour.[62]

Knowing about the incentive problem doesn't mean that scientists should be let off the hook when they engage in malpractice. We all feel the force of the incentives, but we still ought to do our best to defy them, for the sake of the science.[63] It would be better, though, if we didn't have to resist the crushing weight of the publish-or-perish system when trying to make discoveries about the world. It would be better if we could reach a happy medium where scientists are incentivised towards hard work and creativity, but also caution and rigour; where the bias is towards getting it *right* rather than merely getting it published.[64] So how might we achieve that? How might we improve our incentives and in doing so improve the reliability of science? That's what we'll discuss in the next and final chapter.

8

Fixing Science

The process of scientific discovery ... will change more over the
next twenty years than in the past three hundred years.
Michael Nielsen, *The Future of Science* (2008)

The majority of the problems with science that we've seen in
this book are encapsulated in one meta-science paper from
2018. The psychiatry researcher Ymkje Anna de Vries and her
colleagues examined all the steps that typically occur between
the trialling of a new drug and that drug's eventual presentation
to the world.[1] As their sample, they took 105 different trials of
antidepressants that had been approved by the US Food and
Drug Administration. The ratio of positive to negative results
among these trials happened to be almost fifty-fifty, with fifty-
three finding that the antidepressant in question worked better
than the control or placebo, and fifty-two getting results that
were either null (the FDA referred to them as 'negative') or
'questionable'.[2] So far, so predictable: some studies find significant results, some don't. The problem was what happened next.

Looking across all the studies, de Vries and her team
observed a process of literature-laundering that transformed
a messy, varied set of trials into a much cleaner-looking story
of scientific discovery – a process that made the drugs in question look substantially more effective than they really were.
The first step was publication bias. 98 per cent of the positive

trials (fifty-two of fifty-three) were eventually published, compared to only 48 per cent of the negative ones (twenty-five of fifty-two). In reality, the positive-negative ratio was balanced; after publication, the positives outnumbered the negatives two to one. As we know, scientific journals don't favour null results – but for a clear-eyed look at the scientific record, we need to see them.

The second step was *p*-hacking – specifically, outcome-switching, where scientists change the focus of their study upon finding that their main result isn't statistically significant. Once the outcomes being studied were switched, ten more of the negative studies had changed into positive ones (you'll recall from Chapter 4 how outcome-switching increases the chances that our result is nothing more than a false positive). At this point there were still fifteen studies with clear negative results – and this was where the spin came in. Ten of those fifteen studies included some kind of spin in their abstract or the main paper to make their results appear more positive.

After this cycle of bias and spin, a mere five studies were left that were unambiguously negative – that is, ten times fewer than existed at the start. And as a final, rotten cherry on top, the positive trials ended up being cited by later research three times as often as the negative ones. The whole dispiriting step-by-step process is illustrated in Figure 4 below.

This isn't unique to research on antidepressants: the authors found a similar chain of events in trials of new psychotherapies.[3] Indeed, the same thing is happening, to a greater or lesser extent, almost everywhere in scientific research. Any future meta-analyst who wants a full picture of a particular research area is going to get a drastically warped perspective – and unless they look in detail through the trial registries (which, incidentally, aren't required in fields other than medicine), they won't even know about it. De Vries's study didn't investigate the possibility of fraud, or errors in study design or analysis, or the presence of hype in the presentation of the treatments in the media and marketing, but we can safely assume that

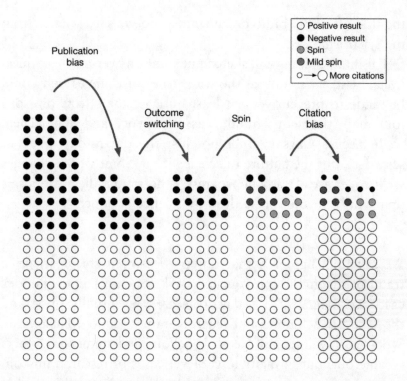

Figure 4. The cycle of bias and spin that hides negative results. Each dot represents a trial of antidepressants; black dots are negative studies. We can watch the negative trials disappear as we move from left to right. Negative studies are: published substantially less than the positive ones; transformed into positive studies by outcome switching; spun into positive-sounding ones, either mildly or substantially; and cited far less than the positive studies, which have their dots inflated in the final column. At the end, almost all you can see are the positive results. Adapted from de Vries et al. (2018), *Psychological Medicine.*

these are also occurring all the time, further obscuring the reality of the research.

Clearly, what de Vries and her colleagues found is a far cry from the ideal of science described at the beginning of this book, in which the social processes of publication and peer review ensure the facts are shared with the world and any accidental errors are filtered out. So how can science be rescued? How do we realign the culture and the incentives of the scientific publication, funding, and ranking processes so that they reward truth and reliability instead of flashy stories? How, to adopt the title of a BBC radio documentary, do we save science from the scientists?

This final chapter will sketch out some answers to these questions. First, we'll look at the wide array of changes that could be made to prevent – or at least mitigate the effects of – our four main problems: fraud, bias, negligence and hype. Then, we'll discuss ways to alter not just the day-to-day work of scientists, but the culture of science itself. Some of the changes we'll cover are already in progress; others are radical proposals that would revolutionise the way science is done.

We saw in Chapter 3 that universities often shield scientific fraudsters from the consequences of their actions. In the famous cases we looked at, the dam of secrecy eventually burst and the fraudsters' names – Diederik Stapel, Paolo Macchiarini, Woo-suk Hwang, Jan Hendrik Schön, and numerous others – came out. But in many lower-level cases of research misconduct, identities are never publicised.[4] It's hardly surprising that so many scientific fraudsters act with impunity if there's both little chance of them being noticed *and* little chance that many people will find out if they do get caught. The first change, then, might simply be to name and shame more of those who have been found to commit scientific misconduct.[5]

Of course, the incentives for universities to do this aren't particularly strong – so another idea is to stop them from marking their own homework. There are some developments on this front. For example, having seen how badly the Karolinska Institute disgraced itself over the Macchiarini trachea-transplant case, in 2019 the Swedish government passed a law that stopped universities from investigating cases of research misconduct themselves, instead handing the responsibility to a new, independent government agency.[6]

That's all very well once fraud has been discovered, but it would be better to prevent it from being published in the journals in the first place. This is where technology could come in.

Researchers are developing increasingly effective algorithms that can spot faked data in scientific papers, and others that can detect problems such as image duplication.[7] It'll be interesting to see how these compare to the skills of a seasoned human duplication-spotter like Elisabeth Bik, whom we met in Chapter 3, but in theory they should make the task of flagging up inappropriate data manipulation far less labour intensive. Journals could require each submitted paper to be assessed by algorithms such as these – as well as others we've encountered, like GRIM and statcheck – to highlight anything suspicious before peer review even begins. They could also use programs to flag up any potential plagiarism or self-plagiarism in the text.[8]

Such algorithms are also ways to combat negligence.[9] A large proportion of the errors picked up by the statcheck algorithm were probably banal mistakes made when researchers were copying and pasting numbers from their statistical analysis software into the word processor they were using to write up the paper. Running the finished product through statcheck could catch these errors before they make it into the literature. But technology could also help keep those errors from happening in the first place. In recent years, software has been developed that combines statistical analysis and word processing into one program, automatically populating all the relevant tables and figures within the paper.[10] The data can bypass the distractible, fallible scientist, and because the whole 'pipeline' from data to paper is there for all to see, intentionally fiddling with the numbers or analyses becomes trickier, too.[11]

It would be easy to get overexcited by these new technologies. Every piece of software is subject to its own bugs. To take one particularly embarrassing example, around 20 per cent of genetics papers that used a Microsoft Excel spreadsheet to list the genes they examined were found to have an autocorrect error, whereby gene names like *SEPT2* and *MARCH1* were mistakenly converted to dates.[12] Automatic software should have its work carefully checked by humans, at least until we're sure we have dealt with the gremlins. In theory, though, it seems that many routine scientific tasks could be done more accurately by non-human intelligences: analysing large amounts of data to find patterns; sifting through numbers in the scientific literature to work out a consensus; even interpreting images of blots and cells and brain scans. Given the sheer number of errors that are found in scientific papers and given how easy it would be to sidestep many of them with a more automated paper-writing process, it might eventually be unethical to rely solely on humans to do these jobs.

One of the concerns that we've repeatedly encountered is scientists' persistent bias towards novelty. Whereas new, exciting results are the engine driving scientific progress, we've seen how an obsession with 'groundbreaking' results has led to entire fields of research being based on flimsy, unreplicable evidence. To paraphrase the biologist Ottoline Leyser, the point of breaking ground is to begin to build something; if all you do is groundbreaking, you end up with a lot of holes in the ground but no buildings.[13] How do we reverse the prioritisation of novel results over solid ones? How do we combat publication bias, ensuring that all results get published, no matter whether they're groundbreaking or null?

One answer has been to create journals that specialise in publishing null results, providing a more attractive alternative to the file drawer. The *Journal of Negative Results in Biomedicine*, for instance, was set up in 2002 for just this purpose. It was a well-meaning idea but, perhaps unsurprisingly, hardly anyone wanted their study to appear in a low-status null-results journal – a journal that was 'defined as publishing articles that no other journal will publish'.[14] The journal shut down in 2017, an unusual fate for a scientific outlet in a world groaning under the weight of new papers.[15]

If journals specifically for null results won't work, what about journals that explicitly accept any result, providing the study that found it is judged methodologically sound? These could also provide a home for replication studies, as they suffer the same prejudice as null results. In recent years, a slew of such journals (often called mega-journals, because not having a requirement for positive or 'exciting' results means that they end up publishing a *lot*) has appeared, including *PLOS ONE*, where my colleagues and I eventually published our null replication of Daryl Bem's psychic study.[16] This is progress, but such journals still risk being relegated to a lower tier in the mind of status-conscious scientists. Ideally, what we want to see is an accurate proportion of null results, and more attempted replications, in the glamorous, high-impact journals too.

There is some good news on this front. Even if it's not an explicit commitment to publishing null results, many big-name journals are now softening their previous attitude towards publishing replications. Take the *Journal of Personality and Social Psychology*, which published Bem's paper and rejected ours due to their blanket 'no replications' policy. Now, post-crisis, their website has a whole section on replication, which notes that the editorial board 'acknowledges the significance of replication in building a cumulative knowledge base in our field. We therefore encourage submissions that attempt to replicate important findings, especially research previously published in the *Journal of Personality and Social Psychology*.'[17] This is a nice instance of a new rule for scientific journals proposed by the psychologist Sanjay Srivastava, based on the 'you break it, you buy it' signage from some pottery shops: if you publish an article, you're at least partly responsible for publishing further work checking whether it replicates.[18]

More and more editors from across different fields are following this lead; over 1,000 journals recently adopted a set of guidelines explicitly announcing, among other things, that replication studies are welcome.[19] Some funders, such as the Dutch Organisation for Scientific Research, are pouring money into replication studies.[20] These are positive moves, but the proof of the pudding will be in whether the journals do, in fact, start routinely publishing more replications. Meta-scientists will be keeping an eye out.

Making it easier for scientists to publish replications and null results might reduce publication bias. But what about the other forms of bias we encountered, having to do with *p*-hacking? Many dozens of papers, and even entire books, have been written on the pitfalls of *p*-values: they're hard to understand, they don't tell us what we really want to know and they're easily abused.[21] There's truth to all these criticisms. In broad terms, what is needed is less focus on *statistical* significance – a *p*-value below the arbitrary threshold of 0.05 – and more on *practical* significance. In a study with a large enough sample size (and high enough statistical power), even very small effects – for example, a pill reducing headache symptoms by one per cent of one point on our 1–5 pain scale – can come up as statistically significant, often with *p*-values far below 0.05, though they could be essentially useless in absolute terms. The economists Stephen Ziliak and Deirdre McCloskey call this the 'sizeless stare of statistical significance', where scientists develop a laser-like focus on *p*-values at the expense of considering, as Ziliak and McCloskey put it, the 'oomph' of their effect.[22]

The remedy proposed most often is simply to abandon the idea of statistical significance. In 2019, over 850 scientists signed an open letter in *Nature* arguing just that. 'It's time,' they wrote, 'for statistical significance to go.'[23] Rather than emphasising significance, they argued, researchers should be clearer about the uncertainty of their findings, reporting instead the margin of error around each number and generally having more humility about what can be derived from often-blurry statistical results.[24] There's a lot to be said for this, although it should be borne in mind that the most common margin of error that's calculated – the so-called 'confidence interval' – just provides a different perspective on the data than that of the *p*-value, rather than much new statistical information.[25] But there's also a baby in the bathwater of statistical significance. By providing scientists with an objective measure, albeit an arbitrary one, it ties their hands. Doing away with *p*-values wouldn't necessarily improve matters; in fact, by introducing another source of subjectivity, it might make the situation a lot worse.[26] With tongue only partly in cheek, John Ioannidis has noted that if we remove all such objective measures we invite a situation where 'all science will become like nutritional epidemiology' – a scary prospect indeed.[27]

The same criticism is often levelled at the other main alternative to *p*-values: Bayesian statistics. Drawing on a probability theorem devised by the eighteenth-century statistician Thomas Bayes, this method allows researchers to take the strength of previous evidence – referred to as a 'prior' – into account when assessing the significance of new findings. For instance, if someone tells you their weather forecast predicts a rainy day in London in the autumn, it won't take too much to convince you that they're right. On the other hand, if their forecast predicts a snowstorm in the Sahara Desert in July, you'd probably want to assess that claim quite sceptically, given all the prior experience we have of scorching Saharan summers. A Bayesian can build all that pre-existing evidence into their initial calculation – in the latter case, they'd require the new forecast to be extraordinarily convincing in order to overturn all the previous meteorological knowledge.[28] This isn't something you can do so easily with *p*-values, since they're almost always calculated independently of any prior evidence. However, the Bayesian 'prior' is inherently subjective: we can all agree that the Sahara is hot and dry, but how strongly we should believe *before a study starts* that a particular drug will reduce depression symptoms, or that a specific government policy will boost economic growth, is wholly debatable.

Aside from taking prior evidence into account, Bayesian statistics also have other differences from *p*-values.[29] They're less affected by sample size, for example: statistical power is not a factor because the Bayesian approach is aimed not at detecting the effect of a particular set of conditions, but simply at weighing up the evidence for and against a hypothesis. Arguably, they're also closer to how people normally reason about statistics. Bayesians say 'what is the probability my hypothesis is true, given these observations?' – a more intuitive approach than that of p-values, which asks 'what is the probability that I'd get these observations, assuming my hypothesis *isn't* true?'[30]

All statistical schemes have their pros and cons.[31] While some pundits in these debates make it sound as if p-values are the root of all evil – a numerical Pied Piper that leads otherwise level-headed scientists astray – it's highly unlikely that all the fraud, bias, negligence and hype we've seen in this book would evaporate if we just got rid of this one statistical tool and adopted another. Statistics alone cannot solve the underlying problem: the crooked timber of human nature and, by extension, that of the scientific system. No matter which statistical perspective became dominant, some scientists would find ways to game it to make their results look more impressive than they really are. As we'll see below, the solutions to these problems will have to be motivational and cultural.

In the meantime, rather than advocating that researchers drop a statistical method entirely – especially one as deeply ingrained as significance testing – it may be better to educate scientists more effectively on what it can and can't show, and reform its use so that the mistakes can be avoided. For instance, it's recently been proposed that we should change the standard criterion for significance from $p < 0.05$ to $p < 0.005$, creating a much higher hurdle for results to clear before we consider them interesting.[32] Given the flaws the replication crisis has revealed, the logic goes, we should be substantially more conservative about what we accept as evidence for our hypotheses. However, the downside of raising the hurdle is that unless we make simultaneous across-the-board increases in our sample sizes, our tests will have much less statistical power. But the advocates of 0.005 are making the case that the problem of false positives, which their method would likely reduce, is a more pressing concern than that of false negatives.

Here's another way to deal with statistical bias and p-hacking: take the analysis completely out of the researchers' hands. In this scenario, upon collecting their data, scientists would hand them over for analysis to independent statisticians or other experts, who would presumably be mostly free of the specific biases and desires of those who designed and performed the experiment.[33] Such a system would be tricky to run and one can imagine it leading to conflict when scientists disagree with the analysis or interpretation that their assigned statistician has imposed on their precious data.[34] As with some of the radical ideas for reforms that we'll see later in the chapter, it could still be worth trying at small scale.

We saw in Chapter 4 that the sheer number of ways a dataset can be analysed also gets scientists into trouble: how do they know that the one analysis they chose isn't the one that gives them fluke results? An alternative to worrying about whether you chose your specific analysis correctly is to embrace the problem of the 'garden of forking paths' and run *all* the analyses that could possibly be run on your dataset. You could include *and* exclude certain participants; combine *and* split certain variables; adjust *and* not adjust for certain confounders – and base your conclusions on what the results tell you as a whole. This idea has been given many names, like 'specification-curve analysis', 'vibration-of-effects analysis' and, my personal favourite, 'multiverse analysis'.[35] If we imagine infinite parallel universes, in each of which you ran the analysis slightly differently, in what proportion of them would you find the same effect? In what proportion would you find the complete opposite? Would all these analyses generally converge on the same overall result?

The Oxford psychologists Amy Orben and Andrew Przybylski, for example, used a multiverse analysis to tackle a hot-button issue: the effects of screen time on young people's mental health.[36] Studies on this are routinely hyped in the press, with many newspaper articles and popular books arguing that today's youngsters are harmed by the amount of time they spend online.[37] Social media is seen as a particular problem, as it purportedly reduces young people's face-to-face contact with others, exposes them to cyber-bullying and hardcore pornography, and reduces their attention span.[38] New psychological diagnoses have even been proposed: 'video gaming disorder', 'online porn addiction', 'iPhone addiction', and the list goes on.[39] Most of the evidence that inspires these tech panics comes from big observational studies looking at correlations between screen time and mental health problems in adolescents. Given the strong potential for p-hacking in such studies (remember how easy that was to do in the big datasets from nutritional research, where essentially all foods could be linked to cancer in some way), they're perfect candidates for a multiverse analysis.

Orben and Przybylski looked at all the possible ways you could analyse three big observational datasets to examine these claims. For example, you could ask about self-esteem and suicidal thoughts rather than just wellbeing, or choose two of them, or all three; you could take ratings from parents, or self-reports, or use both; you could classify just TV watching as 'screen time', or include video games too; you could adjust for factors like gender, or school grades, or any number of other variables that might be important; you could use averages or total scores from the questionnaires; and so on and on. The total number of 'defensible' combinations – that is, ones where you could make a plausible-sounding scientific argument that this was the right way to analyse the data – was in the hundreds in the first dataset, in the tens of thousands in the second, and in the hundreds of millions in the third (since running so many analyses would overwhelm most computers, they pared it down to a 'mere' twenty thousand for that last one).

Having run through all these combinations, Orben and Przybylski found that there were a few analyses showing fairly substantial negative effects of screen time, some that showed no effect at all, and some that showed that screen time was actually beneficial. They took the average. It was negative, but very weak indeed, with screen time accounting for approximately 0.4 per cent of the variation in wellbeing. To put that into perspective, it's around the same-sized correlation one finds between wellbeing and regularly eating potatoes, and smaller than the link between wellbeing and wearing glasses. So much for all the scare stories – the multiverse analysis implied that we'll have to look beyond the easy scapegoat of screen time if we want to explain adolescent mental health problems.[40] The wider relevance is obvious: instead of running just one single analysis, which might suit your individual bias, we instead should take a much broader view of statistics, looking at all the counterfactuals, and asking ourselves what might have happened if we'd decided to run things slightly differently.

The drawback of multiverse analysis is that it usually requires the services of a supercomputer, which aren't available to most researchers. And while analyses of this sort are a great way to bring more clarity to hotly debated questions, they don't rid scientists of the ever-present temptation and pressure to cherry-pick their most impressive-looking results and present them as their initial hypothesis. To tackle this, we might use another tool for fixing science: *pre-registration*.

Pre-registration has been mandatory for US government-funded clinical trials since 2000, and a precondition for publication in most medical journals since 2005.[41] Registering a study involves posting a public, time-stamped document online that details what the researchers are planning to do, in advance of collecting any data. A public repository of experiments that are about to be run provides a baseline by which to check what proportion of these studies actually make it to publication. And it allows us to see what hypotheses the researchers intended to test, so we can check if any of them were switched mid-study.

In addition to pre-registering the fact that a study will take place, researchers can also pre-register a detailed plan for how they intend to analyse the data. We've seen how it's the *unplanned* nature of statistical analysis – the undisclosed flexibility – that can lead scientists down forking paths to results that are statistically significant (and publishable) but don't actually correspond to reality. The idea of pre-registering your analyses is a scientific Ulysses pact: by posting a plan for your analysis somewhere public, you lash yourself to the mast and stop yourself giving in to the Siren's call of *p*-hacking.

Some would rightly object that if scientists permit themselves no wiggle room whatsoever, there's no longer an opportunity for serendipitous findings (penicillin and Viagra are two of the more famous accidental discoveries that are often referenced as part of this argument).[42] But that isn't what pre-registration is about. In a pre-registered study, impromptu analyses to explore interesting patterns in the data are still allowed, they just can't be spun to look as if they'd been planned in advance. These so-called *exploratory* analyses can lead to many important new insights and ideas – for example, you might unexpectedly find that a new drug works better in older than it does in younger participants and then build up a new research line to find out why. But as we've seen again and again in the previous chapters, numbers are noisy and you're guaranteed to find *something* interesting-looking if you slice and dice your data in enough ways. Because you've given yourself more tries to find statistically significant results, positive findings from exploratory analyses are much more likely to be random flukes that won't replicate in a new sample. Yet somewhat scandalously, the majority of science frames *exploratory* results as though they were *confirmatory*; as though they were the results of tests planned before the study started. Pre-registration lets you be clear with your readers whether you were using the data in an exploratory way, to generate hypotheses ('huh, interesting, variable X seems to be linked to variable Y! We'd better check whether this replicates in a new dataset'), or in a confirmatory way, to nail them down ('I predicted that variable X would be related to variable Y in this dataset, and sure enough, it is!').[43]

A study of large-scale trials on heart disease prevention, shown in Figure 5 below, provides a stark illustration of the impact of registration.[44] Before registration was required, journals reported an impressive number of positive effects, represented by the white dots in the lower half of the graph (indicating a lower risk of heart disease), as well as the odd null result. But look what happened after 2000, when pre-registration was brought in. Suddenly there were only a couple of positive results, with the rest of the studies reporting null effects, clustering around zero. The success rate for trials before registration was required was 57 per cent; afterwards, it plummeted to 8 per cent. Just to reiterate: before the advent of trial registration, the research *appeared* hugely successful, with many viable-seeming heart disease interventions. Afterwards, the truth was revealed: the drugs and dietary supplements being tested in these studies were nowhere near as useful as we'd been led to believe.

We mustn't make a correlation-causation error by assuming that the new registration requirements necessarily *caused* the decrease in positive findings – maybe other things changed in that same year, like a shift in focus to different kinds of treatments. But it's plausible to think that the public pre-registration of their plans might have led the clinical triallists to be more transparent and honest about what they found. If these results really do show causality, they serve as a strong argument for pre-registering all our studies – as well as being a powerful indictment of the standard scientific practices from before we started registration.[45]

Not that pre-registration is a silver bullet. Many scientists who pre-register their study still fail to publish it (or at least report the results) in the time period that's mandated by the trial registry. Others, despite pre-registration, still make changes to their analysis after it begins.[46] In the case of clinical trials, the scientists aren't just flouting best practice, they're breaking

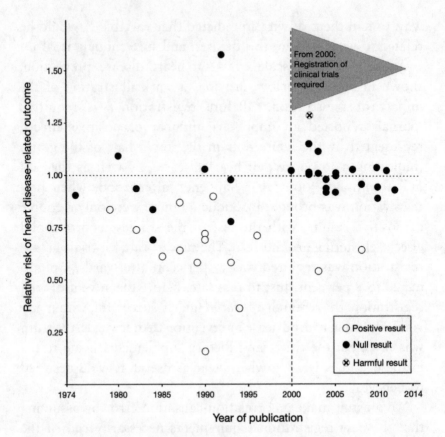

Figure 5. Published studies of heart disease treatment and prevention before and after 2000, when registration of clinical trials, or at least their primary outcomes, began to be required (the vertical dashed line). Studies below the horizontal dotted line found that the prevention or treatment being tested led to a lower risk of heart disease (the lower the dot, the lower the risk). White dots are studies that found a statistically significant effect in favour of the prevention or treatment; black dots are null results. Note that there's also one statistically significant post-2000 study that found the treatment was actually harmful (the dot with a cross inside). Adapted from Kaplan & Irvin (2015), *PLOS ONE*.

the law – and yet one investigation by the journal *Science* in 2020 found that over 55 per cent of trials had their results reported late to the US government's trial registry. Suffice to say, this isn't how it was supposed to work: pre-registration is only beneficial if the registered plan is followed. For clinical trials, we need to tighten up the enforcement and punishments for these 'clinical scofflaws' (for example, banning them from receiving government grants in future, or from publishing in certain journals for a certain period).[47] For other fields, where there's no legal obligation to follow your registration, we need to find other ways to ensure that scientists stick to it. The UK's National Institute for Health Research, for instance, manages to get nearly all its funded projects published, in part because it withholds 10 per cent of the grant money until the report of the trial appears.[48]

Another option is to use an even more rigorous version of pre-registration. In this scenario, a scientist submits the registration itself to peer review and, if it's approved and the reviewers agree that the study design is sound, the journal commits to publishing the eventual paper *no matter how its results come out*. Only then do the scientists start to collect their data.[49] Not only does this type of study, called a 'Registered Report', kill publication bias stone dead, by removing the pernicious link between the statistical significance of the results and the decision to publish, but it reduces *p*-hacking as well, since you have to agree to your analysis with the reviewers beforehand and can't just alter it *post hoc* without making it very clear what you've done. Best of all, it nullifies many of the perverse incentives that lead to bias and fraud in the first place. You know you're going to get a publication in any case, so there's no longer so much pressure to beautify your findings.[50]

Leaving aside the practicalities of enforcing it, the idea of pre-registration is a good one. What it's really all about is *transparency*: not hiding away your plans or your analysis from the world. The related concept of 'Open Science' far predates the replication crisis but might be one of the most powerful antidotes to it.[51] Open Science is the idea that as far as possible, every part of the scientific process should be made freely accessible.[52] The perfect Open Science study would have an associated webpage where you could download all its data, all the statistical code that the scientists used to analyse those data, and all the materials they used to gather the data in the first place.[53] The peer reviews and previous drafts of the article could be published alongside the article (even if the identity of the reviewers isn't revealed), allowing the reader to see the whole publication process.[54]

Providing free access to all your data and methods is an embodiment of the Mertonian norm of communalism: allowing other scientists to re-use your work to make their own research more efficient. It also serves the norm of organised scepticism: it makes it easier for other researchers to check your paper for negligent errors, attempt to replicate your results, and generally dig into your findings in more detail. They can do that far more readily than they could were they using just your study's summary tables and graphs, which are typically all that's included in a journal publication; and they can do it without having to write to you directly to beg for your dataset, which as we know rarely works. Taking the Open Science approach demonstrates to your scientific colleagues that they don't need to have blind faith that everything's been reported accurately: you have nothing to hide.

The corollary is that if you *do* have something to hide – if you faked your data or results – Open Science makes your life far more difficult. As we saw in Chapter 3, it's quite tricky to create a fraudulent dataset that looks like a real one. So far, data sleuths who have uncovered fraudulent results have tended to look at just the data summaries provided in the papers. Having access to the complete underlying set of data makes such sleuthing a lot easier. Mostly, though, open data acts as a deterrent against committing fraud in the first place, since it would take the brassiest of brass necks to post a fake dataset on a public website.[55] The same principle works for *p*-hacking that borders on fraud, and for more innocent errors: allowing other scientists to see your data and how you analysed it means that eagle-eyed peers can spot spreadsheet typos, incorrect statistics, improbable numbers, or undeclared analyses that might affect the way your results should be interpreted.

But not all data can be posted online.[56] Posting the genetic data of participants in a study, for example, would violate the anonymity that's a precondition of their involvement – not to mention their legal entitlement to privacy. Making available some other kinds of data might also be dangerous. In 2011, a controversy erupted when virologists published a few too many details about how they had produced a genetically modified, more contagious version of the H5N1 'bird flu' virus.[57] This was important research into how viruses work, but it was a rare situation where the consequences would be terrible if the information fell into the hands of people with more sinister motives. The researchers were asked to remove some of the relevant methodological details from their paper.[58] These, however, are exceptions: there's nothing in principle that stops the vast majority of scientists from being far more open about their work.

There's another way of opening up science: enlarging it. It's become increasingly clear that to answer the bigger scientific questions, we need to take another step along the pathway that began historically with individual alchemists labouring alone and jealously guarding their secrets, and which has brought us today to scientists sharing their work in international journals. That next step is having scientists from many different labs and universities work together on much larger-scale collaborative projects: so-called 'team science'.[59] Such collaborative studies were pioneered in particle physics, which requires huge teams of researchers to perform the thousands of incredibly complex tasks involved in particle-accelerator experiments.[60] They are also proving crucial in modern genetics, where researchers have been combining their samples in international consortia projects, helping them attain the all-important statistical power needed to identify the tiny correlations between tens of thousands of DNA variations and human traits and diseases.[61] It was this collaborative way of working that exorcised the erroneous 'candidate gene' approach to genetics (as covered in Chapter 5), producing research that, at the very least, replicates.[62]

More and more fields are catching the collaborative bug: the multi-lab attempt to replicate a hundred prominent psychology studies that we encountered back in Chapter 2, for instance, is a great example of how increasing the scale, and therefore the statistical power, of research can correct the mistakes of the past. As well as in physics and genetics, there are now trans-national collaborative projects in neuroscience, cancer epidemiology, psychology, and translational medical research, among others.[63] These large-scale projects can directly address the replicability of their respective fields and because the results are being shared around a larger community of usually very opinionated scientists, they can also, in theory, act as a check on the biases of any individual researcher.

Open Science doesn't stop at the scientific community: a final part of making science more transparent is to open it to the general public. After all, the Mertonian norm of communalism, which emphasises the shared ownership of scientific findings, captures an important truth: a huge chunk of science is funded by taxpayer money. Most journals charge anyone who doesn't have a subscription something close to $35 for access to an individual paper. It seems problematic and even undemocratic that members of the public, who've funded scientific research through their taxes, should be charged to access the results of that research. Of course, there are many things that the tax-payer funds to which they don't have access, and shouldn't: for example, the secret patrol routes of the submarines that form the UK's continuous at-sea nuclear deterrent. Like the sensitive data on viruses we saw above, however, this kind of information is an exception. Most governments have now passed Freedom of Information Acts that allow members of the public to access their documents and statistics; in the United States, the work of all government employees is automatically in the public domain.[64] The branch of Open Science that tries to bring science closer to this free-information default is the Open Access movement. Its impact is already obvious: many journals have begun allowing scientists to pay a fee upon publication that lets their paper be downloaded by anyone, in perpetuity, at no cost.

Paying to have your research made freely-available is optional; now, though, funders are beginning to *demand* that the research that they've paid for gets published with Open Access. The most ambitious strategy, named 'Plan S', comes from Science Europe, the body that represents the funding agencies of European governments. The idea is that by 2021, all research funded by its members will be published in fully Open Access journals.[65] They have the backing of the government funding councils from sixteen countries, as well as major funders like the Wellcome Trust and the Bill & Melinda Gates Foundation. The rules stipulate that scientists who get Science Europe grants won't even be able to submit research to journals that aren't fully Open Access – and that includes *Science*, *Nature*, and indeed 85 per cent of all journals.[66] The research funders also state that they'll act in a coordinated way to ensure that publishers don't charge the scientists too much for Open Access publication. If the journals ramp up the Open Access fees beyond some reasonable amount for covering the costs of publishing, the funders will simply cap the amount that they give for publishing fees. This would guarantee that a huge portion of scientists would have to stop using those journals, possibly putting them out of business. Plan S, and Open Access more generally, is an example of how collective action on the part of funders can force a massive change in research practice – and it should give us hope that such cooperation might in the future help to fix some of the other malfunctioning parts of science.

We've now seen ways of preventing fraud, avoiding negligence, and eliminating bias. What about the fourth of our scientific flaws – hype? Since hype is about how research findings are communicated, the solution might be connected to the publication system we've just been discussing.

From the beginning of the book, we've taken for granted the publishing model that first appeared with *Philosophical Transactions* in the seventeenth century: an editor, with the advice of peer reviewers, decides which studies will get published in the journal, then takes care of getting them into print. In the not-so-distant past, when all this work was done entirely on paper, this process of peer review and journal publication was incredibly laborious: corresponding with the reviewers, sharing the necessary documents amongst them, gathering and collating their feedback, then editing, checking, printing and distributing the paper journal itself. It's understandable that publishers would charge for such a service. But with the advent of email and online journals, the whole process is vastly simplified. In the age of online publication, when for-profit publishers charge exorbitant subscription fees for access to their journals, what exactly are we paying for?

It's worth emphasising just *how* exorbitant those subscription fees are. In 2019 the Nobel Prize-winning biologist and journal editor Randy Schekman compared the fees paid by the University of California to two different publishers: the non-profit National Academy of Sciences and the for-profit Elsevier.[67] Per individual scientific paper downloaded, the National Academy of Science subscription cost the university $0.04. The Elsevier per-paper fee was $1.06. More than *twenty-six times* more expensive. Researchers from across the world are pouring millions into Elsevier's coffers, and indeed billions into those of for-profit publishers in general. What additional service are they getting for all this money? It's not obvious: the National Academy of Sciences offers essentially the same services at its journals as Elsevier, or even better. The time and expertise of peer reviewers is the crux of the whole scientific system, providing the critical analysis that's required for science to progress. Yet those reviewers are volunteers and aren't employed by the journals. They are free labour, so they clearly can't be what justifies the cost. Rather, Elsevier and other for-profit publishers are engaging in what economists call rent-seeking: gaining more profits without providing more value.[68] One highly critical investigation of the scientific publication system noted its absurdities:

> It is as if the *New Yorker* or the *Economist* demanded that journalists write and edit each other's work for free, and asked the government to foot the bill. Outside observers tend to fall into a sort of stunned disbelief when describing this setup ... A 2005 Deutsche Bank report referred to it as a 'bizarre' 'triple-pay' system, in which 'the state funds most research, pays the salaries of most of those checking the quality of research, and then buys most of the published product'.[69]

Once you remember that the fees universities pay to rent-seeking companies such as Elsevier are largely public funds, the whole thing begins to seem immoral: shouldn't universities be making more efficient use of taxpayer money? It's not even that we should be opposed to the profit motive, or to commercial publishers getting involved in science. But the irrational system of prestige that's built up around the journal publishing system represents a market failure, impeding the necessary competition that could drive down its costs. It's hard to argue that journal publishers are adding much value to the papers: they often serve merely as a conduit between the authors and the volunteer peer reviewers, doing barely any editing or other work of their own beyond superficial typesetting.[70]

Big changes can happen to the journal publishing system – and we're witnessing one right now. It's the rise of *preprints*. A preprint is an early draft of a scientific paper, posted online in a free repository for anyone to look at. The intention is that other scientists will read and comment on the study so that the author can make any relevant tweaks before they submit it to a journal for official publication: a new variation on the social processes of science. Economists and physicists have been using the preprint model for decades, but in recent years, preprints have also taken other subjects, like biology, medicine, and psychology, by storm.[71] In a field like genetics, where new techniques, datasets and results are being produced at dizzying speed, preprints allow scientists to keep up with all the latest advancements without the lengthy wait for formal publication. They also allow mistakes, bad ideas and bad papers to be caught and criticised immediately by anyone from the scientific community (or the wider world), rather than the two or three people who'd be involved in normal peer review. These criticisms – as well as any positive comments – are usually posted online as well, making the peer review more informal but also more transparent than the standard model.[72]

Preprints have already quickened the pace and improved the openness of scientific research – and, one hopes, reduced the need to put 'failed' studies in the file drawer, since you don't need to have them accepted by a journal. But how does this address the problem of hype? As we've seen, one of the main reasons for unwarranted exaggeration is the need to convince reviewers and editors that your paper is worth publishing: scientists puff up the implications of their findings and spin away their null results in an effort to get those coveted publications. Journals don't just have the role of disseminating research to the world – they make the decisions on what gets published. What if we were to separate these two roles entirely?[73] One radical proposal goes like this: after completing a study, scientists write a preprint and upload it to an online repository. They then request that it be reviewed by a review service: a new kind of organisation that recruits peer reviewers in the usual way but that is separate from any scientific journal.[74] The service reviews the paper and gives it a grade. The authors, if they wish, can then go back and revise the paper, collect more data, or whatever else is required, and resubmit the improved version for a new grade and, they will hope, a higher one. Journal editors, whose role in this system is more like curators than gatekeepers, can peruse all the graded preprints and choose those that they would like to include in their journal.

With this model, everything gets published in preprint form, and journals become amplifiers for the best or most relevant research – like newspapers that print the most important stories from newswires such as Reuters and Agence France-Presse. The journals that would be worth subscribing to would be the ones that provided the best additional value, sifting through the thousands of new preprints to find the ones most worthy of your attention.[75] This way, you'd still get the competition that drives better-quality papers: scientists would want to have their work picked up by the most renowned curator-editors and would compete based on *where* their work is published. But the decision about *whether* to publish it would already have been made.[76]

This idea isn't the one true answer to science's problems – it might end up no better than the status quo, or even have unknown unknowns that make things worse. It's certainly worth trying.[77] We do need to be careful, however, with these new models of publishing. How do we control the quality of science in a world where anyone can simply post their findings online without any peer review?[78]

A 2016 preprint by the Harvard economist Roland Fryer provides a cautionary tale. It addressed one of the most contentious topics in modern American life: the use of force by the police against black people.[79] Fryer's analysis mainly focused on a detailed dataset of arrests by the police force in Houston, Texas. The results were paradoxical: although black and Hispanic people were over 50 per cent more likely than white people to have non-lethal force used against them during arrests, it was the opposite way round for lethal force: 'blacks [were] 23.8% *less* likely to be shot by police, relative to whites.'[80] The preprint reached public attention via a front-page article in the *New York Times* ('Surprising New Evidence Shows Bias in Police Use of Force but Not in Shootings'), which quoted Fryer saying it was 'the most surprising result of my career'.[81] It quickly became a talking point for conservative commentators displeased with the Black Lives Matter movement, which campaigns against police violence. One wrote that Fryer's study 'proves' that the movement was 'built on lies'.[82]

Fryer's 23.8 per cent figure was based on a straightforward look at the raw data. There was a lot more to the study than just the raw data, but this sadly became lost in the media coverage – especially after Fryer himself had highlighted that very specific number to journalists. After adjustment for other details of the case, such as whether the arrestee had a weapon, the results flipped: in comparable situations, blacks were more likely to be shot than whites. This difference between the raw and the adjusted results is consistent with racial bias on the part of the police: it's the effect you would see if the police just arrested more black people in general, no matter whether they were a threat, and were more likely to shoot black arrestees if the situation escalated. In other words, the number of black arrestees being shot was diluted by the fact that the raw data included lots of non-threatening black people who were arrested for no reason.[83] We shouldn't read too much into any of this, however: Fryer's sample was far too small. The 23.8 per cent figure wasn't even a statistically significant difference; had Fryer stuck to highlighting only statistically significant results, this controversy would never have arisen in the first place.[84]

If the *New York Times* had waited for the published version of the paper, which appeared in April 2019, they'd have found a somewhat different spin: the paper's abstract dropped the 'less likely' claim and just noted that there were no differences in the lethality rates against black people versus white people.[85] Though it still might have been interesting, this way of framing it was quite a climbdown from Fryer's initial claim. So, is the moral of the story to beware scientists bearing preprints? Since a preprint can be posted online without any oversight, we should certainly be extra sceptical about them, while scientists should have the intellectual humility not to publicise their work before it's been at least looked over by their peers.[86]

As the scientific ecosystem changes, journalists will become more aware that there are different 'stages' of scientific publication and that they should be particularly cautious of papers that are still at the earlier ones. Soon after the start of the coronavirus (COVID-19) pandemic in early 2020, preprints appeared on a major biological preprint server that sparked widespread discussion about the origins and effects of the virus. Some of the papers were of obviously low quality, rushed out to capitalise on the media frenzy about the pandemic. Others included phrasing that, whether inadvertently or otherwise, seemed to stoke conspiracy theories about the virus having been designed deliberately as a biological weapon. In response (and in addition to their permanent warning that the preprints they host aren't peer-reviewed), the archive added the following to the top of every page:

A reminder: these are preliminary reports that have not been peer-reviewed. They should not be regarded as conclusive, guide clinical practice/health-related behaviour, or be reported in news media as established information.[87]

We've seen throughout this book, however, that peer review is no guarantee of scientific quality. Think of Andrew Wakefield's MMR vaccine study – perhaps the most damaging scientific fraud of all time. Not only peer-reviewed, it was peer-reviewed and accepted for publication at the *Lancet*, one of the world's most prestigious medical journals. Although preprints can act as a vector for erroneous information, it's unlikely we could design a system that eradicated all mistakes. We need to weigh the downsides of preprints against the upsides that they bring of increased openness, transparency and rapidity. Indeed, for virologists and epidemiologists responding to the coronavirus crisis, the preprinting revolution has brought forth a wave of new data that substantially accelerates science, producing a research culture that's utterly different from those during previous disease outbreaks. Not having to wait for formal peer review, being able to comment instantly on drafts of new findings, and sharing important null results that wouldn't normally survive the publication-bias process has – despite the rare misleading claims

– given us a scientific literature that's months, or maybe years, ahead of where it would be otherwise. Although it remains to be seen whether this accelerated science will save us from the effects of the worst pandemic for a century, which at the time of writing is killing thousands and shutting down the world economy, the lesson for science is clear: in a contest between the old journal-only scheme and one that's supplemented by preprints, the preprints system wins hands down.[88]

The above ideas are promising, but they're mainly aimed at addressing the *symptoms* of science's modern maladies, rather than the causes. If we ignore the ultimate reasons, then any would-be science reformers are in for a Sisyphean task, perpetually rolling back the damage from every new generation of scientists eager to bolster their CVs with attention-grabbing, 'ground-breaking' results. In the previous chapter, we saw how this eagerness is driven by external pressures. Scientists need jobs and grant funding, and their endless chase for results that promise high-impact publications is in many ways a pursuit of those. Witness, for example, how many universities make their scientists sign an agreement about the number of papers they'll publish, and the impact factors of the journals they'll be in, as a step to getting tenure.[89]

The weird thing, though, is that scientists who already have tenure, and who already run well-funded labs, continue regularly to engage in the kinds of bad practices described in this book. The perverse incentives have become so deeply embedded that they've created a system that's self-sustaining. Years of implicit and explicit training to chase publications and citations at any cost leave their mark on trainee scientists, forming new norms, habits, and ways of thinking that are hard to break even once a stable job has been secured. And as we discussed in the previous chapter, the system creates a selection pressure where the only academics who survive are the ones who are naturally good at playing the game. If you're the type whose insatiable ego drives you to do anything to beat your colleagues' *h*-index, the academic system is currently welcoming you with open arms.

Self-sustaining systems are particularly stubborn. Yes, changing the academic incentive structure and its perverse demands to keep publishing and to keep bringing in more grants is crucial to fix science. The three main players responsible for this are the universities, the journals and the funders: they wield the power to lift the lid from the publish-or-perish pressure cooker, and we're about to see how each of them could do so. But if we stop there, we'd be ignoring the fact that many scientists still know no other game to play. To change that, we need to combine the top-down changes from the big players with bottom-up forces that come from the scientific community itself.

To help get the scientists themselves on board with necessary reforms, we should admit that the Mertonian norm of disinterestedness, admirable as it might be, only goes so far. In practice, personal reward is and always has been an important spur towards scientific progress.[90] If we want to encourage researchers to collaborate more with each other, and to share their data online freely, we will need to make doubly sure they get the credit they deserve.

That will take considerable work, and substantial changes in how we evaluate research efforts. But it also offers a welcome opportunity to replace the current opaque and frequently inequitable system for describing scientific collaboration. In most scientific fields currently, the convention for multi-author papers is to list first the name of the scientist who did the majority of the work, ending with the senior researcher who oversaw the project.[91] Everyone else – who might have played a role in advising on the study design, organising the lab, recruiting participants, gathering materials, collecting data or running statistics – is listed somewhere in the middle. As a way of assigning credit, this is far from ideal. On the one hand, many of those listed as 'authors' may have had little or even nothing to do with the actual paper; on the other, the genuine contributions of numerous middle authors go unrecognised, since universities mainly focus on a scientist's first- and last-author papers when assigning jobs and salaries.

Sweeping, collaborative 'team science' efforts will only exacerbate the problem: they often result in papers that have author lists in the hundreds or in some cases thousands.[92] The data-sharing that's required for Open Science also raises concerns about unrewarded effort. What if it generates a new class of scientists – 'research parasites', as they were dubbed by an editorial in the *New England Journal of Medicine* – who never gather data on their own, and just rely on running analyses of the data that others have painstakingly collected?[93] Our current system of vague unwritten rules and personal citation indices is of little use in addressing this issue.

What's required, then, is a new way of apportioning credit and responsibility in science: a system in which scientists are rewarded for their contributions, rather than for merely having their name on a paper.[94] To achieve this, it seems imperative that universities take the first steps. When making hiring and tenure decisions they should consider 'good scientific citizenship' in addition to, or even in place of, measures such as a scientist's h-index.[95] They should recognise not just publication but the complexity of building international collaborations, the arduousness of collecting and sharing data, the honesty of publishing every study whether null or otherwise, and the unglamorous but necessary process of running replication studies. They should, in other words, reward researchers for working towards a more open and transparent scientific literature, expanding the range of what's considered worthwhile.[96] They should explicitly value quality over quantity and focus more on methods ('does this person do rigorous work that produces solid science?') than results ('does this person keep getting p-values below 0.05?'). And they should view grants, in the words of the psychologist Scott Lilienfeld, 'as means to an end rather than ends in and of themselves.'[97]

Funders should follow suit: instead of basing decisions on a scientist's previous numbers of papers or on how big the claims are in the proposal, they should build in criteria that require scientists to do more with the money than just publish papers in high-impact journals. How much data will be created by the proposed study? To what extent will it be shared with the community? The funders could also follow the initiatives we saw above, and withhold money if scientists contribute to publication bias by failing to report all their results. As Plan S shows, when funders band together to push Open Science principles they can be a potent force.

Far more drastic changes could and should be made. Various methods have been proposed to dismantle the arms race of scientists competing for funds with ever-more-overblown claims. One idea, which some funders are already taking on board, is mainly to fund *scientists*, rather than their specific projects. That is, you could give those researchers who are judged to be sufficiently creative enough money to fund their planned research without constraints and hope that the increased freedom will lead to new and exciting advances. This would give the researchers breathing space away from the constant struggle for funding and would allow them to focus on longer-term, higher-quality projects, rather than being forced to submit short, possibly salami-sliced papers to beef up their CVs for their next grant application.[98]

But this wouldn't work so well for larger scale ideas, where the relevant staff, equipment, and subjects have to be planned and budgeted well in advance – and it could be unfair to lesser-known scientists who may have good ideas but not the network or clout that's required to be selected for long-term funding. In a world where we still fund projects, what's the best way to allocate the money? A particularly intriguing idea is to create a shortlist of grant applications that are all above a certain quality level and then allocate the funding by lottery. Given that the scientific system is supposed to be a meritocracy, this perhaps sounds bizarre. As one set of lottery proponents put it, however, the current system is so bad at allocating money that it's 'already in essence a lottery without the benefits of being random'.[99] An analysis in 2016 found that the scores given to potential US

National Institutes of Health grants by reviewers had almost no correlation with the quality of the eventual research produced from the grant (measured by the number of citations it received).[100] If that's the case, a substantial portion of the time scientists spend buffing up their grant proposals is wasted. Indeed, by one calculation, 'the value of the science that researchers forgo while preparing [grant] proposals can approach or exceed the value of the science that the funding program supports'.[101]

It's probably impossible to fairly differentiate between a selection of high-quality grant proposals. Letting random chance do the job saves reviewers the time of attempting the unachievable as well as circumventing biases that might favour certain types of scientists for their seniority, gender, or other characteristics.[102] If scientists knew that their project proposal would be subject to a random lottery so long as it reached a certain level of quality – a bar that could be set quite high, and could include criteria like 'adherence to Open Science principles' – they'd feel far less pressure to hype up the importance of their work.[103]

The journals, too, should adopt new standards that promote openness and replicability: explicitly inviting scientists to submit replications, to pre-register their plans, and to attach their datasets to their papers would be a great start.[104] This extends the list of ways in which scientists can endear their paper to journal editors beyond the narrow focus on positive studies and means that researchers with null results won't give up at the first hurdle.[105] Journals could also institute policies of scanning studies for basic errors or employ data integrity officers to do random spot checks using methods like the GRIM test.[106] They should certainly push authors to present their findings with humility: just as they require scientists to disclose their conflicts of interest, they could require them to write a 'study limitations' section that's displayed in a little box on the final published paper.[107] They could even try to discourage salami-slicing by making it more costly (in terms of time, effort, or money) to submit papers.[108]

All of this is what the universities, funders and journals should or could do. But what's in it for them? After all, university departments want to secure status in an increasingly deafening marketplace that ranks them on how many high-impact papers their scientists produce; funders want to be seen to be supporting the highest-prestige science; journals want to see their impact factors growing each year. There are, however, two reasons that now is the time for change.

The first is the ongoing replication crisis itself. For decades, even centuries, many scientists have harboured serious suspicions about the reliability of results and about the system in which those results are produced. But only in the last few years have they amassed stacks of hard data that back up their case. Now, across many fields, we can place high-profile single instances of science going wrong within the wider context of systemic problems, and point to strong meta-scientific data to back up the anecdotes. It's only recently, in other words, that we've had the data that provides a concrete basis for fixing science.

No university wants their scientists to be known for their work failing to replicate, or to become embroiled in fraud investigations. Nor do funders: they recognise that if they don't fund high-quality, replicable research, their money is being wasted and in the long run, they'll end up looking foolish. Nor do journals: the last thing they or their publishers want is to become a laughing stock, as has happened to some journals that became notorious for publishing flimsy research.[109] Reputational concerns are partly what got us into this mess, but if the scandals and shortcomings become sufficiently well known, those same concerns might also help us out of it.

In fact, worries about the replication crisis among scientists themselves have already produced a groundswell of support for reform in science. Scientists have shown themselves willing to break *en masse* with the standard publication system, as their embrace of preprints demonstrates. The scientific community saw the benefits that preprints offer when it comes to speed and transparency, and took the plunge; the journals followed, changing their policies to say that preprint articles would still be acceptable for submission. Similar demands for Open Access, Open Data and pre-registration are all having their effects on journal policies.

Linking bottom-up demands and top-down policies can result in positive feedback loops – virtuous cycles that create beneficial self-sustaining norms, rather than perverse ones. For instance, the UK Reproducibility Network, formed by academics concerned about the replication crisis and represented by grassroots groups at universities, is now talking with those universities about ways to change their hiring practices to better reward openness and transparency.[110] Such efforts can serve as an antidote to the 'natural selection of bad science' that we saw in the previous chapter. Rewarding scientists who value openness and transparency can create its own virtuous cycle: as the number of Open Science practitioners increases, it'll help to drive further bottom-up changes and encourage other researchers to buy into the movement and its associated reforms. It'll also produce even more meta-science, helping us understand which kinds of studies replicate, which academic structures foster reliable research, and which results we should generally view with suspicion.[111]

Having the approval and respect of one's colleagues is a big motivator; we shouldn't underestimate the power of shame. The fear of being ridiculed online for making silly mistakes, displaying obvious bias, or overhyping one's research should be a strong motivation to double-check the numbers and keep claims within the bounds of reality.[112] But we needn't focus entirely on negative reasons for change. The second big reason why now is the best time to revolutionise our scientific system is technology.

Quite simply, it's never been easier to do Open Science. Algorithms can now check papers for errors; preprints and pre-registrations can be made available instantaneously and even very large datasets can be shared in ways that were impossible just a few years ago; authors' contributions to papers, and to the scientific community more generally, can be logged in minute detail; the whole chain from data collection to publication can be laid out for the world to see. This is a point emphasised by Brian Nosek, the director of the Center for Open Science: before tackling ingrained norms, incentives and policies, the first, deceptively simple step to effecting cultural change is to make it possible and easy for people to go along with your new ideas.[113] A great many academics would like to improve the quality of their work but are held back by the thought of how much effort the changes might require. New and ever-developing technological fixes can eliminate that concern and enable more and more researchers to come into Open Science.

If making it possible and making it easy isn't quite enough, we can always appeal to scientists' self-interest. The cancer biologist Florian Markowetz, in a paper about how scientists should take advantage of new automated tools to make the links between their data, their analysis and their papers crystal clear, gives 'five selfish reasons to work reproducibly':

1. Making your data open and transparent helps you and your co-authors to *spot errors* that might undermine your results, stopping your study from becoming the next Reinhart-Rogoff Excel-typo disaster.
2. The new automated methods make the paper *easier to write*.
3. If anyone can see how you analysed your data, it's easier to *convince reviewers* that you did the right thing (and if the data are available, they can even try running your analysis for themselves).
4. Documenting every step of the analysis openly helps you *continue your work* a few months down the line: 'future you' won't have to rely on memory of what 'past you' did.
5. Making everything available shows the scientific community that you've nothing to hide and that you've done everything in good faith: in other words, it *builds your reputation* as an honest researcher.[114]

We clearly see the problem. The solutions are within our grasp. All we need to fix science is to give people the right motivation.

Henryk Górecki's Third Symphony, the *Symphony of Sorrowful Songs*, has sold more than a million copies, completely unheard of for recordings of modern classical music.[115] Its popularity comes partly from its simplicity: it's slow moving, almost like a film score, with none of the harsh atonality of Górecki's previous works. Despite its eventual incredible sales, its 1977 live premiere apparently didn't go down well with all audience members. As the final movement closed, with its twenty-one repetitions of an A major chord, Górecki overheard 'a prominent French musician' sitting in the front row (most assume it was the irascible avant-gardist Pierre Boulez) hiss 'Merde!'.[116] Even though it's now recognised as a modern masterpiece, the Third Symphony was just not novel or experimental enough for the musical revolutionaries.

Scientists think more like Boulez than Górecki. They've taken excitement about novelty in science too far, producing a perverse neophilia where every study needs to be a massive breakthrough that revolutionises the way we think about the world – perhaps announced, as though in the movies, by a scientist in a white lab coat bursting excitedly into the room waving a sheaf of paper. Scientists want their research to sound as though it's of this Eureka-shouting kind, so they analyse it, write it up, and publicise it accordingly. But although unexpected breakthroughs do occur from time to time, most science is incremental and cumulative, building slowly towards tentative theories rather than suddenly leaping to conclusive truths.[117] Most science, to be perfectly honest, is quite boring.

Cooling down the hype surrounding new results and taking a humbler attitude to what we know might make science duller – but boring-but-reliable should win over exciting-but-insubstantialunder almost all circumstances. Just as publishing more null results and replication studies is a more dependable way to build our knowledge, becoming more aware of the uncertain and preliminary nature of research is, in the long run, a better way to appreciate science fully. Let's work to resist our neophiliac, magpie-like focus on shiny research findings, and instead learn to value results that are solid, even if they're less immediately thrilling. In other words, let's Make Science Boring Again.[118]

But not *too* boring. It's only *most* science that's incremental; in reforming the incentives, we don't want the pendulum to swing too far in the opposite direction, discouraging scientists from taking risks, trying their luck with wild, maverick new ideas, and engaging in the kind of exploration that every so often leads to a major innovation. The trick will be to balance our new, step-by-step attitude towards knowledge with an appreciation that sometimes moonshot research by eccentric, Boulez-style characters really can have enormous payoffs.[119] These perspectives are more easily reconciled than you'd think: not only can it be argued that the dreary focus on publication and citation also holds back the kind of quirky exploratory research that can spark revolutionary advances, but the kinds of reforms I've outlined in this chapter will make novel findings more meaningful, distinguishing the play of chance (or of human bias and spin) from genuine, tantalising leads that could point us towards the Next Big Thing.[120]

At its best, there's a certain nobility in doing scientific research – in following those Mertonian virtues of universalism, communalism, disinterestedness and organised scepticism in honest pursuit of truth and knowledge about the world. But all these virtues are left in tatters in a system that encourages outrageous hyping of results, jealous guarding of data, lazy corner-cutting, brazen prestige-hunting and shameful fraud. If we can train a new generation of researchers to aim for the Mertonian norms, while at the same time holding back the deluge of incentives that push them in the opposite direction, we might be able to save science from itself.

It will, of course, take some time to convince the scientific community that the ideas in this chapter are worth trying. We mustn't be dogmatic about it: not only would that be unscientific, but different fields of science need different reforms. There's no one-science-fits-all solution.[121] We should proceed carefully, experimenting with new ways of working and gathering evidence about them, rather than smashing the current system and imposing reforms by decree.

Ideally, the meta-scientific evidence adduced in this book will convince almost everyone that something has gone very wrong with science, and that there's a dire need for change. But even if you think the talk of a 'crisis' is grandiose or exaggerated, there's one final argument in my quiver.[122] It's this: the reforms we've discussed in this chapter would all be beneficial for science *even if there weren't a replication crisis*. This brings to mind the following classic cartoon about climate change, from the *Lexington Herald-Leader*'s Joel Pett:

With apologies to Pett, let me rewrite his cartoon to respond to a different kind of doubter:

Openness. Transparency. Improved statistics. Pre-registration. Automated error-checking. Clever ways to catch fraudsters. Preprints. Better hiring practices. A new culture of humility. Etc. Etc. What if the replication crisis is a big hoax and we create a better science for nothing?

Epilogue

O, while you live, tell truth, and shame the Devil!
William Shakespeare, *Henry IV*, 1.3.1.59

As I write this, astrophysicists have recently taken the first photograph of a black hole.[1] Medical geneticists have announced that seven children with severe immune deficiencies, who'd been forced to live in isolation lest they catch a common – but for them, deadly – infection, might have been cured by gene therapy; and that gene-based therapies for cystic fibrosis have shown results that imply they could work for 90 per cent of people with the condition.[2] Public health researchers have shown that, for HIV-positive gay men who are taking the latest antiretroviral drugs, the chance of transmitting the virus to a sexual partner is 'effectively zero'.[3] Engineers have teleported information within a diamond using quantum entanglement.[4] Scientists have injected nanoparticles into the eyes of mice, giving them infrared vision.[5] These are all marvels and to see them amid a stream of scientific and medical progress is a regular reminder that science is one of humanity's proudest achievements.

Or, at least, they should be marvels. We should be proud of them. But a disheartening effect of learning about all the scientific flaws we've seen in this book is that we might become suspicious of any and all new results, given our knowledge that the stream of scientific progress is far from pure. Around the same time as all the developments above, the US Office of Research Integrity concluded that a Duke University medical researcher falsified data in thirty-nine published papers and in more than sixty grants worth over $200 million.[6] A genetics professor was found to have run his lab at University College London so 'recklessly' that it produced potentially dozens of fraudulent papers, yet he refused to resign or to accept any responsibility.[7] A hyped psychology study, published in *Science*, claimed that conservatives have stronger physiological reactions (to sudden noises, for example) yet it failed to replicate in a much bigger sample and – in a more or less exact repeat of what happened to our attempted replication of the Bem psychic study – *Science* rejected the replication paper out of hand.[8] A new anti-plagiarism algorithm identified over 70,000 Russian-language journal articles that have been published at least twice, with some of them appearing in up to seventeen different journals.[9] One of the world's most-cited researchers, a US biophysicist, was barred from his position on the board of a biology journal after he was found to have regularly coerced authors whose papers he edited into citing his own publications – sometimes more than fifty of them at a time, apparently giving him a massive spike in his citation count.[10]

We can't simply marvel at new scientific advances, because we know that the solid, replicable findings come to us alongside a glut of erroneous, biased, misleading and falsified research. The nephrologist Drummond Rennie, writing in 1986, put it best:

Anyone who reads journals widely and critically is forced to realize that there are scarcely any bars to eventual publication. There seems to be no study too fragmented, no hypothesis too trivial, no literature citation too biased or too egotistical, no design too warped, no methodology too bungled, no presentation of results too inaccurate, too obscure, and too contradictory, no analysis too self-serving, no argument too circular, no conclusions too trifling or too unjustified, and no grammar and syntax too offensive for a paper to end up in print.[11]

He was hardly the first to notice. In 1830, the mathematician and 'father of the computer' Charles Babbage wrote his remarkable *Reflections on the Decline of Science in England, and on Some of its Causes*, where he offered a taxonomy of scientific ills.[12] These were 'hoaxing' (by those who produce fake findings which they later unmask, to prove a point), 'forging' (by our familiar scientific fraudsters, who have no intention of revealing their deceptions), 'trimming', and 'cooking' (both of which correspond, in our modern understanding, to p-hacking, where scientists manipulate their data and observations to give them the appearance of greater interest or accuracy). So although our modern publication system exacerbates science's problems, it's hardly their ultimate cause: they've been with us for a long time. Imagine how much more progress we could be making, how many more diseases could be rendered extinct or impotent, how much more we'd know about space and evolution and the cell and the brain and human society – not to mention how many false dawns and blind alleys we could avoid – if we finally did something about all these long-running inadequacies.

The idea of this book has been to help that process of change on its way – as Babbage put it at the start of his treatise, to 'do some service to science'. Babbage's friends warned him that to criticise scientific practice would make him enemies (though they may also have been taking into account his rather spiky personality).[13] The usual reaction I received when I told *my* friends about this book was a broader concern regarding trust in science: 'Isn't it irresponsible to write something like that? Won't you encourage a free-for-all, where people use your arguments to justify their disbelief in evolution, or in the safety of vaccines, or in man-made global warming? After all, if mainstream science is so biased, and its results so hyped, why should the average person believe what scientists are telling them?'

But this isn't the right way to think about it. First, trust in science starts from a very high baseline. The Wellcome Global Monitor collects data from across the world about people's attitudes to science and scientists; in the global sample from 2018, an average of 72 per cent of people reported having either medium or high levels of trust in science, with numbers reaching as high as 92 per cent for Australia and New Zealand.[14] Admittedly, some regions reported much lower trust: only 48 per cent in Central Africa, for example, and 65 per cent in South America. But the average is high and in some Western countries, such as the UK, there's evidence that people's views toward science have become *more* trusting over time.[15] People across the world respect science to a high degree and although these levels might drop a little when they hear about some of its systemic problems, it's hardly likely that there'll be a sudden catastrophic decline.[16] My view is that scientists need to work harder to deserve trust. Although I'm not aware of representative surveys on this, it could certainly be argued that the constant media drumbeat of overblown claims about scientific and medical breakthroughs, as well as the contradictory conveyor belt of fly-by-night findings from fields like nutritional epidemiology, serve to do more damage to trust in science than any amount of discussion about the replication crisis ever could.

More importantly, though, a sophisticated view on science isn't one of unquestioning trust. It's one that's pithily summed up by the motto of the UK's Royal Society: *nullius in verba*, or 'take nobody's word for it'. (Almost the same notion is expressed by the Russian proverb favoured by Ronald Reagan during Cold War negotiations: *doveryai, no proveryai* – 'trust, but verify'.) That's the idea of Open Science, and of the Mertonian norms of communalism and organised scepticism in a nutshell: reduce our reliance on unthinking trust as far as possible, and share as much checkable, testable, verifiable evidence with the world as we can. It's been said that 'there is really no such thing as alternative medicine, just medicine that works and medicine that doesn't'.[17] In the same way, there isn't really such a thing as Open Science: there's science, and then there's an inscrutable, closed-off, unverifiable activity academics engage in where your only option is to have blind faith that they're getting it right.

We can go further: it's actively dangerous to encourage people to think of science as a body of facts that can never be doubted. Not only is this view contrary to the norm of organised scepticism, but it can backfire very badly. If you're convinced that science is an implacable wall of truth that you have no choice but to believe in, what do you do when it becomes clear that something has gone wrong? After all, if we've learned one thing from this book, science *will*, quite often, go wrong. The science historian Alex Csiszar discusses the case of climate change, where sceptics

> have invoked a fairy-tale image of scientific publishing as the bedrock of legitimate consensus only to profess outrage when it turns out not to live up to this fantasy. This reaction was exemplified by the reaction to the leak of thousands of emails and documents from the Climate Research Unit at the University of East Anglia in November 2009. The emails seemed to reveal climate scientists engaging in secretive behavior and politicking with peer review. Evidence that scientific life behind the printed pages of journals was not a precise reflection of its better-behaved public face was seized upon by commentators to argue that the bottom had fallen out.[18]

Climate science, as it happens, is an apt example. In recent years, the field has come under a particularly subtle kind of attack, where the language of science reform is co-opted as a political move. In 2019, the US Department of Agriculture announced to its researchers that they had to add a statement to every piece of research noting that it was 'preliminary'.[19] At face value, this appears to be exactly what I've been recommending in this book: treating each study as a tentative step towards an answer, rather than as the answer in itself. But nobody thinks this policy was driven by an innocent desire to improve people's interpretation of research. It was announced because much of the work done at the Department of Agriculture involves climate change, and the results are often inconvenient for an administration that's as pro-fossil-fuels that of as Donald Trump.

The new regulations caused an uproar and the guidelines on the use of 'preliminary' were rolled back a month after they were introduced. What was particularly striking about this case, though, was how some scientists overcompensated in their response to the political attack. The editor of the *Journal of Environmental Quality* was quoted in the *Washington Post* arguing that a published paper is 'the end product to your research … It is now finalised. There's nothing preliminary about it'.[20] This is a naïve statement, informed by the kind of idealised, prettified view of science we've seen and dismissed throughout this book. Even if politicians use concerns about replicability as a disingenuous pretext for their scepticism of climate change, it doesn't justify scientists overstating how much confidence we should have in our results. As the editors of *Retraction Watch*, Ivan Oransky and Adam Marcus, put it, 'scientists and policymakers need to take a long-term view, even if politicians on the fossil fuel dole are eager to embrace every opportunity to cast doubt on global warming … Significant efforts are afoot to improve reproducibility, and … the threat of bad-faith attempts to exploit science reform shouldn't be allowed to derail these attempts.'[21]

Politicians have long suppressed science that's inconvenient for their policies. History's most extreme example might be the forced adherence to the baroque pseudoscience of Trofim Lysenko, whose cultish, genetics-denying ideas contributed to famines that killed millions in Stalin's USSR and Mao's China.[22] But a government needn't be a totalitarian dictatorship to have an anti-science attitude. Politicians in democracies regularly pander to their voters by denying or distorting scientific evidence, whether it's American politicians who back the teaching of creationism, Italian populists who campaign against vaccines, the South African government's disastrous denial of the link between HIV and AIDS, or Indian Prime Minister Narendra Modi's bizarre theories about stem cell technology having been available in ancient India.[23] Even the comparatively liberal Scottish government announced in 2015 a ban on the commercial growing of genetically modified crops, a decision that will hamper research and force Scottish farmers to miss out on technological advances in, for example, pest resistance. The policy, which was intended to protect the 'integrity' of Scotland's 'clean and green ... brand' (whatever that means), was derided as 'cheap populism' by a political commentator and described as 'extremely concern[ing]' in an open letter signed by twenty-eight scientific societies.[24] What all this tells us is that, regardless of our discussion of the replication crisis and its associated failings, politicians will still trample all over science if they think it'll lead them towards votes.

The worry that the arguments in this book might be mis-appropriated to make selective, insincere attacks on research shouldn't stop us from publicly discussing the replication crisis and its associated problems. We mustn't make science suck in its stomach whenever a member of the public or a politician is watching. In fact, a frank admission of science's weaknesses is the best way to pre-empt attacks by science's critics and to be honest more generally about how the uncertainty-filled process of science really works.

The argument that airing our dirty laundry in public will reduce trust in science seems even more misconceived when we look at the sheer amount of worthless, misleading and fundamentally untrustworthy research we're putting out into the world. Every time we allow a flawed or obviously biased study to be published; every time we write another boy-who-cried-wolf press release that can't be backed up by the data; every time a scientist writes a popular book full of feel-good-but-flimsy advice, we hand science's critics another round of ammunition. Fix the science, I'd suggest, and the trust will follow.

In spite of the perverse incentives, in spite of the publication system, in spite of academia and in spite of scientists, science does actually contain the tools to heal itself. It's with more science that we can discover where our research has gone wrong and work out how to fix it. The ideals of the scientific process aren't the problem: the problem is the betrayal of those ideals by the way we do research in practice. If we can only begin to align the practice with the values, we can regain any wavering trust – and stand back to marvel at all those wondrous discoveries with a clear conscience.

Émile Zola defined art as 'a corner of nature seen through a temperament'.[25] As we've witnessed over and over throughout the book, this definition could equally well apply to science – or, at least, to the way science is currently done. The corners of nature with which science deals are seen through all-too-human temperaments, with their attendant prejudice, arrogance, care-lessness and dishonesty. You needn't believe that science is just one among many equal 'truths' to agree that it's definitely a human activity, and thus that it bears the imprint of human failings.

A task as grand as revolutionising science was never going to be straightforward. There'll be trial, error and, appropriately, experimentation along the way. This isn't just about discarding some faulty theory, like geocentrism, or phlogiston, or alchemy, or any of the menagerie of incorrect ideas that litter the history of science. This is about root-and-branch (or lab-and-journal) reform of the way we do research, and of scientific culture – an attempt to master the faults and biases that have crept in, largely unnoticed. The world is rightly proud of where science has brought us. To retain that pride, we owe it something far better than the product of our flawed human temperaments.

We owe it the truth.

Appendix
How to Read a Scientific Paper

Remember Daniel Kahneman's imperative from Chapter 2 that 'you have no choice' but to believe in behavioural 'priming' effects – effects that were subsequently discredited in the replication crisis? The message of this book is that when faced with a scientific finding, you *do* have a choice: you can choose to suspend your judgement until you've properly evaluated the science.

How, though? There's no denying that fully understanding the strengths and weaknesses of a given study usually takes years of training in the relevant scientific discipline. However, if you're willing to do a little searching online, you can usually get a decent impression of a study's pros and cons. And if you're willing to check the paper itself, there are some red flags that might stand out even above the impenetrable technical jargon.

Of course, checking the paper means being able to download it and in a world where full Open Access isn't yet a thing, the article you're interested in could be behind a paywall. There are a few things you can do about that, assuming you don't want just to shell out the money for article access, which you absolutely shouldn't do unless all the alternatives that I'm about to describe fail and you're still desperate to read it. First, you can check the authors' personal or work websites to see if they've put up a free version, sometimes in the form of a non-typeset file, for anyone to download. Just searching for the article using Google Scholar and clicking on the 'versions' link under each item can often point you to these free manuscripts. Second, you can check whether there's a preprint version of the article – these are always free, and although they might differ a little from the final published version that's been through peer review, they'll likely be similar in most respects. (As we've seen, sometimes the preprint *is* the version that gets through into the mainstream media, which might be where you saw the study in the first place.) Third, you could email the authors: although we've seen depressingly low rates of response when would-be replicators or reproducers contact authors to ask for their data, asking for the paper is much less likely to be a problem (in fact, many scientists will be delighted to hear that someone's interested enough to want to read their study). Fourth, I'm told there are also ways to use online piracy to access papers illegally, but I wouldn't know anything about that.[1]

Let's assume that you get your hands on the full version of the paper. When reading it, you can put together almost everything we've learned in the book using the following top-ten list of questions:

1. *Is everything above board?* First, the basics. Are the authors from what look like reputable universities, companies or labs? Does the journal where the study is published seem professional? If it has a shoddy-looking website that appears as if it's from the 1990s, it's probably one of the depressingly common 'predatory' journals that we encountered in Chapter 7. Anything published in such an outlet shouldn't be trusted, since they don't even *attempt* peer review.[2]

2. *How transparent is it?* How closely, in other words, does the study fit with the ideas of Open Science, outlined in Chapter 8? Was it pre-registered? An answer in the affirmative doesn't make the findings true by any means – and a negative answer doesn't make them false. But if you

can find an online registration of the study, it should at least modestly increase your confidence that the results aren't just due to *p*-hacking.[3] Tracking down the pre-registration document can also help you spot whether the main analysis is different from the one the scientists pre-registered: that is, whether they've engaged in outcome-switching. Also, are the data and other materials online? As we discussed, not every dataset can just be made public, for instance if it contains information that might identify the individual participants in a study. These are rare cases, though; if there's a link to the full dataset that's easy to find, it's a compelling piece of evidence that the scientists are being open with the reader.[4]

3. *Is the study well designed?* You'll recall from Chapter 5 that a worrying proportion of animal-research studies don't even mention blinding or randomisation. These are crucial aspects of experimental design and if you can't find any discussion of them in a paper – at least, for studies like clinical trials where these aspects would be important – it should increase your scepticism. Likewise, for many study designs, an adequate control group is necessary: when seeing the headline claim of an article, you should always ask 'compared to what'? If the answer is 'compared to a control group that differed from the treatment group in important ways before the experiment began', you've got a poorly designed study on your hands.

4. *How big is the sample?* Sample size matters, mainly for reasons of statistical power. It's true that statistical power can be increased in other ways, so sample size isn't the only consideration; for some kinds of studies, such as ones that expect large effects, or that test their participants again and again, a small sample is perfectly adequate. And even massive samples can be hopelessly biased if they're not random or representative. But it's a common error for studies in areas like neuroscience, ecology and psychology to search in tiny samples for what are generally small effects – a

strategy that tends to be worse than useless. Another thing to look out for is how many subjects were excluded from the final sample. Some exclusions are perfectly normal and usually inevitable. For example, human participants rarely all follow the instructions as they should. However, if the exclusion rate gets extremely high, say, perhaps excluding more than half of the sample, you might start to wonder if the results generalise to the studied population or if the authors cherry-picked subjects that showed the desired effect, while excluding those that didn't.

5. *How big is the effect?* The first thing to check is whether the effect reported in the study is statistically significant, and at what level. Are there lots of p-values that are just below the 0.05 significance threshold? Do the authors use ambiguous phrases such as 'trending towards significance' to excuse the fact that their results didn't quite make the grade? That, however, is just the beginning – you also need to ask how big an effect the study found. How does it compare to other studies, or other relevant effects? For example, if the study is analysing a new educational or medical intervention, how does it compare to other, already established ones? Are media reports, or the scientists themselves, interpreting something that has a small effect as if it were the only thing that matters? Since we know that there might be several null studies on a similar topic hiding in the file drawer, it could be useful mentally to revise the size of the effect downwards somewhat. The other side of the coin is that effects that are implausibly large – basically, too good to be true – should also raise suspicion that something in the study has gone wrong. The same goes for p-values: a study reporting exclusively, or nearly exclusively, significant results should cause you to raise an eyebrow. This is because, as we've seen, studies never have perfect statistical power and often have very low power indeed. So even if their effects are true, we'd still expect some p-values not to make the 0.05 threshold. A perfect line-up of significant

results in a study with many p-values is likely the result of p-hacking (or worse).

6. *Are the inferences appropriate?* As we've seen, scientists regularly slip into causal-sounding language even if they've only run a correlational study. If scientists are using an observational study to talk about how variable X affects, impacts or influences variable Y, they're going beyond their data. In an observational study, there isn't any randomised intervention, so causal conclusions can't normally be drawn. Similarly, if an experiment is run in rats or mice, or is a computer simulation, the inference that such an experiment necessarily tells us something about 'how humans work' simply isn't valid. Ditto for a study only carried out on a small, selected subset of people that's presented as telling us something about humanity in general.

7. *Is there bias?* Does the study have obvious political or social implications, and do the scientists write about these in a way that seems less than impartial? We've seen that hype and spin are often blatant, even in peer-reviewed papers. Was the study funded, in whole or in part, by a group or company that would favour one particular outcome? You can look at the Funding statements and the Conflict of Interest section, required in virtually all journals, to get some idea (though note that at present these aren't required to mention things like book contracts or lecture tours that might relate directly to the results in the paper – you might want to check the authors' websites for ancillary activities of this kind). If the scientists are appropriately tentative when discussing their findings and can't be found in the media telling journalists about how these resoundingly validate some political perspective or specific policy, it's a good sign that they're keeping a lid on their biases. Incidentally, checking for bias is even more important when the study suits your *own* ideological preconceptions. You might ask yourself whether you're scrutinising a study disproportionately harshly because

you happen to disagree with its conclusions or giving a free pass to a weak study that reinforces your preferred beliefs.

8. *How plausible is it, really?* For studies that involve human participants, a useful strategy is to imagine that you yourself have taken part.[5] For example, in a study of nutritional epidemiology, think about how accurate your memory would've been when you filled in the Food Frequency Questionnaire about your snack habits for the past decade, or even the past few weeks. The answer is probably 'not very accurate'. Would you have been exhausted by the end of all the tests in a behavioural experiment, and did the researchers account for that? Did the environment of the study (say, a lab in a university) even approximate the setting that the scientists really want to know about (such as a high-stakes job interview)? In other words, did this study really answer the question that was being asked? Putting yourself in the participants' shoes can help to reveal these basic questions about the study's plausibility.

9. *Has it been replicated?* We need to stop relying so heavily on individual studies. It's reassuring when scientists have replicated their own results; it's even better if others, from completely independent labs, have also done so. The first thing to do is to search to see if there are any published replications.[6] There might also be a review or meta-analysis of the main result, or of similar results, that can tell you if this study is just an outlier, and also whether its result fits into a broader theory (remember that the particular study you're looking at may itself be an attempted replication of an earlier result). Of course, reviews and meta-analyses can themselves be corrupted by poor research and publication bias in their source material; if you find a meta-analysis of studies that were all themselves pre-registered, then you've hit the jackpot, but I don't think I've ever encountered such a gem myself (although now that pre-registration is becoming more popular, this should change in future). Naturally,

you can't expect a replication to exist for studies that are entirely new, but you can withhold judgement on their validity until one appears.

10. *What do other scientists think about it?* The best news reports about scientific studies will quote an independent scientist to give their views, so it's worth looking to see if there are any instant reactions. There are some groups that do this more systematically: for example, the Science Media Centre is a UK charity that solicits comments and reactions from a range of independent experts each time a new paper is press released, publishing them on its website.[7] It's a nice example of how peer review can still occur even after a paper has officially been published. You can also search websites like Pubpeer, the anonymous science-commentary site where the faked stem-cell figures of Haruko Obokata, along with attempted deceptions by many other fraudsters, were first identified.[8] Googling to see if there are any blogs or other websites discussing the paper, or even searching Twitter, can be a good idea – although you should beware that this will give you both informed *and* uninformed, serious *and* unserious, unbiased *and* biased discussion of a study.[9] If the study has been around for a while, you can use Google Scholar's 'cited by' function to look for citations of it and check whether they tend to be positive or negative.[10]

None of these rather general methods are perfect, nor are they all applicable to every form of research. Obviously, it's always better to have prior knowledge and experience about the research area to get a better insight into the strengths and weaknesses of a specific study. Anything, though, is better than just accepting claims at face value.

It's also essential to bear in mind what we learned from the saga of Stephen Jay Gould, Samuel Morton and the apparently never-ending debate over skull sizes: even if you read a seemingly devastating critique of a piece of research, the critique itself might be mistaken, and so might the critiques of the critique. That also goes for everything I've written in this book.

The fundamental lesson is to *be humbler about what we do and do not know*. At first this might appear to be antithetical to the idea of scientific research, which is surely about uncovering new facts about the world and always adding to our knowledge. But if you think about it for longer, it turns out to be the very essence of science itself.

Acknowledgements

It all starts with Will Francis, my literary agent, whose idea it was for me to write a book on this topic. Will, PJ Mark, and the whole team at Janklow & Nesbit have been enormously helpful at every stage of conceiving, proposing and writing the book.

Working with my editors, Will Hammond and Grigory Tovbis, frequently brought to mind the old cliché about editing being a process of turning an ugly block of marble into an attractive sculpture. Trust me: you would not have wanted to read this book (if you could call it that) before they gave it their incredibly clarifying, thoughtful and detailed input. Thanks are also due to Will and Grigory's respective teams at The Bodley Head and Metropolitan Books (in particular Alison Davies and Sarah Fitts), for all their efforts to help make the book happen. I'm also grateful for a meticulous and good-humoured copyedit from Marigold Atkey, and reassuring legal advice from Henry Kaufman.

Several friends read and commented on various drafts, and I'm greatly obliged for their input. They are Nick Brown, Iva Čukić, Jeremy Driver, Stacy Shaw, Chris Snowdon, and Katie Young. There are two readers to thank most of all: Saloni Dattani, who instantly read the first drafts of each chapter as soon as I wrote them, giving me immediate feedback about how they looked and what did and didn't work; and Anne Scheel, who went far beyond the call of duty, using her expertise in statistics, her immense knowledge of Open Science, and her terrifyingly insightful eye for phrasing and nuance to save me from numerous statistical (and other) elephant traps.

I'm also indebted to all of those who either pointed me towards new stories or references, had interesting conversations or arguments with me about science and its problems, or just provided much-needed encouragement during the book-writing process. They include: all the members of Best Picture (Bobby Bluebell, Kenny Farquharson, Euan McColm, and Ian Rankin), all the not-previously-mentioned members of Fat Cops (Chris Ayre, Chris Deerin, Al Murray, and Neil Murray), Mhorag & Nigel Atkinson, Mike Bird, Ewan Birney, Robin Bisson, Sam Bowman (who, it pains me to admit, gave me the initial idea for the book's title), Chris Chabris, Tom Chivers, Simon Cox, Gail Davies, Ian Deary, Rory Ellwood, Alasdair Ferguson, Patrick Forscher, Anna Fürtjes, Roger Giner-Sorolla, Niall Gooch, Saskia Hagenaars, Sarah Haider, Lewis Halsey, Paige Harden, Kirsty Johnson, Mike Jones, Mustafa Latif-Aramesh, Riccardo Marioni, Damien Morris, Nick Partington, Robert Plomin, Jennifer Raff, Jo Rowling, Adam Rutherford, Aylwyn Scally, Adrian Smith, Ben Southwood, Michael Story (and Laska the dog), Elliot Tucker-Drob, Simine Vazire, Rachael Wagner, Ed West, Sam Westwood, Tal Yarkoni, and the members of the Science Media Centre. My colleagues at the SGDP Centre at King's College London made it a real pleasure to work each day before returning home to get back to the book. They're too numerous to name individually, but special mention goes to my wonderfully supportive Heads of Department, Franky Happé and Cathryn Lewis.

Of course, none of the people named above (or below) are necessarily in agreement with all (or any) of the points I make in the book, and none of them are responsible for any errors caused by my own biases or negligence. Incidentally, I'd encourage anyone who might discover such an error to get in touch via the website sciencefictions.org, and point me towards the accurate information. I'll post any corrections on that same site.

Many, many people told me (often shocking) stories of fraudulent, biased, negligent or hyped science that they'd uncovered – or sometimes experienced first-hand as students or research assistants in labs where things were going badly wrong. I'm sorry I could only fit a small fraction of these into the book, and can only hope that its publication inspires anyone who knows of an example of research misconduct, or even honest-but-flawed research, to do their best to expose it to the world.

It's tricky, writing a book, and I can only apologise to all the colleagues who wondered why on Earth I was taking so long to get back to them on this project or that and to friends who must've thought I'd disappeared off the planet while I was in the busiest writing phases. My parents were, as ever, magnificently kind and uplifting throughout the whole process, and I'm more grateful to them than I can say. But the person who was most directly affected by the book was Katharine Atkinson, who nevertheless showed only patience, and never frustration, with my constant refrain of '… I need to get back to the book …'. It was definitely more patience than I deserved. It's to her that the book is dedicated.

Stuart Ritchie
March 2020

The author gratefully acknowledges permission to reproduce images from the following sources: (p. 36) Sidney Harris, ScienceCartoonsPlus.com; (p. 197) the Getty's Open Content Program; (p. 201) Cambridge University Press; (p. 236) Joel Pett. The figures on p. 61 and p. 214 are reproduced under CC-BY and CC-0 Creative Commons licences respectively.

Notes

Opening Epigraph: Chris Morris et al., 'Paedogeddon!', *Brass Eye*, Tristram Shapeero, dir. (Series 2, Episode 1, 26 July 2001).

Preface

Epigraph: Francis Bacon, *Novum Organum*, ed. Joseph Devey (New York: P. F. Collier & Son, 1620/1902).

1   Daryl J. Bem, 'Feeling the Future: Experimental Evidence for Anomalous Retroactive Influences on Cognition and Affect', *Journal of Personality and Social Psychology* 100, no. 3 (2011): pp. 407–25; https://doi.org/10.1037/a0021524
2   They also had the reverse ability: when a violent picture was behind one of the curtains, participants psychically shifted away from it, only choosing it 48.3 per cent of the time – again, a statistically significant difference from what pure chance would predict.
3   Peter Aldhous, 'Journal Rejects Studies Contradicting Precognition', *New Scientist*, 5 May 2011; https://www.newscientist.com/article/dn20447-journal-rejects-studies-contradicting-precognition/
4   The Colbert Report, *Time Travelling Porn – Daryl Bem*, 2011; http://www.cc.com/video-clips/bhf8jv/the-colbert-report-time-traveling-porn–daryl-bem
5   After a few more frustrations, we did eventually publish the paper in a different scientific journal: Stuart J. Ritchie et al., 'Failing the Future: Three Unsuccessful Attempts to Replicate Bem's "Retroactive Facilitation of Recall" Effect', *PLOS ONE* 7, no. 3 (14 Mar. 2012): e33423; https://doi.org/10.1371/journal.pone.0033423. Note that the journal did publish a statistical critique of Bem's study (Eric-Jan Wagenmakers et al., 'Why psychologists must change the way they analyze their data: the case of psi: comment on Bem (2011)', *Journal of Personality and Social Psychology* 100, no. 3 (2011): pp. 426-432; https://doi.org/10.1037/a0022790), as well as a response from Bem and his colleagues (Daryl J. Bem et al., 'Must psychologists change the way they analyze their data?', *Journal of Personality and Social Psychology* 101, no.4 (2011): pp. 716-719; https://doi.org/10.1037/a0024777). But they still wouldn't consider publishing a replication attempt. Later in the book, we'll see that the editors of this journal have since changed their mind on this crucial issue.

6   D. A. Stapel & S. Lindenberg, 'Coping with Chaos: How Disordered Contexts Promote Stereotyping and Discrimination', *Science* 332, no. 6026 (8 April 2011): pp. 251–53; https://doi.org/10.1126/science.1201068

7   Philip Ball, 'Chaos Promotes Stereotyping', *Nature*, 7 April 2011; https://doi.org/10.1038/news.2011.217 and Nicky Phillips, 'Where There's Rubbish There's Racism', *Sunday Morning Herald*, 11 April 2011; https://www.smh.com.au/world/where-theres-rubbish-theres-racism-20110410-1d9df.html

8   Stapel and Lindenberg, 'Coping with Chaos', p. 251.

9   Levelt Committee et al., 'Flawed Science: The Fraudulent Research Practices of Social Psychologist Diederik Stapel [English Translation]', 28 Nov. 2012; https://osf.io/eup6d

10 D. A. Stapel, *Derailment: Faking Science*, tr. Nicholas J. L. Brown (Strasbourg, France, 2014,2016): p. 119; http://nick.brown.free.fr/stapel

11 Stapel, *Derailment*, p. 124.

12 Indeed, scientific progress depends on us finding missteps in previous work. Around the beginning of the twentieth century, for example, physicists realised that Newton's theory of classical mechanics, long regarded as true, wasn't compatible with the behaviour of very small and very fast particles, and replaced it with the theory of quantum mechanics. For a discussion of this issue from the perspective of measuring quantities like the speed of light and the Planck constant, see Martin J. T. Milton and Antonio Possolo, 'Trustworthy Data Underpin Reproducible Research', *Nature Physics* 16, no. 2 (Feb. 2020): pp. 117–19; https://doi.org/10.1038/s41567-019-0780-5

13 Quoted in Daniel Engber, 'Daryl Bem Proved ESP Is Real: Which Means Science Is Broken', *Slate*, 17 May 2017; https://slate.com/health-and-science/2017/06/daryl-bem-proved-esp-is-real-showed-science-is-broken.html

14 A classic example of such a book is Carl Sagan, *The Demon-Haunted World: Science as a Candle in the Dark*, reprint. ed. (New York: Ballantine Books, 1997).

15 This book considers numerous motes in the eyes of other scientists, so – if you'll indulge me a moment of self-reflection – it's only right that I check for any potential beams in my own. In the years since my attempted replication of Bem, I've published many papers on various topics, though mainly on my primary interest: human intelligence. The first thing to say is that I haven't deliberately faked any results. But it would be silly to suggest that I'm immune to biases. They are often, or perhaps usually, unconscious, and the history of a study is easily rewritten so it seems as though you'd intended to do it that way all along. On the plus side, I've published a fair few null results – papers that didn't find support for their main hypothesis. For example, see Stuart J. Ritchie et al., 'Polygenic Predictors of Age-Related Decline in Cognitive Ability', *Molecular Psychiatry* (13 Feb. 2019); https://doi.org/10.1038/s41380-019-0372-x; as well as my first ever scientific publication: S. J. Ritchie et al., 'Irlen Colored Overlays Do Not Alleviate Reading Difficulties', *Pediatrics* 128, no. 4 (1 Oct. 2011): pp. e932–38; https://doi.org/10.1542/peds.2011-0314. Then again, one could easily argue that this initial null paper had too small a sample and might have missed real effects (see the discussion of statistical power in Chapter 5). Some of my work has been justifiably criticised by other scientists, such as when I naïvely stumbled into overfitting (a phenomenon covered in Chapter 4). See Drew H. Bailey & Andrew K. Littlefield, 'Does Reading Cause Later Intelligence? Accounting for Stability in Models of Change', *Child Development* 88, no. 6 (Nov. 2017):

pp. 1913–21; https://doi.org/10.1111/cdev.12669. I've even published a 'candidate gene' study, using a method that I'll lay into in Chapter 5. See Stuart J. Ritchie et al., 'Alcohol Consumption and Lifetime Change in Cognitive Ability: A Gene × Environment Interaction Study', *AGE* 36, no. 3 (June 2014): 9638; https://doi.org/10.1007/s11357-014-9638-z. I've also almost definitely engaged in hype: I've had several conversations about science with journalists where I've been too loose with language or, in *l'esprit d'escalier* fashion, I've later regretted not adding important cautions and caveats. And I've made the mistake of arguing that 'there are hundreds of published, peer-reviewed papers on this topic', as if that was some indicator of truth. And talking of peer review, there have certainly been occasions where I haven't given a paper that I'm reviewing the time it deserves and might inadvertently have allowed errors to slip through. I've no doubt that other errors or regrets will come to light in the future.

1: *How Science Works*

Epigraph: David Hume, 'Of Essay-Writing', *Essays: Moral, Political, and Literary*, ed. Eugene F. Miller (Indianapolis: Liberty Fund, 1777).

1   Alan Sokal & Jean Bricmont, *Intellectual Impostures*, tr. Sokal & Bricmont (London: Profile Books, 1998, 2003).

2   John Stuart Mill, *On Liberty* (London: Dover Press, 1859) p. 29.

3   Helen E. Longino, *Science as Social Knowledge* (Princeton: Princeton University Press, 1990). See also Helen Longino, 'The Social Dimensions of Scientific Knowledge', *The Stanford Encyclopedia of Philosophy*, ed. Edward N. Zalta (Summer 2019); https://plato.stanford.edu/archives/sum2019/entries/scientific-knowledge-social; and Julian Reiss & Jan Sprenger, Jan, 'Scientific Objectivity', *The Stanford Encyclopedia of Philosophy*, ed. Edward N. Zalta (Winter 2017); https://plato.stanford.edu/archives/win2017/entries/scientific-objectivity

4   In making this argument, I'm influenced by the idea from the evolutionary theorists Hugo Mercier and Dan Sperber that the basic function of human reasoning itself is to work out how best to convince other people. Hugo Mercier & Dan Sperber, 'Why Do Humans Reason? Arguments for an Argumentative Theory'. *Behavioral and Brain Sciences* 34, no. 2 (April 2011): pp. 57–74; https://doi.org/10.1017/S0140525X10000968

5   Julie McDougall-Waters, Noah Moxham, and Aileen Fyfe. *Philosophical Transactions: 350 Years of Publishing at the Royal Society (1665 – 2015)* (London: Royal Society, 2015); https://royalsociety.org/~/media/publishing350/publishing350-exhibition-catalogue.pdf. Some historians would argue that a French publication, the *Journal des sçavans*, which was launched just two months before *Philosophical Transactions* in 1665, should really be considered the first scientific journal. However, the *Journal* published articles on a huge number of different learned topics and, at the beginning, mainly consisted of book reviews and book extracts. *Philosophical Transactions*, on the other hand, was focused on publishing scientific news and observations from the start. It might be fairer to regard the *Journal* as the first academic publication and *Philosophical Transactions* as the first scientific one. See Roger Philip McCutcheon, 'The "Journal Des Scavans" and the "Philosophical Transactions of the Royal Society"', *Studies in Philology* 21, no. 4 (1924): pp. 626–28; https://www.jstor.org/stable/4171899; and David Banks, 'Thoughts on Publishing the Research Article over the Centuries', *Publications* 6, no. 1 (8 Mar. 2018): 10; https://doi.org/10.3390/publications6010010

6   Paul A. David, 'The Historical Origins of "Open Science": An Essay on Patronage, Reputation and Common Agency Contracting in the Scientific Revolution', *Capitalism and Society* 3, no. 2 (2008): 5; https://papers.ssrn.com/sol3/papers.cfm?abstract_id=2209188

7   Robert Hooke, 'A Spot in One of the Belts of Jupiter', *Philosophical Transactions*, Vol. 1, Issue 1, 30 May 1665; https://doi.org/10.1098/rstl.1665.0005. Italics, capitalisation and variant spelling of the name as 'Hook' in the original. I've changed the long s's in the original to modern ones.

8   In 1900 it was split into two sub-journals, one for mathematics and physical sciences, and one for biological sciences. See: https://royalsocietypublishing.org/journal/rstl

9   Mark Ware & Michael Mabe, 'The STM Report: An Overview of Scientific and Scholarly Journal Publishing'. The Hague, Netherlands: International Association of Scientific, Technical and Medical Publishers, March 2015; https://www.stm-assoc.org/2015_02_20_STM_Report_2015.pdf

10 Note that it took until the mid-eighteenth century before *Philosophical Transactions*, up until then run by various individual researchers and compilers, was officially run by the Royal Society.

11 Most journal articles are reports of new studies, known as 'empirical papers', but some are 'review papers', synthesising everything that's known so far on a given scientific question.

12 https://www.nih.gov/ and https://www.nsf.gov/. Similar organisations in other countries include UK Research and Innovation (https://www.ukri.org/), the National Natural Science Foundation of China (http://www.nsfc.gov.cn/english/site_1/index.html), and the Japan Society for the Promotion of Science (https://www.jsps.go.jp/english/). See also https://wellcome.ac.uk/ and https://www.gatesfoundation.org/

13 Some scientific journals, for instance, format their papers so that the Method section is at the very end, as if this critical information was a mere afterthought.

14 https://www.sciencemag.org/site/feature/contribinfo/faq/index.xhtml#pct_faq

15 Alex Csiszar, 'Peer Review: Troubled from the Start', *Nature* 532, no. 7599 (April 2016): pp. 306–8; https://doi.org/10.1038/532306a

16 Quoted in Melinda Baldwin, 'Scientific Autonomy, Public Accountability and the Rise of "Peer Review" in the Cold War United States', *Isis* 109, no. 3 (Sept. 2018): pp. 538–58; https://doi.org/10.1086/700070

17 Ibid.

18 https://shitmyreviewerssay.tumblr.com/

19 I should note that these are norms for scientific investigation and analysis, and are separate from the ethical concerns that all scientists also have to consider. These are perhaps especially important for those who work with human (or other animal) subjects, but also for those who work with potentially dangerous technologies or whose experiments might cause environmental or other harm.

20 Robert K. Merton, 'The Normative Structure of Science' (1942), *The Sociology of Science: Empirical and Theoretical Investigations* (Chicago and London: University of Chicago Press, 1973): pp. 267–278.

21 Darwin Correspondence Project, 'Letter no. 2122', 9 July 1857; https://www.darwinproject.ac.uk/letter/DCP-LETT-2122.xml

22 Merton actually called communality 'communism', but that term has some, shall we say, different connotations. Some subsequent work altered the name to 'communality', and I'm following it here. See e.g. Melissa S. Anderson et al., 'Extending the Mertonian Norms: Scientists' Subscription to Norms of Research', *Journal of Higher Education* 81, no. 3 (May 2010): pp. 366–93; https://doi.org/10.1080/00221546.2010.11779057

23 Merton mentions the highly introverted Henry Cavendish, the eighteenth-century physicist and chemist, as a historical violator of this norm – he hid many of his important experiments and theories from the world out of pure shyness, and they weren't rediscovered until well after his death.

24 Nicholas W. Best, 'Lavoisier's "Reflections on Phlogiston" I: Against Phlogiston Theory', *Foundations of Chemistry* 17, no. 2 (July 2015): pp. 137–51; https://doi.org/10.1007/s10698-015-9220-5

25 Richard Dawkins, *The God Delusion* (London: Bantam Books, 2006): pp. 320–21.

26 Max Planck, *Scientific Autobiography and Other Papers*, tr. Frank Gaynor (London: Williams & Norgate, Ltd., 1949): pp. 33–34.

27 Karl Popper, *The Logic of Scientific Discovery* (London & New York: Routledge Classics, 1959/2002): p. 23.

28 Though set around a century after Boyle, one such replication is depicted, with dramatic lighting, in the painting *An Experiment on a Bird in the Air Pump* by Joseph Wright of Derby, currently in the National Gallery in London.

29 Robert Boyle, *The New Experiments Physico-Mechanicall, Touching the Spring of the Air and Its Effects* (London: Miles Flesher, 1682): p. 2; quoted in Steven Shapin & Simon Schaffer, *Leviathan and the Air-Pump: Hobbes, Boyle, and the Experimental Life* (Princeton: Princeton University Press, 1985).

30 Shapin & Schaffer, *Leviathan*.

2: *The Replication Crisis*

Epigraph: Brian A. Nosek et al., 'Scientific Utopia: II. Restructuring Incentives and Practices to Promote Truth Over Publishability', *Perspectives on Psychological Science* 7, no. 6 (Nov. 2012): pp. 615-631; https://doi.org/10.1177/1745691612459058, p. 616.

1 Daniel Kahneman, *Thinking, Fast and Slow* (New York: Farrar, Straus and Giroux, 2011).

2 James Neely, 'Semantic Priming Effects in Visual Word Recognition: A Selective Review of Current Findings and Theories', in *Basic Processes in Reading: Visual Word Recognition*, ed. Derek Besner, 1st ed. (Abingdon: Routledge, 2012); https://doi.org/10.4324/9780203052242

3 C. B. Zhong & K. Liljenquist, 'Washing Away Your Sins: Threatened Morality and Physical Cleansing', *Science* 313, no. 5792 (8 Sept. 2006): pp. 1451–52; https://doi.org/10.1126/science.1130726

4 K. D. Vohs et al., 'The Psychological Consequences of Money', *Science* 314, no. 5802 (17 Nov. 2006): pp. 1154–56; https://doi.org/10.1126/science.1132491

5 Ibid. p. 1154.

6 Kahneman, *Thinking, Fast and Slow*, pp. 55, 57.

7 As far as I can tell, the term originates in a paper by Pashler & Wagenmakers, who didn't use the exact phrase 'replication crisis' directly but talked of 'a crisis of confidence' in psychological research after a string of failed replications. Nelson, Simmons and Simonsohn discuss the triggers of the crisis. Harold Pashler & Eric-Jan Wagenmakers, 'Editors' Introduction to the Special Section on Replicability in Psychological Science: A Crisis of Confidence?', *Perspectives on Psychological Science* 7, no. 6 (Nov. 2012): pp. 528–30; https://doi.org/10.1177/1745691612465253 and: Leif D. Nelson et al., 'Psychology's Renaissance', *Annual Review of Psychology* 69, no. 1 (4 Jan. 2018): pp. 511–34; https://doi.org/10.1146/annurev-psych-122216-011836

8 John A. Bargh et al., 'Automaticity of Social Behavior: Direct Effects of Trait Construct and Stereotype Activation on Action', *Journal of Personality and*

   Social Psychology 71, no. 2 (1996): pp. 230–44; https://doi.org/10.1037/0022-3514.71.2.230; citation numbers (precisely 5,208 citations) come from Google Scholar as of January 2020.

9   Stéphane Doyen et al., 'Behavioral Priming: It's All in the Mind, but Whose Mind?', *PLOS ONE* 7, no. 1 (18 Jan. 2012): e29081; https://doi.org/10.1371/journal.pone.0029081

10   Brian D. Earp et al., 'Out, Damned Spot: Can the "Macbeth Effect" Be Replicated?' *Basic and Applied Social Psychology* 36, no. 1 (Jan. 2014): pp. 91–98; https://doi.org/10.1080/01973533.2013.856792; Money-priming effect: Richard A. Klein et al., 'Investigating Variation in Replicability: A "Many Labs" Replication Project', *Social Psychology* 45, no. 3 (May 2014): pp. 142–52; https://doi.org/10.1027/1864-9335/a000178

11   Original study: Lawrence E. Williams & John A. Bargh, 'Keeping One's Distance: The Influence of Spatial Distance Cues on Affect and Evaluation', *Psychological Science* 19, no. 3 (Mar. 2008): pp. 302–8; https://doi.org/10.1111/j.1467-9280.2008.02084.x; Replication: Harold Pashler et al., 'Priming of Social Distance? Failure to Replicate Effects on Social and Food Judgments', *PLOS ONE* 7, no. 8 (29 Aug. 2012): e42510; https://doi.org/10.1371/journal.pone.0042510

12   Original study: Theodora Zarkadi & Simone Schnall, '"Black and White" Thinking: Visual Contrast Polarizes Moral Judgment', *Journal of Experimental Social Psychology* 49, no. 3 (May 2013): pp. 355–59; https://doi.org/10.1016/j.jesp.2012.11.012; Replication: Hans IJzerman & Pierre-Jean Laine, 'Does Background Color Affect Moral Judgment? Three Pre-Registered Replications of Zarkadi and Schnall's (2012) Study 1', Preprint, *PsyArXiv* (30 July 2018); https://doi.org/10.31234/osf.io/ktfxq

13   Priming disgust was often done by filling a room with a bad smell. The studies on this topic are thus particularly notable for the many papers where straight-faced psychologists had to talk about the effects of 'fart spray', including one that involved a deadpan discussion of 'a proprietary odorant called Liquid Ass®. For 'Liquid Ass' see T. G. Adams et al., 'The Effects of Cognitive and Affective Priming on Law of Contagion Appraisals', *Journal of Experimental Psychopathology* 3, no. 3 (July 2012): p. 473; https://doi.org/10.5127/jep.025911. For the review of that line of research, see Justin F. Landy & Geoffrey P. Goodwin, 'Does Incidental Disgust Amplify Moral Judgment? A Meta-Analytic Review of Experimental Evidence', *Perspectives on Psychological Science* 10, no. 4 (July 2015): pp. 518–36; https://doi.org/10.1177/1745691615583128

14   Alison McCook, '"I Placed Too Much Faith in Underpowered Studies:" Nobel Prize Winner Admits Mistakes', *Retraction Watch*, 20 Feb. 2017; https://retractionwatch.com/2017/02/20/placed-much-faith-underpowered-studies-nobel-prize-winner-admits-mistakes/. Kahneman also wrote an open letter to social psychologists, telling them that he saw a 'train wreck looming' and urged them to change the way they went about their research. A copy can be found at the following link: https://go.nature.com/2T7A2NV

15   Dana R. Carney et al., 'Power Posing: Brief Nonverbal Displays Affect Neuroendocrine Levels and Risk Tolerance', *Psychological Science* 21, no. 10 (Oct. 2010): pp. 1363–68; https://doi.org/10.1177/0956797610383437

16   The total of the 56 million views on the TED website and the additional 17.6 million views on YouTube – numbers at the time of writing in February 2020. The talk was originally titled 'Your Body Language Shapes Who You Are' but at some point, post-replication crisis, it has been renamed 'Your Body Language May Shape Who You Are'. Amy Cuddy, 'Your Body Language May Shape Who

You Are', presented at *TEDGlobal 2012*, June 2012; https://www.ted.com/talks/amy_cuddy_your_body_language_may_shape_who_you_are

17 Amy J. C. Cuddy, *Presence: Bringing Your Boldest Self to Your Biggest Challenges* (New York: Little, Brown and Company, 2015). The quotation is from the publisher page at the following link: https://www.littlebrown.com/titles/amy-cuddy/presence/9780316256575/

18 Homa Khaleeli, 'A Body Language Lesson Gone Wrong: Why is George Osborne Standing like Beyoncé?' *Guardian*, 7 Oct. 2015; https://www.theguardian.com/politics/shortcuts/2015/oct/07/who-told

19 Eva Ranehill et al., 'Assessing the Robustness of Power Posing: No Effect on Hormones and Risk Tolerance in a Large Sample of Men and Women', *Psychological Science* 26, no. 5 (May 2015): pp. 653-56; https://doi.org/10.1177/0956797614553946, p. 655. The power-posing debate has gone on and on since then. A 2017 review concluded that power-posing effects are 'hypotheses currently lacking in empirical support'. See Joseph P. Simmons & Uri Simonsohn, 'Power Posing: P-Curving the Evidence', *Psychological Science* 28, no. 5 (May 2017): pp. 687–93; https://doi.org/10.1177/0956797616658563. Then Cuddy hit back with her own review that did find an overall effect, though it has since been pointed out that, among other problems with the research, most of the effects in the cited studies were probably due to the negative effects of slouching, rather than the beneficial effects of power posing. See Amy J. C. Cuddy et al., 'P-Curving a More Comprehensive Body of Research on Postural Feedback Reveals Clear Evidential Value for Power-Posing Effects: Reply to Simmons and Simonsohn (2017)', *Psychological Science* 29, no. 4 (April 2018): pp. 656–66; https://doi.org/10.1177/0956797617746749. Then for slouching, see Marcus Credé, 'A Negative Effect of a Contractive Pose is not Evidence for the Positive Effect of an Expansive Pose: Commentary on Cuddy, Schultz, and Fosse (2018)', *SSRN*: https://doi.org/10.2139/ssrn.3198470

20 Philip Zimbardo, *The Lucifer Effect: How Good People Turn Evil* (London: Rider, 2007).

21 Stanley Milgram, 'Behavioral Study of Obedience', *Journal of Abnormal and Social Psychology* 67, no. 4 (1963): pp. 371–78; https://doi.org/10.1037/h0040525. The Milgram experiments have also seen their fair share of criticism – for evidence that the more the participants believed they were really shocking the 'learners', the less likely they were to give more powerful shocks, see e.g. Gina Perry et al., 'Credibility and Incredulity in Milgram's Obedience Experiments: A Reanalysis of an Unpublished Test', *Social Psychology Quarterly*, 22 Aug. 2019; https://doi.org/10.1177/0190272519861952

22 Philip Zimbardo, 'Our inner heroes could stop another Abu Ghraib', *Guardian*, 29 Feb. 2008; https://www.theguardian.com/commentisfree/2008/feb/29/iraq.usa

23 Erich Fromm, *The Anatomy of Human Destructiveness* (New York: Holt, Rinehart and Winston, 1975).

24 Thibault Le Texier, 'Debunking the Stanford Prison Experiment', *American Psychologist* 74, no. 7 (Oct. 2019): pp. 823–39; https://doi.org/10.1037/amp0000401

25 The debate continues and Zimbardo has responded to the criticisms. For example, see Philip Zimbardo, 'Philip Zimbardo's Response to Recent Criticisms of the Stanford Prison Experiment', 23 June 2018; https://static1.squarespace.com/static/557a07d5e4b05fe7bf112c19/t/5dee52149d16d153cba11712/1575899668862/Zimbardo2018-06-23.pdf. See also Le Texier's reply to a more recent (at the time of writing unpublished) version: Thibault Le Texier, 'The SPE Remains Debunked: A Reply to Zimbardo and Haney (2020)', Preprint, *PsyArXiv* (24 Jan. 2020); https://doi.org/10.31234/osf.io/9a2er

26 Open Science Collaboration, 'Estimating the Reproducibility of Psychological Science', *Science* 349, no. 6251 (28 Aug. 2015): aac4716; https://doi.org/10.1126/science.aac4716

27 77 per cent: Colin F. Camerer et al., 'Evaluating the Replicability of Social Science Experiments in Nature and Science between 2010 and 2015', *Nature Human Behaviour* 2, no. 9 (Sept. 2018): pp. 637–44; https://doi.org/10.1038/s41562-018-0399-z

28 This number is derived from six out of sixteen studies showing successful replications. Charles R. Ebersole et al., 'Many Labs 3: Evaluating Participant Pool Quality across the Academic Semester via Replication', *Journal of Experimental Social Psychology* 67 (Nov. 2016): pp. 68–82; https://doi.org/10.1016/j.jesp.2015.10.012

29 At this point, some critics might argue I've been hoist by my own petard. I've been stressing the importance of robust results, but in making the case that there's a replication crisis, I'm relying on multi-study replication attempts that weren't representative samples of all the scientific literature. The conclusion of 'only about half of published results replicate' might not generalise to all science. This was a point made in a critique of one of the replication survey studies: D. T. Gilbert et al., 'Comment on "Estimating the Reproducibility of Psychological Science"', *Science* 351, no. 6277 (4 Mar. 2016): p. 1037; https://doi.org/10.1126/science.aad7243. Whereas I disagree with many of the arguments made in this rejoinder (for some reasons to be sceptical of it, see Daniël Lakens, 'The Statistical Conclusions in Gilbert et al (2016) Are Completely Invalid', *The 20% Statistician*, 6 March 2016; https://daniellakens.blogspot.com/2016/03/the-statistical-conclusions-in-gilbert.html), the criticism about representativeness was fair. We're still largely in the dark about exactly what proportion of findings across *whole subjects* are replicable, even in fields like psychology where these large-scale replication attempts have been done – the truth might be better, or perhaps worse, than those studies indicate. But the very fact that we don't know – along with the fact that so many high-profile, puffed-up findings have fallen apart upon closer inspection – is, I'd argue, cause for enough concern. For responses to other criticisms of the idea that there's a crisis, see Harold Pashler & Christine R. Harris, 'Is the Replicability Crisis Overblown? Three Arguments Examined', *Perspectives on Psychological Science* 7, no. 6 (Nov. 2012): pp. 531–36; https://doi.org/10.1177/1745691612463401

30 Alexander Bird, 'Understanding the Replication Crisis as a Base Rate Fallacy', *British Journal for the Philosophy of Science*, 13 Aug. 2018; https://doi.org/10.1093/bjps/axy051

31 Of course, the argument of the original authors (those whose findings failed to replicate) has often been that the modifications aren't, in fact, slight, and break the experiment in important ways. Every case should be taken on its merits, but this sort of argument does rather often sound like a form of special pleading.

32 Another area that's doing OK is personality psychology. The psychologist Christopher Soto ran a large replication study of effects from personality research: correlations of personality traits, as measured by questionnaires, with outcomes such as life and romantic satisfaction, religious and political views, and job success. He found a replication rate of 87 per cent, which is pretty solid compared to the other fields that we've seen. Christopher J. Soto, 'How Replicable Are Links Between Personality Traits and Consequential Life Outcomes? The Life Outcomes of Personality Replication Project', *Psychological Science* 30, no. 5 (May 2019): pp. 711–27; https://doi.org/10.1177/0956797619831612

33 C. F. Camerer et al., 'Evaluating Replicability of Laboratory Experiments in Economics', *Science* 351, no. 6280 (25 Mar. 2016): pp. 1433–36; https://doi.org/10.1126/science.aaf0918

34 Benjamin O. Turner et al., 'Small Sample Sizes Reduce the Replicability of Task-Based fMRI Studies', *Communications Biology* 1, no. 1 (Dec. 2018): 62; https://doi.org/10.1038/s42003-018-0073-z

35 Anders Eklund et al., 'Cluster Failure: Why fMRI Inferences for Spatial Extent Have Inflated False-Positive Rates', *Proceedings of the National Academy of Sciences* 113, no. 28 (12 July 2016): pp. 7900–5; https://doi.org/10.1073/pnas.1602413113 and Anders Eklund et al., 'Cluster Failure Revisited: Impact of First Level Design and Physiological Noise on Cluster False Positive Rates', *Human Brain Mapping* 40, no. 7 (May 2019): 2017–32; https://doi.org/10.1002/hbm.24350

36 Kathryn A. Lord et al., 'The History of Farm Foxes Undermines the Animal Domestication Syndrome', *Trends in Ecology & Evolution* 35, no. 2 (Feb. 2020): pp. 125–36; https://doi.org/10.1016/j.tree.2019.10.011

37 Finches: Daiping Wang et al., 'Irreproducible Text-Book "Knowledge": The Effects of Color Bands on Zebra Finch Fitness: Color Bands Have No Effect on Fitness in Zebra Finches', *Evolution* 72, no. 4 (April 2018): pp. 961–76; https://doi.org/10.1111/evo.13459. See also Yao-Hua Law, 'Replication Failures Highlight Biases in Ecology and Evolution Science', *The Scientist*, 31 July 2018; https://www.the-scientist.com/features/replication-failures-highlight-biases-in-ecology-and-evolution-science-64475. Sparrows: Alfredo Sánchez-Tójar et al., 'Meta-analysis challenges a textbook example of status signalling and demonstrates publication bias', *eLife* 7 (13 Nov. 2008): e37385; https://doi.org/10.7554/eLife.37385.001. Blue tits: Timothy H. Parker, 'What Do We Really Know about the Signalling Role of Plumage Colour in Blue Tits? A Case Study of Impediments to Progress in Evolutionary Biology: Case Study of Impediments to Progress', *Biological Reviews* 88, no. 3 (Aug. 2013): pp. 511–36; https://doi.org/10.1111/brv.12013

38 Timothy D. Clark et al., 'Ocean Acidification Does Not Impair the Behaviour of Coral Reef Fishes', *Nature* 577, no. 7790 (Jan. 2020): pp. 370–75; https://doi.org/10.1038/s41586-019-1903-y. See also Martin Enserink, 'Analysis Challenges Slew of Studies Claiming Ocean Acidification Alters Fish Behavior', *Science*, 8 Jan. 2020; https://doi.org/10.1126/science.aba8254. As the latter article notes, the fact that fish behaviour seems to be unaffected is not a reason to stop worrying about ocean acidification, which has many other negative effects.

39 http://www.orgsyn.org/instructions.aspx; see also Dalmeet Singh Chawla, 'Taking on Chemistry's Reproducibility Problem', *Chemistry World*, 20 March 2017; https://www.chemistryworld.com/news/taking-on-chemistrys-reproducibility-problem/3006991.article

40 Because of the way the search was done, these numbers might miss research that doesn't explicitly brand itself as a replication attempt, so the true numbers might be slightly higher. For economics, see: Frank Mueller-Langer et al., 'Replication Studies in Economics – How Many and Which Papers Are Chosen for Replication and Why?', *Research Policy* 48, no. 1 (Feb. 2019): pp. 62–83; https://doi.org/10.1016/j.respol.2018.07.019. For psychology: Matthew C. Makel et al., 'Replications in Psychology Research: How Often Do They Really Occur?', *Perspectives on Psychological Science* 7, no. 6 (Nov. 2012): pp. 537–42; https://doi.org/10.1177/1745691612460688

41 Board of Governors of the Federal Reserve System, Andrew C. Chang & Phillip Li, 'Is Economics Research Replicable? Sixty Published Papers from Thirteen

Journals say "Usually Not"', *Finance and Economics Discussion Series* 2015, no. 83 (Oct. 2015): pp. 1–26; https://doi.org/10.17016/FEDS.2015.083. For a detailed review of replicability in economics, see Garret Christensen & Edward Miguel, 'Transparency, Reproducibility, and the Credibility of Economics Research' (Cambridge, MA: National Bureau of Economic Research, Dec. 2016); https://doi.org/10.3386/w22989

42   Markus Konkol et al., 'Computational Reproducibility in Geoscientific Papers: Insights from a Series of Studies with Geoscientists and a Reproduction Study', *International Journal of Geographical Information Science* 33, no. 2 (Feb. 2019): pp. 408–29; https://doi.org/10.1080/13658816.2018.1508687

43   Even worse: of that seven, fully six were redundant compared to much simpler methods that had been known about for years before these new algorithms appeared. Maurizio Ferrari Dacrema et al., 'Are We Really Making Much Progress?: A Worrying Analysis of Recent Neural Recommendation Approaches', in *Proceedings of the 13th ACM Conference on Recommender Systems – RecSys 2019* (Copenhagen, Denmark: ACM Press, 2019): pp. 101–9; https://doi.org/10.1145/3298689.3347058. See also this report from computer science, which hints that new researchers are having trouble reproducing the performance of several classic algorithms – something of a ticking time bomb, since 'young researchers don't want to be seen as criticising senior researchers' by publishing failures to reproduce the performance of algorithms the senior researchers had developed and on which they'd staked their reputations: Matthew Hutson, 'Artificial Intelligence Faces Reproducibility Crisis', *Science* 359, no. 6377 (16 Feb. 2018): pp. 725–26; https://doi.org/10.1126/science.359.6377.725, p. 726.

44   C. Glenn Begley & Lee M. Ellis, 'Raise Standards for Preclinical Cancer Research', *Nature* 483, no. 7391 (Mar. 2012): pp. 531–33; https://doi.org/10.1038/483531a

45   Florian Prinz et al., 'Believe It or Not: How Much Can We Rely on Published Data on Potential Drug Targets?', *Nature Reviews Drug Discovery* 10 (Sept. 2011): 712; https://doi.org/10.1038/nrd3439-c1. Note that the Bayer figure only included 70 per cent cancer research studies – the remaining 30 per cent were on women's health or cardiovascular research.

46   Chi Heem Wong et al., 'Estimation of Clinical Trial Success Rates and Related Parameters', *Biostatistics* 20, no. 2 (1 April 2019): pp. 273–86; https://doi.org/10.1093/biostatistics/kxx069. Across all different drugs, the percentage that make it from preclinical trials to human use was estimated in this study to be 13.8 per cent, so cancer research was doing particularly badly.

47   Brian A. Nosek & Timothy M. Errington, 'Reproducibility in Cancer Biology: Making Sense of Replications', *eLife* 6 (19 Jan. 2017): e23383; https://doi.org/10.7554/eLife.23383. The project is called the 'Reproducibility Project: Cancer Biology', where the term 'reproducibility' is used in the way that I've been using 'replicability' (that is, trying to get the same results in different samples). I've chosen my definitions to try and reflect the broad consensus, but you should be aware that not everyone sticks to the same terminology.

48   John Repass et al., 'Replication Study: Fusobacterium Nucleatum Infection is Prevalent in Human Colorectal Carcinoma', *eLife* 7 (13 Mar. 2018): e25801; https://doi.org/10.7554/eLife.25801

49   Tim Errington, 'Reproducibility Project: Cancer Biology – Barriers to Replicability in the Process of Research' (2019); https://osf.io/x9p5s/

50   Monya Baker & Elie Dolgin, 'Cancer Reproducibility Project Releases First Results', *Nature* 541, no. 7637 (Jan. 2017): pp. 269–70; https://doi.org/10.1038/541269a; Daniel Engber, 'Cancer Research Is Broken', *Slate*, 19 April 2016; https://slate.

com/technology/2016/04/biomedicine-facing-a-worse-replication-crisis-than-the-one-plaguing-psychology.html

51 Errington, 'Reproducibility Project', slide 11.

52 J. Kaiser, 'The Cancer Test', *Science* 348, no. 6242 (26 June 2015): pp. 1411–13; https://doi.org/10.1126/science.348.6242.1411

53 Shareen A. Iqbal et al., 'Reproducible Research Practices and Transparency across the Biomedical Literature', *PLOS Biology* 14, no. 1 (4 Jan. 2016): e1002333; https://doi.org/10.1371/journal.pbio.1002333. Note that 441 studies were included in the full sample, but only 268 reported empirical data.

54 Nicole A. Vasilevsky et al., 'On the Reproducibility of Science: Unique Identification of Research Resources in the Biomedical Literature', *PeerJ* 1 (2013): e148; https://doi.org/10.7717/peerj.148. Problems with poor reporting go far beyond biomedicine. For example, for a perspective from political science, see Alexander Wuttke, 'Why Too Many Political Science Findings Cannot Be Trusted and What We Can Do About It: A Review of Meta-Scientific Research and a Call for Academic Reform', *Politische Vierteljahresschrift* 60, no. 1 (Mar. 2019): pp. 1–19; https://doi.org/10.1007/s11615-018-0131-7. For a perspective from ecology, see Timothy H. Parker et al., 'Transparency in Ecology and Evolution: Real Problems, Real Solutions', *Trends in Ecology & Evolution* 31, no. 9 (Sept. 2016): pp. 711–19; https://doi.org/10.1016/j.tree.2016.07.002

55 Jocelyn Kaiser, 'Plan to Replicate 50 High-Impact Cancer Papers Shrinks to Just 18', *Science*, 31 July 2018; https://doi.org/10.1126/science.aau9619. Note that the reference in endnote 49, above (Errington, 'Reproducibility Project') discusses fifty-one studies, not fifty.

56 All of the Reproducibility Project: Cancer Biology studies can be found at the following link from the journal *eLife*: https://elifesciences.org/collections/9b1e83d1/reproducibility-project-cancer-biology

57 Vinayak K. Prasad & Adam S. Cifu, *Ending Medical Reversal: Improving Outcomes, Saving Lives* (Baltimore: Johns Hopkins University Press, 2015).

58 Joshua Lang, 'Awakening', *The Atlantic*, Feb. 2013; https://www.theatlantic.com/magazine/archive/2013/01/awakening/309188/

59 Michael S. Avidan et al., 'Anesthesia Awareness and the Bispectral Index', *New England Journal of Medicine* 358, no. 11 (13 Mar. 2008): 1097; https://doi.org/10.1056/NEJMoa0707361

60 Diana Herrera-Perez et al., 'A Comprehensive Review of Randomized Clinical Trials in Three Medical Journals Reveals 396 Medical Reversals', *eLife* 8 (11 June 2019): e45183; https://doi.org/10.7554/eLife.45183.This was a follow-up to a similar study they'd done before, which revealed 146 reversals: Vinay Prasad et al., 'A Decade of Reversal: An Analysis of 146 Contradicted Medical Practices', *Mayo Clinic Proceedings* 88, no. 8 (Aug. 2013): pp. 790–98; https://doi.org/10.1016/j.mayocp.2013.05.012

61 Jon F. R. Barrett et al., 'A Randomized Trial of Planned Cesarean or Vaginal Delivery for Twin Pregnancy', *New England Journal of Medicine* 369, no. 14 (3 Oct. 2013): pp. 1295–1305; https://doi.org/10.1056/NEJMoa1214939

62 George Du Toit et al., 'Randomized Trial of Peanut Consumption in Infants at Risk for Peanut Allergy', *New England Journal of Medicine* 372, no. 9 (26 Feb. 2015): pp. 803–13; https://doi.org/10.1056/NEJMoa1414850

63 Francis Kim et al., 'Effect of Prehospital Induction of Mild Hypothermia on Survival and Neurological Status Among Adults with Cardiac Arrest: A Randomized Clinical Trial', *JAMA* 311, no. 1 (1 Jan. 2014): pp. 45–52; https://doi.org/10.1001/jama.2013.282173

64 AVERT Collaboration, 'Efficacy and Safety of Very Early Mobilisation within 24 h of Stroke Onset: A Randomised Controlled Trial', *Lancet* 386, no. 9988 (July 2015): pp. 46–55; https://doi.org/10.1016/S0140-6736(15)60690-0

65 M. Irem Baharoglu et al., 'Platelet Transfusion versus Standard Care after Acute Stroke Due to Spontaneous Cerebral Haemorrhage Associated with Antiplatelet Therapy (PATCH): A Randomised, Open-Label, Phase 3 Trial', *Lancet* 387, no. 10038 (June 2016): pp. 2605–13; https://doi.org/10.1016/S0140-6736(16)30392-0

66 Paolo José Fortes Villas Boas et al., 'Systematic Reviews Showed Insufficient Evidence for Clinical Practice in 2004: What about in 2011? The Next Appeal for the Evidence-Based Medicine Age: The Next Appeal for EBM Age', *Journal of Evaluation in Clinical Practice* 19, no. 4 (Aug. 2013): pp. 633–37; https://doi.org/10.1111/j.1365-2753.2012.01877.x

67 Leonard P. Freedman et al., 'The Economics of Reproducibility in Preclinical Research', *PLOS Biology* 13, no. 6 (9 June 2015): e1002165; https://doi.org/10.1371/journal.pbio.1002165

68 Iain Chalmers & Paul Glasziou, 'Avoidable Waste in the Production and Reporting of Research Evidence', *Lancet* 374, no. 9683 (July 2009): pp. 86–89; https://doi.org/10.1016/S0140-6736(09)60329-9. See also Malcolm R. Macleod et al., 'Biomedical Research: Increasing Value, Reducing Waste', *Lancet* 383, no. 9912 (Jan. 2014): pp. 101–4; https://doi.org/10.1016/S0140-6736(13)62329-6

69 Monya Baker, '1,500 Scientists Lift the Lid on Reproducibility', *Nature* 533, no. 7604 (May 2016): pp. 452–54; https://doi.org/10.1038/533452a

70 John P. A. Ioannidis, 'Why Most Published Research Findings Are False', *PLOS Medicine* 2, no. 8 (30 Aug. 2005): e124; https://doi.org/10.1371/journal.pmed.0020124

71 Citation numbers from Google Scholar.

72 One critique of Ioannidis is by Jeffrey T. Leek & Leah R. Jager, 'Is Most Published Research Really False?', *Annual Review of Statistics and Its Application* 4, no. 1 (7 Mar. 2017): pp. 109–22; https://doi.org/10.1146/annurev-statistics-060116-054104

3: *Fraud*

Epigraph: Norman MacDonald, *Maxims and Moral Reflections* (New York: 1827).

1 Cochlear implants: Vivien Williams, 'Baby Hears for First Time with Cochlear Implants', *Mayo Clinic News Network*, 13 Nov. 2018; https://newsnetwork.mayoclinic.org/discussion/baby-hears-for-first-time-with-cochlear-implants/; cataracts: National Geographic, 'Two Blind Sisters See for the First Time', 26 Sept. 2014; https://youtu.be/EltIpB4EtYU; prosthetic limbs: Victoria Smith, 'Video Of Rick Clement Walking On New Legs Goes Viral', *Forces Network*, 23 July 2015; https://www.forces.net/services/tri-service/video-rick-clement-walking-new-legs-goes-viral; see also 'Boy, 5, given Prosthetic Arm That Lets Him Hug Brother', *BBC News*, 14 Dec. 2019; https://www.bbc.co.uk/news/uk-wales-50762563

2 The reattachment is a procedure called anastomosis, which often has good results. Advances in surgical techniques across the early to mid-twentieth century meant that anastomosis remained possible after ever longer parts of the trachea were removed. These efforts eventually reached their limit: if over half of the trachea had to be removed, which happens in cases where tracheal tumours have grown particularly large, anastomosis is no longer an option.

3 Hermes C. Grillo, 'Tracheal Replacement: A Critical Review', *The Annals of Thoracic Surgery* 73, no. 6 (June 2002): 1995–2004; https://doi.org/10.1016/S0003-4975(02)03564-6

4   Paolo Macchiarini et al., 'Clinical Transplantation of a Tissue-Engineered Airway', *Lancet* 372, no. 9655 (Dec. 2008): 2023–30; https://doi.org/10.1016/S0140-6736(08)61598-6

5   Karolinska Institute, 'First Successful Transplantation of a Synthetic Tissue Engineered Windpipe' (news release), 29 July 2011; https://ki.se/en/news/first-successful-transplantation-of-a-synthetic-tissue-engineered-windpipe

6   Philipp Jungebluth et al., 'Tracheobronchial Transplantation with a Stem-Cell-Seeded Bioartificial Nanocomposite: A Proof-of-Concept Study', *Lancet* 378, no. 9808 (Dec. 2011): pp. 1997–2004; https://doi.org/10.1016/S0140-6736(11)61715-7

7   Christian Berggren & Solmaz Filiz Karabag, 'Scientific Misconduct at an Elite Medical Institute: The Role of Competing Institutional Logics and Fragmented Control', *Research Policy* 48, no. 2 (Mar. 2019): pp. 428–43; https://doi.org/10.1016/j.respol.2018.03.020

8   Ibid. p. 432.

9   Madeleine Svärd Huss, 'The Macchiarini Case: Timeline' (Karolinska Institute, 26 June 2018); https://ki.se/en/news/the-macchiarini-case-timeline

10 AFP Newswire, 'Macchiarini's Seventh Transplant Patient Dies', *Local*, 20 March 2017; https://www.thelocal.it/20170320/macchiarinis-seventh-transplant-patient-dies-sweden-italy

11 Translated by Berggren & Karabag, 'Scientific Misconduct', p. 432; originally quoted in Johannes Wahlström, 'Den Bortglömda Patienten', *Filter*, 18 May 2016; https://magasinetfilter.se/granskning/den-bortglomda-patienten/ [Swedish]

12 William Kremer, 'Paolo Macchiarini: A Surgeon's Downfall', *BBC News Magazine*, 10 Sept. 2016; https://www.bbc.co.uk/news/magazine-37311038

13 'Although she was healthy, a car accident had left her with a tracheostomy, and in order to speak, she had to cover the opening with her hand. She hoped the surgery would let her sing to her son.' Carl Elliott, 'Knifed with a Smile', *New York Review of Books*, 5 April 2018; https://www.nybooks.com/articles/2018/04/05/experiments-knifed-with-smile/

14 A British patient, who had previously been operated upon by Macchiarini, received a Macchiarini-inspired synthetic trachea from other surgeons in London in 2011, but died the next year: Kremer, 'Paolo Macchiarini'.

15 It's heart-breaking to compare the articles in the same Canadian publication before and after her death: AP Newswire, '"We Feel like She's Reborn": Toddler Born without Windpipe Gets New One Grown from Her Own Stem Cells', *National Post*, 30 April 2013; https://nationalpost.com/news/south-korean-2-year-old-youngest-ever-to-get-lab-made-windpipe-from-her-own-stem-cells; and Joseph Brean, 'Swashbuckling Surgeon's Collapsing Reputation Threatens Canadian Girl's Legacy as "pioneer" Patient', *National Post*, 18 Feb. 2016; https://nationalpost.com/news/canada/swashbuckling-surgeons-collapsing-reputation-threatens-canadian-girls-legacy-as-pioneer-patient

16 Eve Herold, 'A Star Surgeon Left a Trail of Dead Patients – and His Whistleblowers were Punished', *leapsmag*, 8 Oct. 2018; https://leapsmag.com/a-star-surgeon-left-a-trail-of-dead-patients-and-his-whistleblowers-were-punished/

17 The report is hosted by the website Retraction Watch: http://retractionwatch.com/wp-content/uploads/2015/05/Translation-investigation.doc (see p. 36).

18 David Cyranoski, 'Artificial-Windpipe Surgeon Committed Misconduct', *Nature* 521, no. 7553 (May 2015): 406–7; https://doi.org/10.1038/nature.2015.17605. See also Alison McCook, 'Misconduct Found in 7 Papers by Macchiarini, Says English Write-up of Investigation', *Retraction Watch*, 28 May 2015; https://retractionwatch.com/2015/05/28/misconduct-found-in-7-papers-by-macchiarini-says-english-write-up-of-investigation/

19 Kremer, 'Paolo Macchiarini'.

20 'Paolo Macchiarini Is Not Guilty of Scientific Misconduct', *Lancet* 386, no. 9997 (Sept. 2015): 932; https://doi.org/10.1016/S0140-6736(15)00118-X

21 Adam Ciralsky, 'The Celebrity Surgeon Who Used Love, Money, and the Pope to Scam an NBC News Producer', *Vanity Fair* (Feb. 2016); https://www.vanityfair.com/news/2016/01/celebrity-surgeon-nbc-news-producer-scam

22 Ibid.

23 *Vanity Fair*'s investigation also alleged that Macchiarini had invented many of the qualifications and affiliations he'd claimed on his CV, and that the findings of the Italian investigation into him had never been publicly released.

24 Kremer, 'Paolo Macchiarini'.

25 Huss, 'The Macchiarini Case'. See also David Cyranoski, 'Nobel Official Resigns over Karolinska Surgeon Controversy', *Nature*, 8 Feb. 2016; https://doi.org/10.1038/nature.2016.19332

26 Macchiarini didn't act alone: the Karolinska Institute report also accused several of his co-authors of scientific misconduct.

27 'The Final Verdict on Paolo Macchiarini: Guilty of Misconduct', *Lancet* 392, no. 10141 (July 2018): 2; https://doi.org/10.1016/S0140-6736(18)31484-3

28 Karolinska Institute, 'Seven Researchers Responsible for Scientific Misconduct in Macchiarini Case', 28 June 2015; https://news.ki.se/seven-researchers-responsible-for-scientific-misconduct-in-macchiarini-case

29 Matt Warren, 'Disgraced Surgeon is Still Publishing on Stem Cell Therapies', *Science*, 27 April 2018; https://doi.org/10.1126/science.aau0038

30 Margarita Zhuravleva et al., 'In Vitro Assessment of Electrospun Polyamide-6 Scaffolds for Esophageal Tissue Engineering: Polyamide-6 Scaffolds for Esophageal Tissue Engineering', *Journal of Biomedical Materials Research Part B: Applied Biomaterials* 107, no. 2 (Feb. 2019): pp. 253–68; https://doi.org/10.1002/jbm.b.34116

31 Alla Astakhova, 'Superstar Surgeon Fired, Again, This Time in Russia', *Science*, 16 May 2017; https://doi.org/10.1126/science.aal1201

32 Swedish Prosecution Authority, 'Investigation Concerning Surgeries Resumed after Review', 11 Dec. 2018; https://via.tt.se/pressmeddelande/investigation-concerning-surgeries-resumed-after-review?publisherId=3235541&releaseId=3252745

33 Herold, 'A Star Surgeon'.

34 Berggren & Karabag, 'Scientific Misconduct', p. 432.

35 Summerlin had also falsified the results of a cornea transplant he'd claimed in rabbits. Jane E. Brody, 'Inquiry at Cancer Center Finds Fraud in Research', *New York Times*, 25 May 1974; https://www.nytimes.com/1974/05/25/archives/article-5-no-title-fraud-is-charged-at-cancer-center-premature.html; see also the titular essay in Peter Medawar, *The Strange Case of the Spotted Mice: And Other Classic Essays on Science* (Oxford: Oxford University Press, 1996).

36 P. G. Pande et al., 'Toxoplasma from the Eggs of the Domestic Fowl (*Gallus gallus*)', *Science* 133, no. 3453 (3 March 1961): pp. 648–648; https://doi.org/10.1126/science.133.3453.648

37 G. DuShane et al., 'An Unfortunate Event', *Science* 134, no. 3483 (29 Sept. 1961): pp. 945–46; https://doi.org/10.1126/science.134.3483.945-a; see also J. L. Kavanau & K. S. Norris, 'Letter to the Editor', *Science* 136, no. 3511 (13 April 1962): p. 199; https://doi.org/10.1126/science.136.3511.199; and Nicholas B. Wade & William Broad, *Betrayers of the Truth: Fraud and Deceit in the Halls of Science* (New York: Simon & Schuster, 1982).

38 A portmanteau of 'SNU' (Seoul National University) and 'puppy'. The achievement was particularly impressive because dog ova are relatively fragile and unstable compared to other mammals; even though by 2005 animals such as

sheep, cats, pigs and horses had been cloned, nobody until Hwang had managed to do the same for dogs. Byeong Chun Lee et al., 'Dogs Cloned from Adult Somatic Cells', *Nature* 436, no. 7051 (Aug. 2005): p. 641; https://doi.org/10.1038/436641a

39 Jaeyung Park et al., 'The Korean Press and Hwang's Fraud', *Public Understanding of Science* 18, no. 6 (Nov. 2009): pp. 653–69; https://doi.org/10.1177/0963662508096779

40 R. Saunders & J. Savulescu, 'Research Ethics and Lessons from Hwanggate: What Can We Learn from the Korean Cloning Fraud?', *Journal of Medical Ethics* 34, no. 3 (1 Mar. 2008): pp. 214–21; https://doi.org/10.1136/jme.2007.023721

41 Constance Holden, 'Bank on These Stamps', *Science* 308, no. 5729 (17 June 2005): p. 1738a; https://doi.org/10.1126/science.308.5729.1738a

42 Supreme Scientist: Jongyoung Kim & Kibeom Park, 'Ethical Modernization: Research Misconduct and Research Ethics Reforms in Korea Following the Hwang Affair', *Science and Engineering Ethics* 19, no. 2 (June 2013): p. 358; https://doi.org/10.1007/s11948-011-9341-8. Egg donation: Saunders & Savulescu, 'Research Ethics', p. 217.

43 Jennifer Couzin, 'STEM CELLS: ... And How the Problems Eluded Peer Reviewers and Editors', *Science* 311, no. 5757 (6 Jan. 2006): pp. 23–24; https://doi.org/10.1126/science.311.5757.23; Mike Rossner, 'Hwang Case Review Committee Misses the Mark', *Journal of Cell Biology* 176, no. 2 (15 Jan. 2007): pp. 131–32; https://doi.org/10.1083/jcb.200612154

44 Saunders & Savulescu, 'Research Ethics', p. 215.

45 Kim & Park, 'Ethical Modernization', pp. 360–361.

46 Ibid. p. 361.

47 Sei Chong & Dennis Normile, 'STEM CELLS: How Young Korean Researchers Helped Unearth a Scandal ...', *Science* 311, no. 5757 (6 Jan. 2006): pp. 22–25; https://doi.org/10.1126/science.311.5757.22

48 Mi-Young Ahn & Dennis Normile, 'Korean Supreme Court Upholds Disgraced Cloner's Criminal Sentence', *Science*, 27 Feb. 2014; https://www.sciencemag.org/news/2014/02/korean-supreme-court-upholds-disgraced-cloners-criminal-sentence

49 I feel sure that, unlike the man who cloned him, Snuppy was a very good boy. Min Jung Kim et al., 'Birth of Clones of the World's First Cloned Dog', *Scientific Reports* 7, no. 1 (Dec. 2017): 15235; https://doi.org/10.1038/s41598-017-15328-2

50 I have a completely irrational sense of pride that this now-routine laboratory technique was invented by Southern at my alma mater, the University of Edinburgh. E. M. Southern, 'Detection of Specific Sequences among DNA Fragments Separated by Gel Electrophoresis', *Journal of Molecular Biology* 98, no. 3 (Nov. 1975): pp. 503–17; https://doi.org/10.1016/S0022-2836(75)80083-0

51 The full process of Southern blotting goes something like this: take a DNA molecule, break apart its double helix into single strands using an enzyme, then force it through a gel with an electric current (a process called electrophoresis), separating out all its fragments. The clever part is that fragments of different size move through the gel at different speeds, so you can get some idea of a protein's size by its position in the gel after the current is applied for a time. Then, mix the gel on some filter paper with another DNA strand, which has previously been tagged with radioactivity. This new strand binds to the relevant parts of the first one and when you expose the paper to a piece of X-ray film, it acts as an indicator: you see the different parts of the DNA lined up as blurs of varying size and darkness. You can then work out which blur corresponds to which DNA strand. You can also do blotting using coloured dyes as tags instead of radioactivity.

52 Northern blots detect RNA, and western blots detect proteins. There are also eastern blots (for detecting protein modification) and even far-eastern blots,

these latter so called because they're related to eastern blots but were developed in Japan.

53 Haruko Obokata et al., 'Stimulus-Triggered Fate Conversion of Somatic Cells into Pluripotency', Nature 505, no. 7485 (Jan. 2014): pp. 641–47; https://doi.org/10.1038/nature12968; Haruko Obokata et al., 'Bidirectional Developmental Potential in Reprogrammed Cells with Acquired Pluripotency', Nature 505, no. 7485 (Jan. 2014): pp. 676–80; https://doi.org/10.1038/nature12969

54 This is something of an oversimplification: the induced cells don't have quite the same properties as the embryonic ones. This might end up being very important in the medical context – only more research will tell. Incidentally, the process of making induced pluripotent stem cells hadn't been discovered back in 2004–2005, hence Hwang's focus on creating stem cells from embryos.

55 Nobel Media, 'The Nobel Prize in Physiology or Medicine 2012' (Oct. 2012); https://www.nobelprize.org/prizes/medicine/2012/summary/

56 According to the Japan Times, kappogi sales shot up following Obokata's rise to fame. Rowan Hooper, 'Stem-Cell Leap Defied Japanese Norms', Japan Times, 14 Feb. 2014; https://www.japantimes.co.jp/news/2014/02/15/national/science-health/stem-cell-leap-defied-japanese-norms/

57 Shunsuke Ishii et al., 'Report on STAP Cell Research Paper Investigation' (31 March 2014); http://www3.riken.jp/stap/e/f1document1.pdf

58 'red for failures': https://ipscell.com/stap-new-data/; 'all were in red': Mianna Meskus et al., 'Research Misconduct in the Age of Open Science: The Case of STAP Stem Cells', Science as Culture 27, no. 1 (2 Jan. 2018): pp. 1–23; https://doi.org/10.1080/09505431.2017.1316975. This piece also provides an interesting discussion of how the internet – anonymous comments on the faked images, and the blog cataloguing replication attempts – was used to bring down the STAP research.

59 James Gallagher, 'Stem Cell Scandal Scientist Haruko Obokata Resigns', BBC News, 19 Dec. 2014; https://www.bbc.co.uk/news/health-30534674

60 Isao Katsura et al., 'Report on STAP Cell Research Paper Investigation' (25 Dec. 2014); http://www3.riken.jp/stap/e/c13document52.pdf; Masaaki Kameda, '"STAP Cells" Claimed by Obokata Were Likely Embryonic Stem Cells', Japan Times, 26 Dec. 2014; https://www.japantimes.co.jp/news/2014/12/26/national/stap-cells-claimed-by-obokata-were-likely-embryonic-stem-cells/

61 David Cyranoski, 'Collateral Damage: How One Misconduct Case Brought a Biology Institute to Its Knees', Nature 520, no. 7549 (April 2015): pp. 600–3; https://doi.org/10.1038/520600a

62 David Cyranoski, 'Stem-Cell Pioneer Blamed Media "Bashing" in Suicide Note', Nature, 13 Aug. 2014, https://doi.org/10.1038/nature.2014.15715

63 Elisabeth M. Bik et al., 'The Prevalence of Inappropriate Image Duplication in Biomedical Research Publications', MBio 7, no. 3 (6 July 2016): e00809-16; https://doi.org/10.1128/mBio.00809-16. For a profile of Bik, see Tom Bartlett, 'Hunting for Fraud Full Time', Chronicle of Higher Education, 8 Dec. 2019; https://www.chronicle.com/article/Hunting-for-Fraud-Full-Time/247666

64 Bik et al., 'The Prevalence of Inappropriate Image Duplication'.

65 For example, it's been observed that if you ask people to choose a random number between 1 and 10, they're far more likely to choose 7 than any other number. Seeing a disproportionate number of 7s in a dataset is a big 'tell' that humans have been involved in making it. See, e.g.: https://www.reddit.com/r/dataisbeautiful/comments/acow6y/asking_over_8500_students_to_pick_a_random_number/

66 It's important to note that statisticians don't use 'error' in a pejorative sense – it just means the difference between the value you have in your data and the true value of whatever you measured.

67 Indeed, the whole point of many statistical tests is to separate out real effects – say, of the new drug you're testing – from this random sampling error.

68 J. B. S. Haldane, 'The Faking of Genetical Results', *Eureka* 27 (1964): pp. 21–24. Quoted in J. J. Pandit, 'On Statistical Methods to Test If Sampling in Trials Is Genuinely Random: Editorial', *Anaesthesia* 67, no. 5 (May 2012): pp. 456–62; https://doi.org/10.1111/j.1365-2044.2012.07114.x

69 Lawrence J. Sanna et al., 'Rising up to Higher Virtues: Experiencing Elevated Physical Height Uplifts Prosocial Actions', *Journal of Experimental Social Psychology* 47, no. 2 (Mar. 2011): pp. 472–76; https://doi.org/10.1016/j.jesp.2010.12.013. Dirk Smeesters & Jia (Elke) Liu, 'The Effect of Color (Red versus Blue) on Assimilation versus Contrast in Prime-to-Behavior Effects', *Journal of Experimental Social Psychology* 47, no. 3 (May 2011): pp. 653–56; https://doi.org/10.1016/j.jesp.2011.02.010

70 Uri Simonsohn, 'Just Post It: The Lesson from Two Cases of Fabricated Data Detected by Statistics Alone', *Psychological Science* 24, no. 10 (Oct. 2013): pp. 1875–88; https://doi.org/10.1177/0956797613480366

71 Ed Yong, 'Uncertainty Shrouds Psychologist's Resignation', *Nature*, 12 July 2012; https://doi.org/10.1038/nature.2012.10968. See also Jules Seegers, 'Ontslag Hoogleraar Erasmus Na Plegen Wetenschapsfraude', *NRC Handelsblad*, 25 June 2012; https://www.nrc.nl/nieuws/2012/06/25/erasmus-trekt-artikelen-terug-hoogleraar-ontslagen-om-schenden-integriteit-a1443819 [Dutch].

72 And just like the banks' automatic fraud-detection systems, automatic data-checking algorithms are being developed to assess papers for problematic data – see Chapter 5.

73 For example, there's Benford's law (Frank Benford, 'The Law of Anomalous Numbers', *Proceedings of the American Philosophical Society* 78, no. 4 (22 April 1937): pp. 551–72; https://www.jstor.org/stable/984802; though the 'law' was first noticed in 1881 by the mathematician Simon Newcomb), a mathematical phenomenon seen in many different collections of numbers. Benford's law states that the first significant digit of the numbers in many datasets is far more likely to be low than high: it's about 30 per cent likely to be a 1; 18 per cent likely to be a 2; 13 per cent likely to be a 3, and so on until 9, which appears as the first significant digit only 5 per cent of the time. This is seen in datasets as diverse as population counts from different countries or regions, house and stock prices, surface areas of world rivers, and numbers in the Fibonacci sequence. Even the citations of scientific papers on Benford's Law are distributed according to Benford's Law (Tariq Ahmad Mir, 'Citations to Articles Citing Benford's Law: A Benford Analysis', *ArXiv* (19 Mar. 2016): 1602.01205; http://arxiv.org/abs/1602.01205). If you find Benford's law quite bizarre and counter-intuitive, you're not alone: mathematicians haven't satisfyingly worked out why it occurs. Nonetheless, it's been well established in practice, and datasets that don't follow this distribution even though they'd normally be expected to might have been tampered with. There is substantial controversy about how reliable the law is as an indicator of fraud, though (Andreas Diekmann & Ben Jann, 'Benford's Law and Fraud Detection: Facts and Legends', *German Economic Review* 11, no. 3 (1 Aug. 2010): pp. 397–401; https://doi.org/10.1111/j.1468-0475.2010.00510.x), so you should only rely on it as part of a balanced diet of fraud-detecting techniques.

74 A useful scheme is provided in this paper: Rutger M. van den Bor et al., 'A Computationally Simple Central Monitoring Procedure, Effectively Applied to Empirical Trial Data with Known Fraud', *Journal of Clinical Epidemiology* 87 (July 2017): pp. 59–69; https://doi.org/10.1016/j.jclinepi.2017.03.018

75 M. J. LaCour & D. P. Green, 'When Contact Changes Minds: An Experiment on Transmission of Support for Gay Equality', *Science* 346, no. 6215 (12 Dec. 2014): 1366–69; https://doi.org/10.1126/science.1256151

76 Harry McGee, 'Personal Route to Reach Public Central to Yes Campaign', *Irish Times*, 14 May 2015; https://www.irishtimes.com/news/politics/marriage-referendum/personal-route-to-reach-public-central-to-yes-campaign-1.2211282

77 Quoted by Michael C. Munger, 'L'Affaire LaCour: What It Can Teach Us about Academic Integrity and "Truthiness"', *Chronicle of Higher Education*, 15 June 2015; https://www.chronicle.com/article/LAffaire-LaCour/230905

78 David Broockman et al., 'Irregularities in LaCour (2014)', 19 May 2015; https://stanford.edu/~dbroock/broockman_kalla_aronow_lg_irregularities.pdf

79 Tom Bartlett, 'The Unraveling of Michael LaCour', *Chronicle of Higher Education*, 2 June 2015; https://www.chronicle.com/article/The-Unraveling-of-Michael/230587. It should be noted that LaCour offered a (very weak, in my view) rebuttal to the charges (David Malakoff, 'Gay Marriage Study Author LaCour Issues Defense, but Critics Aren't Budging', *Science*, 30 May 2015; https://www.sciencemag.org/news/2015/05/gay-marriage-study-author-lacour-issues-defense-critics-arent-budging). Finally, Broockman and Kalla published their own, genuine survey study that tested some of the same hypotheses as LaCour's fake one, except targeted at trans rights rather than gay rights. They concluded that face-to-face canvassing did work in reducing prejudice, but that it didn't matter whether the canvasser was themselves transgender. D. Broockman & J. Kalla, 'Durably Reducing Transphobia: A Field Experiment on Door-to-Door Canvassing', *Science* 352, no. 6282 (8 April 2016): pp. 220–24; https://doi.org/10.1126/science.aad9713

80 Jeffrey Brainard, 'What a Massive Database of Retracted Papers Reveals about Science Publishing's "Death Penalty"', *Science*, 25 Oct. 2018; https://doi.org/10.1126/science.aav8384

81 https://retractionwatch.com/retraction-watch-database-user-guide/

82 Inha Cho et al., 'Retraction', *Science* 367, no. 6474 (2 Jan. 2020): p. 155; https://doi.org/10.1126/science.aba6100

83 https://twitter.com/francesarnold/status/1212796266494607360

84 Fraud: Michael L. Grieneisen & Minghua Zhang, 'A Comprehensive Survey of Retracted Articles from the Scholarly Literature', *PLOS ONE* 7, no. 10 (24 Oct. 2012): e44118; https://doi.org/10.1371/journal.pone.0044118. A review of retractions of psychology research found similar figures: Johannes Stricker & Armin Günther, 'Scientific Misconduct in Psychology: A Systematic Review of Prevalence Estimates and New Empirical Data', *Zeitschrift Für Psychologie* 227, no. 1 (Jan. 2019): pp. 53–63; https://doi.org/10.1027/2151-2604/a000356. Plagiarism: this is broadly the same finding as other investigations of retractions – for example: Anthony Bozzo et al.,' Retractions in Cancer Research: A Systematic Survey', *Research Integrity and Peer Review* 2, no. 1 (Dec. 2017): 5; https://doi.org/10.1186/s41073-017-0031-1; Zoë Corbyn, 'Misconduct Is the Main Cause of Life-Sciences Retractions', *Nature* 490, no. 7418 (Oct. 2012): p. 21; https://doi.org/10.1038/490021a; Guowei Li et al., 'Exploring the Characteristics, Global Distribution and Reasons for Retraction of Published Articles Involving Human Research Participants: A Literature Survey', *Journal of*

Multidisciplinary Healthcare 11 (Jan. 2018): pp. 39–47; https://doi.org/10.2147/ JMDH.S151745. For a review, see Charles Gross, 'Scientific Misconduct', *Annual Review of Psychology* 67, no. 1 (4 Jan. 2016): pp. 693–711; https://doi.org/10.1146/ annurev-psych-122414-033437

85 Daniele Fanelli, 'Why Growing Retractions Are (Mostly) a Good Sign', *PLOS Medicine* 10, no. 12 (3 Dec. 2013): e1001563; https://doi.org/10.1371/journal. pmed.1001563

86 Crimes in society: Avshalom Caspi et al., 'Childhood Forecasting of a Small Segment of the Population with Large Economic Burden', *Nature Human Behaviour* 1, no. 1 (Jan. 2017): p. 0005; https://doi.org/10.1038/s41562-016-0005. Individual scientists: Jeffrey Brainard, 'What a Massive Database of Retracted Papers Reveals about Science Publishing's "Death Penalty"', *Science* (25 Oct. 2018); https://doi.org/10.1126/science.aav8384

87 https://retractionwatch.com/the-retraction-watch-leaderboard/. At present, you need a minimum of twenty-one retractions to be listed on the Leaderboard.

88 Peter Kranke et al., 'Reported Data on Granisetron and Postoperative Nausea and Vomiting by Fujii et al. Are Incredibly Nice!', *Anesthesia & Analgesia* 90, no. 4 (April 2000): pp. 1004–6; https://doi.org/10.1213/00000539-200004000-00053

89 Adam Marcus & Ivan Oransky, 'How the Biggest Fabricator in Science Got Caught', *Nautilus*, 21 May 2015; http://nautil.us/issue/24/error/how-the-biggest-fabricator-in-science-got-caught

90 Improbable data: J. B. Carlisle, 'The Analysis of 168 Randomised Controlled Trials to Test Data Integrity: Analysis of 168 Randomised Controlled Trials to Test Data Integrity', *Anaesthesia* 67, no. 5 (May 2012): pp. 521–37; https://doi. org/10.1111/j.1365-2044.2012.07128.x. End of career: Dennis Normile, 'A New Record for Retractions? (Part 2)', *Science*, 2 July 2012; https://www.sciencemag. org/news/2012/07/new-record-retractions-part-2

91 Adam Marcus, 'Does Anesthesiology Have a Problem? Final Version of Report Suggests Fujii Will Take Retraction Record, with 172', *Retraction Watch*, 2 July 2012; https://retractionwatch.com/2012/07/02/does-anesthesiology-have-a-problem-final-version-of-report-suggests-fujii-will-take-retraction-record-with-172/

92 Daniele Fanelli, 'How Many Scientists Fabricate and Falsify Research? A Systematic Review and Meta-Analysis of Survey Data', *PLOS ONE* 4, no. 5 (29 May 2009): e5738; https://doi.org/10.1371/journal.pone.0005738

93 Ibid.

94 Gross, 'Scientific Misconduct', p. 700.

95 It's Rashmi Madhuri, an Indian nanomaterials researcher who has notched up twenty-four retractions, mainly for image duplication. Alison McCook, 'Author under Fire Has Eight Papers Retracted, Including Seven from One Journal', *Retraction Watch*, 25 April 2018; https://retractionwatch.com/2018/04/25/ author-under-fire-has-six-papers-retracted-including-five-from-one-journal/

96 Ferric C. Fang et al., 'Males Are Overrepresented among Life Science Researchers Committing Scientific Misconduct', *MBio* 4, no. 1 (22 Jan. 2013): e00640-12; https://doi.org/10.1128/mBio.00640-12

97 Daniele Fanelli et al., 'Misconduct Policies, Academic Culture and Career Stage, Not Gender or Pressures to Publish, Affect Scientific Integrity', *PLOS ONE* 10, no. 6 (17 June 2015): e0127556; https://doi.org/10.1371/journal. pone.0127556. It does seem plausible that men would be responsible for more scientific fraud than women – in FBI data on arrests for all kinds of fraudulent activity in 2017, men made up 65.5 per cent of arrests for forgery

and counterfeiting, and 62.5 per cent of arrests for fraud (though only 50.9 per cent of arrests for embezzlement; Criminal Justice Information Services, 'Crime in the United States: 2017'; https://ucr.fbi.gov/crime-in-the-u.s/2017/crime-in-the-u.s.-2017/topic-pages/tables/table-42, Table 42). Of course, it's not just fraud and related crimes: not a single one of the crime categories in the FBI data shows a higher percentage of women (embezzlement, with just a 2 percentage-point difference between the sexes, is by far the closest). Overall, men represented 73 per cent of arrests for all categories of crime in the data.

98   Daniele Fanelli et al., 'Testing Hypotheses on Risk Factors for Scientific Misconduct via Matched-Control Analysis of Papers Containing Problematic Image Duplications', *Science and Engineering Ethics* 25, no. 3 (June 2019): pp. 771–89; https://doi.org/10.1007/s11948-018-0023-7

99   Although this may change: a report in 2017 indicated that some courts in China were calling for the death penalty (not metaphorically, as with the notion of retraction as 'science's death penalty' for a paper, but literal execution) for cases of scientific fraud. The founders of Retraction Watch explain why that's a bad idea: Ivan Oransky & Adam Marcus, 'Chinese Courts Call for Death Penalty for Researchers Who Commit Fraud', *STAT News*, 23 June 2017; https://www.statnews.com/2017/06/23/china-death-penalty-research-fraud/

100   Wang et al., 'Positive Results in Randomized Controlled Trials on Acupuncture Published in Chinese Journals: A Systematic Literature Review', *The Journal of Alternative and Complementary Medicine* 20, no. 5 (May 2014): A129–A129; https://doi.org/10.1089/acm.2014.5346.abstract cited in Stephen Novella, 'Scientific Fraud in China', *Science-Based Medicine*, 27 Nov. 2019; https://sciencebasedmedicine.org/scientific-fraud-in-china/https://www.liebertpub.com/doi/abs/10.1089/acm.2014.5346.abstract

101   Qing-Jiao Liao et al., 'Perceptions of Chinese Biomedical Researchers Towards Academic Misconduct: A Comparison Between 2015 and 2010', *Science and Engineering Ethics*, 10 April 2017; https://doi.org/10.1007/s11948-017-9913-3

102   Andrew M. Stern et al., 'Financial Costs and Personal Consequences of Research Misconduct Resulting in Retracted Publications', *eLife* 3 (14 Aug. 2014): e02956; https://doi.org/10.7554/eLife.02956. The 'desperate for funding' interpretation is given by Nicolas Chevassus-au-Louis, *Fraud in the Lab: The High Stakes of Scientific Research*, tr. Nicholas Elliott (Cambridge, MA: Harvard University Press, 2019).

103   Medawar, *The Strange Case of the Spotted Mice*, p. 197.

104   David Goodstein, *On Fact and Fraud: Cautionary Tales from the Front Lines of Science* (Princeton: Princeton University Press, 2010): p. 2.

105   J. H. Schön et al., 'Field-Effect Modulation of the Conductance of Single Molecules', *Science* 294, no. 5549 (7 Dec. 2001): pp. 2138–40; https://doi.org/10.1126/science.1066171

106   Stanford professor: quoted in 'World's Smallest Transistor', *Engineer*, 9 Nov. 2001; https://www.theengineer.co.uk/worlds-smallest-transistor/. For the full story of the Schön affair, see Eugenie Samuel Reich, *Plastic Fantastic: How the Biggest Fraud in Physics Shook the Scientific World* (Basingstoke, Hampshire: Palgrave Macmillan, 2009).

107   Leonard Cassuto, 'Big Trouble in the World of "Big Physics"', *Guardian*, 18 Sept. 2002; https://www.theguardian.com/education/2002/sep/18/science.highereducation

108   American Physical Society, 'Report of the Investigation Committee on the Possibility of Scientific Misconduct in the Work of Hendrick and Coauthors' (2002); https://media-bell-labs-com.s3.amazonaws.com/pages/20170403_1709/

misconduct-revew-report-lucent.pdf, p. 3. Schön also simply hadn't recorded a huge amount of his lab work. See Reich, *Plastic Fantastic*.

109 American Physical Society, 'Report of the Investigation Committee on the Possibility of Scientific Misconduct in the Work of Hendrik Schön and Coauthors' (Sept. 2002): pp. E-5–E-6; https://media-bell-labs-com.s3.amazonaws.com/pages/20170403_1709/misconduct-revew-report-lucent.pdf

110 Ibid. p. H-1.

111 Schön's position on the Retraction Watch Leaderboard may not be that high, but he may have the highest number of retractions from top-tier journals like *Science* and *Nature*.

112 Diederik A. Stapel, *Derailment: Faking Science*, tr. Nicholas J. L. Brown (Strasbourg, France, 2014,2016): p. 103; http://nick.brown.free.fr/stapel

113 For an investigation of over 146 reports of scientific misconduct from the US Office of Research Integrity, and discussion of some common themes, see Donald S. Kornfeld, 'Perspective: Research Misconduct', *Academic Medicine* 87, no. 7 (July 2012): pp. 877–82; https://doi.org/10.1097/ACM.0b013e318257ee6a

114 Jeneen Interlandi, 'An Unwelcome Discovery', *New York Times*, 22 Oct. 2006; https://www.nytimes.com/2006/10/22/magazine/22sciencefraud.html

115 Levelt Committee et al., 'Flawed Science: The Fraudulent Research Practices of Social Psychologist Diederik Stapel [English Translation]', 28 Nov. 2012; https://osf.io/eup6d

116 One analysis from biomedicine showed that within a year of retraction, a paper receives 45 per cent of the citations of comparable non-retracted papers and that this number declines as time goes on. See Jeffrey L. Furman et al., 'Governing Knowledge in the Scientific Community: Exploring the Role of Retractions in Biomedicine', *Research Policy* 41, no. 2 (Mar. 2012): pp. 276–90; https://doi.org/10.1016/j.respol.2011.11.001

117 Helmar Bornemann-Cimenti et al., 'Perpetuation of Retracted Publications Using the Example of the Scott S. Reuben Case: Incidences, Reasons and Possible Improvements', *Science and Engineering Ethics* 22, no. 4 (Aug. 2016): pp. 1063–72; https://doi.org/10.1007/s11948-015-9680-y

118 Judit Bar-Ilan & Gali Halevi, 'Post Retraction Citations in Context: A Case Study', *Scientometrics* 113, no. 1 (Oct. 2017): pp. 547–65; https://doi.org/10.1007/s11192-017-2242-0. For even more depressing figures, see also Anne Victoria Neale et al., 'Analysis of Citations to Biomedical Articles Affected by Scientific Misconduct', *Science and Engineering Ethics* 16, no. 2 (June 2010): pp. 251–61; https://doi.org/10.1007/s11948-009-9151-4

119 It's also possible that scientists saved the papers to their computers or owned a literal paper copy of the journal before the retraction happened, and never checked back. See Jaime A. Teixeira da Silva & Helmar Bornemann-Cimenti, 'Why Do Some Retracted Papers Continue to Be Cited?', *Scientometrics* 110, no. 1 (Jan. 2017): pp. 365–70; https://doi.org/10.1007/s11192-016-2178-9. See also Jaime A. Teixeira da Silva et al., 'Citing Retracted Papers Has a Negative Domino Effect on Science, Education, and Society', *Impact of Social Sciences*, 6 Dec. 2016; https://blogs.lse.ac.uk/impactofsocialsciences/2016/12/06/citing-retracted-papers-has-a-negative-domino-effect-on-science-education-and-society/

120 Joachim Boldt has 100 retractions. You may have noticed that there seem to be rather a lot of anaesthesiologists involved in scientific fraud cases. If it isn't just purely a coincidence, the most plausible reason I can think of is that anaesthesia is still generally rather mysterious and poorly understood. This leaves a lot of space for new 'discoveries' to be made by fraudsters, with no

super-strong evidence in the field with which to compare them. Note that this is pure speculation on my part.

121   Ryan Zarychanski et al., 'Association of Hydroxyethyl Starch Administration With Mortality and Acute Kidney Injury in Critically Ill Patients Requiring Volume Resuscitation: A Systematic Review and Meta-Analysis', *JAMA* 309, no. 7 (20 Feb. 2013): pp. 678–88; https://doi.org/10.1001/jama.2013.430

122   And yet, even after 100 papers by Boldt have been retracted, at the time of writing roughly another 100 of his articles remain in the literature. Journal editors are in an invidious position: for many of those individual papers, fraud hasn't explicitly been proven, but since Boldt was such a prolific faker we all know it's likely to be there. One suggestion that's been made is that editors should add a note to the articles of known fraudsters (an 'expression of concern', in the publishing jargon), indicating that they should be cited only with extreme care. See Christian J. Wiedermann, 'Inaction over Retractions of Identified Fraudulent Publications: Ongoing Weakness in the System of Scientific Self-Correction', *Accountability in Research* 25, no. 4 (19 May 2018): pp. 239–53; https://doi.org/ 10.1080/08989621.2018.1450143. See also Christian J. Wiedermann & Michael Joannidis, 'The Boldt Scandal Still in Need of Action: The Example of Colloids 10 Years after Initial Suspicion of Fraud', *Intensive Care Medicine* 44, no. 10 (Oct. 2018): pp. 1735–37; https://doi.org/10.1007/s00134-018-5289-3

123   A. J. Wakefield et al., 'Ileal-Lymphoid-Nodular Hyperplasia, Non-Specific Colitis, and Pervasive Developmental Disorder in Children', *Lancet* 351, no. 9103 (Feb. 1998): pp. 637–41; https://doi.org/10.1016/S0140-6736(97) 11096-0

124   A. J. Wakefield et al., 'Enterocolitis in Children with Developmental Disorders', *The American Journal of Gastroenterology* 95, no. 9 (Sept. 2000): pp. 2285–95; https://doi.org/10.1111/j.1572-0241.2000.03248.x. There is actually some evidence that children with autism spectrum disorder have higher levels of some gastrointestinal symptoms (B. O. McElhanon et al., 'Gastrointestinal Symptoms in Autism Spectrum Disorder: A Meta-Analysis', *Pediatrics* 133, no. 5 (1 May 2014): pp. 872–83; https://doi.org/10.1542/peds.2013-3995), but there's no evidence that this is caused by vaccines.

125   *MMR: What They Didn't Tell You*, Brian Deer, dir. (Twenty Twenty Television, 2004); https://youtu.be/7UbL80pM6TM

126   Luke E. Taylor et al., 'Vaccines Are Not Associated with Autism: An Evidence-Based Meta-Analysis of Case-Control and Cohort Studies', *Vaccine* 32, no. 29 (June 2014): pp. 3623–29; https://doi.org/10.1016/j.vaccine.2014.04.085; Jean Golding et al., 'Prenatal Mercury Exposure and Features of Autism: A Prospective Population Study', *Molecular Autism* 9, no. 1 (Dec. 2018): 30; https://doi.org/10.1186/s13229-018-0215-7; Matthew Z. Dudley et al., *The Clinician's Vaccine Safety Resource Guide: Optimizing Prevention of Vaccine-Preventable Diseases Across the Lifespan* (Cham: Springer International Publishing, 2018); https://doi.org/10.1007/978-3-319-94694-8; Anders Hviid et al., 'Measles, Mumps, Rubella Vaccination and Autism: A Nationwide Cohort Study', *Annals of Internal Medicine* 170, no. 8 (16 April 2019): pp. 513–520; https://doi.org/10.7326/M18-2101

127   Dudley et al., *The Clinician's Vaccine Safety Resource Guide*, pp. 157–165.

128   F. Godlee et al., 'Wakefield's Article Linking MMR Vaccine and Autism Was Fraudulent', *BMJ* 342 (5 Jan. 2011): c7452; https://doi.org/10.1136/bmj.c7452

129   B. Deer, 'How the Case against the MMR Vaccine Was Fixed', *BMJ* 342 (5 Jan. 2011): c5347; https://doi.org/10.1136/bmj.c5347

130   As any first-year-philosophy student could tell you, the 'MMR-then-symptoms' argument was hardly a clinching one anyway: just because X happens shortly after Y, it's not necessarily evidence that X *caused* Y.

131   B. Deer, 'How the Vaccine Crisis Was Meant to Make Money', *BMJ* 342 (14 Jan. 2011): c5258; https://doi.org/10.1136/bmj.c5258

132   Deer reports that Wakefield was being paid £150 per hour by the lawyer, to a total of £435,643 plus expenses, and that this was taxpayer money, from the UK's Legal Aid fund; https://briandeer.com/wakefield/legal-aid.htm

133   http://briandeer.com/wakefield/vaccine-patent.htm

134   A. J. Wakefield et al., 'Ileal-Lymphoid-Nodular Hyperplasia', p. 641.

135   B. Deer, 'The Lancet's Two Days to Bury Bad News', *BMJ* 342, (18 Jan. 2011): c7001; https://doi.org/10.1136/bmj.c7001

136   The report from the General Medical Council's Fitness to Practice Hearing, held for Wakefield and two of his colleague on 16 July 2007, is hosted at the following link: http://www.channel4.com/news/media/2010/01/day28/GMC_Charge_sheet.pdf

137   'Ruling on Doctor in MMR Scare', *NHS News*, 29 Jan. 2010; https://www.nhs.uk/news/medical-practice/ruling-on-doctor-in-mmr-scare/

138   'Vaxxed: Tribeca Festival Withdraws MMR Film', *BBC News*, 27 March 2016; https://www.bbc.co.uk/news/entertainment-arts-35906470

139   *Daily Mail* as ringleader: Ben Goldacre, 'The MMR Sceptic Who Just Doesn't Understand Science', *Bad Science* (blog), 2 Nov. 2005; https://www.badscience.net/2005/11/comment-the-mmr-sceptic-who-just-doesnt-understand-science/; *Private Eye*: David Elliman & Helen Bedford, 'Press: *Private Eye* Special Report on MMR', *BMJ* 324, no. 7347 (18 May 2002): p. 1224; https://doi.org/10.1136/bmj.324.7347.1224

140   There's good reason to specifically blame the media coverage here. A 2019 study showed that a surge in media misinformation about the MMR was directly (and, under certain assumptions, causally) related to a decline in vaccination rates. Of course, the *ultimate* cause was the publication of Wakefield's paper. See Meradee Tangvatcharapong, 'The Impact of Fake News: Evidence from the Anti-Vaccination Movement in the US', Oct. 2019; https://meradeetang.files.wordpress.com/2019/11/meradee_jmp_oct31_2.pdf

141   80%: NHS Digital, 'Childhood Vaccination Coverage Statistics: England 2017-18', 18 Sept. 2018; https://files.digital.nhs.uk/55/D9C4C2/child-vacc-stat-eng-2017-18-report.pdf, Figure 6. Rates jumping: Vaccine Knowledge Project, 'Measles', University of Oxford, 25 June 2019; https://vk.ovg.ox.ac.uk/vk/measles

142   WHO: 'More than 140,000 Die from Measles as Cases Surge Worldwide', World Health Organisation, 5 Dec. 2019; https://www.who.int/news-room/detail/05-12-2019-more-than-140-000-die-from-measles-as-cases-surge-worldwide. Measles surge: Sarah Boseley, 'Resurgence of Deadly Measles Blamed on Low MMR Vaccination Rates', *Guardian*, 21 Aug. 2018; https://www.theguardian.com/society/2018/aug/20/low-mmr-uptake-blamed-for-surge-in-measles-cases-across-europe

143   This argument has been made by some journalists who were, at the time, on the MMR-scare bandwagon, such as the *Daily Mail*'s Peter Hitchens (Peter Hitchens, 'Some Reflections on Measles and the MMR', *Peter Hitchens's Blog*, 11 April 2013; https://hitchensblog.mailonsunday.co.uk/2013/04/some-reflections-on-measles-and-the-mmr-.html). Moreover, as many writers on anti-vaccinationism have argued, vaccines are a victim of their own success. Measles, mumps and rubella are all highly unpleasant diseases that in rare cases have serious, life-altering complications like deafness. But vaccines have

done such a good job of wiping them out, we've generally forgotten how bad they can be, increasing our complacency.

144 Simon Chaplin et al.,'Wellcome Trust Global Monitor 2018', Wellcome Trust, 19 June 2019; https://wellcome.ac.uk/reports/wellcome-global-monitor/2018, Chapter 5.

145 Another similar story is the recent Theranos scandal, where the company's CEO, Elizabeth Holmes (at the time of writing, on trial for wire fraud), managed to bilk unbelievable amounts of money from investors such as Rupert Murdoch, the Walton family (of Walmart fame), and many more, becoming America's youngest and richest self-made female billionaire. Her company's devices, which ostensibly could diagnose many health conditions from a tiny drop of blood, never actually worked. But the investors, who wanted to get in at the start of what might've been the next Facebook or Uber in terms of its transformative technological effect, managed to miss or ignore the obvious flaws. The story is told by the investigative journalist John Carreyrou in his unputdownable *Bad Blood*, John Carreyrou, *Bad Blood: Secrets and Lies in a Silicon Valley Startup* (New York: Alfred A. Knopf, 2018).

146 Another similar story can be found in: Alison McCook, 'Two Researchers Challenged a Scientific Study About Violent Video Games – and Took a Hit for Being Right', *Vice*, 25 July 2018; https://www.vice.com/en_us/article/8xb89b/two-researchers-challenged-a-scientific-study-about-violent-video-games-and-took-a-hit-for-being-right

147 For further discussion, see Joe Hilgard, 'Are Frauds Incompetent?', *Crystal Prison Zone*, 1 Feb. 2020; http://crystalprisonzone.blogspot.com/2020/01/are-frauds-incompetent.html

4: *Bias*

Epigraph: Arthur Schopenhauer, *The World as Will and Presentation: Vol II*, tr. David Carus and Richard E. Aquila (New York: Routledge, 2011) and T. H. Huxley, 'The Darwin Memorial' (1885).

1 See e.g. Samuel George Morton, *Crania Americana* (London: Simkin, Marshall & Co., 1839); https://archive.org/details/Craniaamericana00Mort

2 Or possibly peppercorns – see Paul Wolff Mitchell, 'The Fault in His Seeds: Lost Notes to the Case of Bias in Samuel George Morton's Cranial Race Science', *PLOS Biology* 16, no. 10 (4 Oct. 2018): e2007008; https://doi.org/10.1371/journal.pbio.2007008

3 Samuel George Morton, 'Aug. 8th, 1848, Vice President Morton in the Chair', *Proceedings of the Academy of Natural Sciences of Philadelphia* 4 (1848): pp. 75–76.

4 Stephen Jay Gould, *The Mismeasure of Man*, Rev. and Expanded (New York: Norton, 1996): p. 97.

5 S. J. Gould, 'Morton's Ranking of Races by Cranial Capacity. Unconscious Manipulation of Data May Be a Scientific Norm', *Science* 200, no. 4341 (5 May 1978): pp. 503–9; https://doi.org/10.1126/science.347573

6 Ibid. p. 504.

7 The appendix of the following article gives a useful taxonomy of these biases: David L. Sackett, 'Bias in Analytic Research', *The Case-Control Study Consensus and Controversy*, Elsevier (1979): pp. 51–63; https://doi.org/10.1016/B978-0-08-024907-0.50013-4

8 In technical terms, a bias is anything that systematically moves results away from the truth. That 'systematically' is important: unlike the random errors (of measurement and sampling) we met in the previous chapter, biases have a direction. Random errors are like a car with a broken steering wheel that causes it to swerve haphazardly all over the street. Bias, on the other hand, is like a car with a misaligned axle that constantly pulls it towards one particular side of the road. Some bias can come from non-human factors, such as malfunctioning instruments or glitches in computer software. But the kind of biases we're interested in here are the ones held by the scientists themselves.

9 Daniele Fanelli, '"Positive" Results Increase Down the Hierarchy of the Sciences', *PLOS ONE* 5, no. 4 (7 April 2010): e10068; https://doi.org/10.1371/journal.pone.0010068

10 There's some disagreement over whether the numbers of positive and negative results are decreasing or increasing over time. For one perspective, see Daniele Fanelli, 'Negative Results Are Disappearing from Most Disciplines and Countries', *Scientometrics* 90, no. 3 (Mar. 2011): pp. 891–904; https://doi.org/10.1007/s11192-011-0494-7. For the other, see Joost C. F. de Winder & Dimitra Dodou, 'A Surge of p-Values between 0.041 and 0.049 in Recent Decades (but Negative Results Are Increasing Rapidly Too)', *PeerJ* 3 (22 Jan. 2015): e733; https://doi.org/10.7717/peerj.733

11 There's another reason why a success rate above 90 per cent wouldn't be a good sign even if it was accurate and nothing fishy was going on: it would mean that scientists are so good at picking true hypotheses that they *already know* what's true before they test it. Scientists in this world of near-perfect success would be shying away from studying truly new, cutting-edge questions, for which the answers are much more uncertain and the studies are more risky. In doing so, they'd be neglecting the important role of science to explore the unknown and advance our knowledge of the world.

12 Robert Rosenthal, 'The File Drawer Problem and Tolerance for Null Results', *Psychological Bulletin* 86, no. 3 (1979): pp. 638–41; https://doi.org/10.1037/0033-2909.86.3.638

13 Since height varies across different countries, it turns out that Austrian women are on average taller than Peruvian men (although the sex difference within each of these countries is preserved – Peruvian women are shorter than their male counterparts, and Austrian men taller than their female counterparts); https://en.wikipedia.org/wiki/Average_human_height_by_country#Table_of_Heights

14 This would be an underestimate of the true effect: according to Wikipedia, the average height difference between men and women in Scotland was 13.7cm (or 5.5 inches) in 2008; https://en.wikipedia.org/wiki/Average_human_height_by_country#Table_of_Heights

15 The specific details of the calculation of the *p*-value aren't strictly necessary for understanding how it works. For a crystal-clear introduction to statistics in general, I'd recommend David Spiegelhalter, *The Art of Statistics: Learning from Data* (London: Penguin, 2019). For an accessible discussion of the more philosophical issues surrounding statistics, see Zoltan Dienes, *Understanding Psychology as a Science: An Introduction to Scientific and Statistical Inference* (New York: Palgrave Macmillan, 2008).

16 Scott A. Cassidy et al., 'Failing Grade: 89% of Introduction-to-Psychology Textbooks That Define or Explain Statistical Significance Do So Incorrectly', *Advances in Methods and Practices in Psychological Science* 2, no. 3 (Sept. 2019): pp. 233–39; https://doi.org/10.1177/2515245919858072. See also Raymond Hubbard & M. J. Bayarri, 'Confusion Over Measures of Evidence (p's) Versus

Errors (α's) in Classical Statistical Testing', *American Statistician* 57, no. 3 (Aug. 2003): pp. 171–78; https://doi.org/10.1198/0003130031856

17 For the American Statistical Association's consensus position on *p*-values, written surprisingly comprehensibly, see Ronald L. Wasserstein & Nicole A. Lazar, 'The ASA Statement on *p*-Values: Context, Process, and Purpose', *The American Statistician* 70, no. 2 (2 April 2016): pp. 129–33; https://doi.org/10.1080/0003130 5.2016.1154108. It defines the *p*-value like this: 'the probability under a specified statistical model that a statistical summary of the data (e.g., the sample mean difference between two compared groups) would be equal to or more extreme than its observed value: p. 131.

18 Why does the definition of the *p*-value ('how likely is it that pure noise would give you results like the ones you have, or ones with an even larger effect') have that 'or an even larger effect' clause in it? (The 'or more extreme' part of the American Statistical Association's definition, given in the previous endnote, serves the same purpose.) It's necessary because any one specific pattern is extremely unlikely: think, for example, how rarely we would find a height difference of *precisely* (say) 10.00144983823 cm in a sample if we were to repeat our study on Scottish men and women an infinite number of times. This very specific number would be extremely improbable regardless of whether there's really a height difference at the population level, so a *p*-value that just tells us how unlikely this very specific number is would hardly be very helpful. The 'or more extreme' clause solves that. In our pretend example – a single study that found a 10 cm difference between the sampled ten men and ten women – a *p*-value of 0.03 means that the chance of finding a difference of 10 cm *or more* would be 3 per cent if there were no 'real' effect in the Scottish population.

19 You might immediately respond that we'd surely want a *zero* per cent chance of making a false-positive error, or something very close to zero. But there's a trade-off. If we're extremely conservative about which results we accept, there's a higher chance that we'll miss real effects in our dataset (that is, that we'll make a false-*negative* error).

20 David Salsburg, *The Lady Tasting Tea: How Statistics Revolutionized Science in the Twentieth Century* (New York: Holt, 2002): p. 98.

21 Ronald A. Fisher, 'The Arrangement of Field Experiments', *Journal of the Ministry of Agriculture of Great Britain* 33 (1926): pp. 503-513, p. 504.

22 https://www.taps-aff.co.uk/. The creator of this wonderful service, which actually takes into account a lot more information than just the temperature, is Colin Waddell.

23 This was also the suggestion of a paper that was part of a big debate on significance levels. See: Daniël Lakens et al., 'Justify Your Alpha', *Nature Human Behaviour* 2, no. 3 (Mar. 2018): pp. 168–71; https://doi.org/10.1038/s41562-018-0311-x

24 David Spiegelhalter, 'Explaining 5-Sigma for the Higgs: How Well Did They Do?', *Understanding Uncertainty*, 8 July 2012; https://understandinguncertainty.org/explaining-5-sigma-higgs-how-well-did-they-do

25 Richard Dawkins, 'The Tyranny of the Discontinuous Mind', *New Statesman*, 19 Dec. 2011; https://www.newstatesman.com/blogs/the-staggers/2011/12/issue-essay-line-dawkins. There's more discussion along these lines in Dawkins's wonderful book *The Ancestor's Tale*. Richard Dawkins & Yan Wong, *The Ancestor's Tale: A Pilgrimage to the Dawn of Life*, rev. ed. (London: Weidenfeld & Nicolson, 2016).

26 The impact a vaccine has on reducing mortality from a disease was the subject of the first ever medical meta-analysis, carried out by the statistician Karl

Pearson in 1904 (the disease being typhoid) – though the technique hadn't yet been named 'meta-analysis'. Karl Pearson, 'Report on Certain Enteric Fever Inoculation Statistics', *BMJ* 2, no. 2288 (5 Nov. 1904): pp. 1243–46; https://doi.org/10.1136/bmj.2.2288.1243. A useful history and summary of meta-analysis is provided in: Jessica Gurevitch et al., 'Meta-Analysis and the Science of Research Synthesis', *Nature* 555, no. 7695 (Mar. 2018): pp. 175–82; https://doi.org/10.1038/nature25753. Climate change: A. J. Challinor et al., 'A Meta-Analysis of Crop Yield under Climate Change and Adaptation', *Nature Climate Change* 4, no. 4 (April 2014): pp. 287–91; https://doi.org/10.1038/nclimate2153

27 That's all else held equal. But there are lots of other things that affect a study's precision other than sample size, such as the quality of its measures. Although sample size is usually a decent proxy, these days most meta-analysts use a more direct measure of the precision of effects, something called the 'standard error'. It's this that you'll normally see on the y-axis of funnel plots.

28 Calculations of *p*-values take the greater fluctuation of small samples into account. If there genuinely was no height difference between men and women in Scotland, finding a difference as large as 10 cm in a sample would be unlikely, but it could still happen occasionally when measuring a small sample. Recall that the *p*-value for a 10 cm difference in our sample of ten males and ten females was 0.03. The chance of finding that same 10 cm difference in a sample with 1,000 males and 1,000 females would be exceedingly small, giving us a tiny *p*-value (say, 0.0000001, or lower). In that case, we'd have better evidence that there really is a 'true' effect in the population. Incidentally, this shows why the *p*-value is not a measure of the size or importance of a finding: the exact same effect size can lead to different *p*-values, depending on the sample size.

29 David R. Shanks et al., 'Romance, Risk, and Replication: Can Consumer Choices and Risk-Taking Be Primed by Mating Motives?', *Journal of Experimental Psychology: General* 144, no. 6 (2015): e142–58; https://doi.org/10.1037/xge0000116. For another example, this time from priming research, with similar results, see Paul Lodder et al., 'A Comprehensive Meta-Analysis of Money Priming', *Journal of Experimental Psychology: General* 148, no. 4 (April 2019): pp. 688–712; https://doi.org/10.1037/xge0000570

30 Panayiotis A. Kyzas et al., 'Almost All Articles on Cancer Prognostic Markers Report Statistically Significant Results', *European Journal of Cancer* 43, no. 17 (Nov. 2007): pp. 2559–79; https://doi.org/10.1016/j.ejca.2007.08.030

31 Ioanna Tzoulaki et al., 'Bias in Associations of Emerging Biomarkers with Cardiovascular Disease', *JAMA Internal Medicine* 173, no. 8 (22 April 2013): p. 664; https://doi.org/10.1001/jamainternmed.2013.3018

32 See Erick H. Turner et al., 'Selective Publication of Antidepressant Trials and Its Influence on Apparent Efficacy', *New England Journal of Medicine* 358, no. 3 (17 Jan. 2008): pp. 252–60; https://doi.org/10.1056/NEJMsa065779. At the time of writing, the most recent meta-analysis of antidepressants does show a (modest) effect on depression symptoms: Andrea Cipriani et al., 'Comparative Efficacy and Acceptability of 21 Antidepressant Drugs for the Acute Treatment of Adults with Major Depressive Disorder: A Systematic Review and Network Meta-Analysis', *Lancet* 391, no. 10128 (April 2018): pp. 1357–66; https://doi.org/10.1016/S0140-6736(17)32802-7

33 Akira Onishi & Toshi A. Furukawa, 'Publication Bias Is Underreported in Systematic Reviews Published in High-Impact-Factor Journals: Metaepidemiologic Study', *Journal of Clinical Epidemiology* 67, no. 12 (Dec. 2014): pp. 1320–26, https://doi.org/10.1016/j.jclinepi.2014.07.002

34   D. Herrmann et al., 'Statistical Controversies in Clinical Research: Publication Bias Evaluations Are Not Routinely Conducted in Clinical Oncology Systematic Reviews', *Annals of Oncology* 28, no. 5 (May 2017): pp. 931–37; https://doi.org/10.1093/annonc/mdw691

35   There's a whole set of techniques to adjust the effect size in your meta-analysis when you discover that there's publication bias. Since these are guesswork (about how much you should reduce the effect size) stacked on guesswork (about how much publication bias there is), I always feel a bit nervous about using them. For details see e.g. Evan C. Carter et al., 'Correcting for Bias in Psychology: A Comparison of Meta-Analytic Methods', *Advances in Methods and Practices in Psychological Science* 2, no. 2 (June 2019): pp. 115–44; https://doi.org/10.1177/2515245919847196

36   Daniel Cressey, 'Tool for Detecting Publication Bias Goes under Spotlight', *Nature*, 31 March 2017; https://doi.org/10.1038/nature.2017.21728; Richard Morey, 'Asymmetric Funnel Plots without Publication Bias', *BayesFactor*, 9 Jan. 2016; https://bayesfactor.blogspot.com/2016/01/asymmetric-funnel-plots-without.html

37   A. Franco et al., 'Publication Bias in the Social Sciences: Unlocking the File Drawer', *Science* 345, no. 6203 (19 Sept. 2014): pp. 1502–5; https://doi.org/10.1126/science.1255484

38   http://www.tessexperiments.org; an additional nice factor about this programme is that all the research applications were peer reviewed with specific criteria in mind, so the ones selected to be turned into actual studies were all high quality and well-powered (see Chapter 5 for a discussion of statistical power and why it matters).

39   These numbers are calculated by dividing the values in the 'Published' column of Franco et al.'s Table 2 by the total in the bottom row.

40   All quotations from Franco et al.'s 'Publication Bias', Supplementary Table S6.

41   The conclusion of the Franco publication bias study is backed up by: Kerry Dwan et al., 'Systematic Review of the Empirical Evidence of Study Publication Bias and Outcome Reporting Bias', *PLOS ONE* 3, no. 8 (28 Aug. 2008): e3081; https://doi.org/10.1371/journal.pone.0003081. Another classic paper on this question is: An-Wen Chan et al., 'Empirical Evidence for Selective Reporting of Outcomes in Randomized Trials: Comparison of Protocols to Published Articles', *JAMA* 291, no. 20 (26 May 2004): pp. 2457–65; https://doi.org/10.1001/jama.291.20.2457

42   Winston Churchill, *The World Crisis*, Vol III, Part 1, abridged and rev. ed. Penguin Classics (London: Penguin, 2007): p.193. Quoted in Andrew Roberts, *Churchill: Walking with Destiny* (London: Allen Lane, 2018).

43   Bush promotion: Susan S. Lang, 'Wansink Accepts 14-Month Appointment as Executive Director of USDA Center for Nutrition Policy and Promotion', *Cornell Chronicle*, 20 Nov. 2007; http://news.cornell.edu/stories/2007/11/wansink-head-usda-center-nutrition-policy-and-promotion. Smarter Lunchrooms: see e.g. https://snapedtoolkit.org/interventions/programs/smarter-lunchrooms-movement-sml/

44   'The 2007 Ig Nobel Prize Winners', 4 Oct. 2007; https://www.improbable.com/ig/winners/#ig2007. The soup bowl paper is: Brian Wansink and Matthew M. Cheney, 'Super Bowls: Serving Bowl Size and Food Consumption', *JAMA* 293, no. 14 (13 April 2005): pp. 1727–28; https://doi.org/10.1001/jama.293.14.1727. It was featured and described as 'another Wansink ... masterpiece' in Richard Thaler and Cass Sunstein's influential 2008 book *Nudge*. Sunstein has since won a real Nobel Prize for economics. Richard H. Thaler & Cass R. Sunstein, *Nudge: Improving Decisions about Health, Wealth and Happiness* (New Haven: Yale University Press, 2008): p. 43.

45 Portion-size research: Wansink & Cheney, 'Super Bowls'. Shopping when hungry: Aner Tal & Brian Wansink, 'Fattening Fasting: Hungry Grocery Shoppers Buy More Calories, Not More Food', *JAMA Internal Medicine* 173, no. 12 (June 24, 2013): 1146–48; https://doi.org/10.1001/jamainternmed.2013.650. Cereal characters' eyes: Aviva Musicus et al., 'Eyes in the Aisles: Why Is Cap'n Crunch Looking Down at My Child?, *Environment and Behavior* 47, no. 7 (Aug. 2015): 715–33; https://doi.org/10.1177/0013916514528793. Wansink also made a series of videos to promote his work, some of which are available on YouTube; the following video explains the cereal study: https://www.youtube.com/watch?v=8u6xdGCIq6o. And this response explains in great detail – perhaps greater than was deserved – why the cereal study is absurd: Donald E. Simanek, 'Debunking a Shoddy "Research" Study', *Donald Simanek's Skeptical Documents and Links*, April 2014; https://www.lockhaven.edu/~dsimanek/pseudo/cartoon_eyes.htm. Elmo on apples: Brian Wansink et al., 'Can Branding Improve School Lunches?', *Archives of Pediatrics & Adolescent Medicine* 166, no. 10 (1 Oct. 2012): 967–68; https://doi.org/10.1001/archpediatrics.2012.999

46 The post has since been deleted, but the internet never forgets, and you can still access it using the Wayback Machine: http://web.archive.org/web/20170312041524/http://www.brianwansink.com/phd-advice/the-grad-student-who-never-said-no

47 Christie Aschwanden, 'We're All "P-Hacking" Now', *Wired*, 26 Nov. 2019; https://www.wired.com/story/were-all-p-hacking-now/

48 Joseph P. Simmons et al., 'False-Positive Psychology: Undisclosed Flexibility in Data Collection and Analysis Allows Presenting Anything as Significant', *Psychological Science* 22, no. 11 (Nov. 2011): pp. 1359–66; https://doi.org/10.1177/0956797611417632

49 Norbert L. Kerr, 'HARKing: Hypothesizing After the Results Are Known', *Personality and Social Psychology Review* 2, no. 3 (Aug. 1998): pp. 196–217; https://doi.org/10.1207/s15327957pspr0203_4

50 For discussion of where the idea of the Texas Sharpshooter came from, see: Barry Popik, 'Texas Sharpshooter Fallacy', *The Big Apple*, 9 March 2013; https://www.barrypopik.com/index.php/texas/entry/texas_sharpshooter_fallacy/

51 If the probability of making a false-positive error is 0.05, then the probability that we *don't* make that error (that is, we correctly declare that there's no effect) is 1 minus 0.05. The probability we don't make an error in n tests is this to the power of n, so $(1 - 0.05)^n$. It follows that the probability that we make at least one false-positive error in n tests is $1 - (1 - 0.05)^n$. So if we run five tests, we get $1 - (1 - 0.05)^5$, which is 0.226, or 22.6%. Technically, this is only true for *independent* tests – situations where the variables involved in each test aren't related to each other at all. In practice, and especially in many cases of *p*-hacking, where the same variables are being used over and over again, the increase in the false-positive rate as a function of the number of tests won't be quite as severe – but it'll still get higher and higher, so the same principle applies.

52 I should also say that there are a whole host of ways to adjust your *p*-value threshold if you've calculated a lot of them – you might only accept *p*-values that fall below 0.01 as significant instead of 0.05, for example. The problem is that most researchers forget to do this – or when they're *p*-hacking, they don't feel like they've *really* done so many tests, even if they have. There's also the interesting philosophical question of how many *p*-values a scientist should be correcting for. Every *p*-value they've calculated in that specific paper? Every *p*-value they've calculated while researching that topic? Every *p*-value they've calculated in their entire career? What about all the *p*-values they might calculate in future?

As with all interesting philosophical questions, there's no simple answer. Here's one perspective: Daniël Lakens, 'Why You Don't Need to Adjust Your Alpha Level for All Tests You'll Do in Your Lifetime', *The 20% Statistician*, 14 Feb. 2016; https://daniellakens.blogspot.com/2016/02/why-you-dont-need-to-adjust-you-alpha.html

53   This analogy for *p*-hacking is from Lee McIntyre, *The Scientific Attitude: Defending Science from Denial, Fraud, and Pseudoscience* (Cambridge, Massachusetts: The MIT Press, 2019).

54   This is hardly a new observation: here's a paper from 1969 that showed that this was the case: P. Armitage et al., 'Repeated Significance Tests on Accumulating Data', *Journal of the Royal Statistical Society*, Series A (General) 132, no. 2 (1969): pp. 235–44; https://doi.org/10.2307/2343787

55   Full disclosure: one of the researchers who looked into Wansink's papers, Nick Brown, is a colleague and friend of mine.

56   Tim van der Zee et al., 'Statistical Heartburn: An Attempt to Digest Four Pizza Publications from the Cornell Food and Brand Lab', *BMC Nutrition* 3, no. 1 (Dec. 2017): 54; https://doi.org/10.1186/s40795-017-0167-x

57   'Notice of Retraction: The Joy of Cooking Too Much: 70 Years of Calorie Increases in Classic Recipes, *Annals of Internal Medicine* 170, no. 2 (Jan. 15, 2019): p. 138; https://doi.org/10.7326/L18-0647

58   Brian Wansink et al., 'Notice of Retraction and Replacement. Wansink B, Just DR, Payne CR. Can Branding Improve School Lunches? *Arch Pediatr Adolesc Med.* 2012;166(10):967-968. Doi:10.1001/Archpediatrics.2012.999', *JAMA Pediatrics*, 21 Sept. 2017; https://doi.org/10.1001/jamapediatrics.2017.3136. The errors were first described by Nicholas J. L. Brown, 'A Different Set of Problems in an Article from the Cornell Food and Brand Lab', *Nick Brown's Blog*, 15 Feb. 2017; http://steamtraen.blogspot.com/2017/02/a-different-set-of-problems-in-article. html. Another one of Wansink's original critics has put together a dossier of all of the articles where errors have been found: Tim van der Zee, 'The Wansink Dossier: An Overview', *The Skeptical Scientist*, March 21, 2017; http://www. timvanderzee.com/the-wansink-dossier-an-overview/

59   Many others beyond the eighteen retracted have been corrected or have been given an 'expression of concern'. You can find them by searching for 'Brian Wansink' on the trusty Retraction Watch database: http://retractiondatabase. org. The Elmo paper came in for a particular kicking: after it was retracted, the journal editors allowed Wansink to publish a corrected version, but no sooner had he done so than another major error was found in the replacement (the children in the study were described as being eight to eleven years old; in fact they had been three to five). The replacement was retracted – making it the only scientific article of which I've ever heard that's been given 'science's death penalty' *twice*. Brian Wansink et al., 'Notice of Retraction. Wansink B, Just DR, Payne CR. Can Branding Improve School Lunches? *Arch Pediatr Adolesc Med.* 2012;166(10):967-968, *JAMA Pediatrics* 171, no. 12 (1 Dec. 2017): 1230; https:// doi.org/10.1001/jamapediatrics.2017.4603

60   Cornell's investigation found that Wansink had 'committed academic misconduct in his research and scholarship, including misreporting of research data, problematic statistical techniques, failure to properly document and preserve research results, and inappropriate authorship.' Michael Kotlikoff, 'Provost Issues Statement on Wansink Academic Misconduct Investigation', *Cornell Chronicle*, 20 Sept. 2018; http://news.cornell.edu/stories/2018/09/provost-issues-statement-wansink-academic-misconduct-investigation

61 Stephanie M. Lee, 'Here's How Cornell Scientist Brian Wansink Turned Shoddy Data Into Viral Studies About How We Eat', *BuzzFeed News*, 25 Feb. 2018; https://www.buzzfeednews.com/article/stephaniemlee/brian-wansink-cornell-p-hacking

62 Alas, in many cases they aren't delicate about making their requests at all, as the researchers behind the Bullied into Bad Science initiative will attest; http://bulliedintobadscience.org/

63 Dana R. Carney, 'My Position on "Power Poses"', 26 Sept. 2016; http://faculty.haas.berkeley.edu/dana_carney/pdf_My%20position%20on%20power%20poses.pdf

64 https://twitter.com/nicebread303/status/780395235268501504; https://twitter.com/PeteEtchells/status/780425109077106692; https://twitter.com/cragcrest/status/780447545126293504; https://twitter.com/timothycbates/status/780386384276230144; https://twitter.com/MichelleNMeyer/status/780437722393698305; https://twitter.com/eblissmoreau/status/780594280377176064

65 Quoted in Jesse Singal & Melissa Dahl, 'Here Is Amy Cuddy's Response to Critiques of Her Power-Posing Research', *The Cut*, 30 Sept. 2016; https://www.thecut.com/2016/09/read-amy-cuddys-response-to-power-posing-critiques.html

66 Leslie K. John et al., 'Measuring the Prevalence of Questionable Research Practices with Incentives for Truth Telling', *Psychological Science* 23, no. 5 (May 2012): pp. 524–32; https://doi.org/10.1177/0956797611430953. I've used numbers that are halfway between the 'self-admission rate' in the control group and the experimental group (see their Table 1), the latter of whom were told money would be donated to charity if they told the truth, and showed higher admission rates for the dodgy research practices. That study was based on psychologists in the US, but similar results have been found in Italy. See Franca Agnoli et al., 'Questionable Research Practices among Italian Research Psychologists', *PLOS ONE* 12, no. 3 (15 Mar. 2017): e0172792; https://doi.org/10.1371/journal.pone.0172792. See also this survey from Germany, which took issue with some of the wording of the questions in the original US study: Klaus Fiedler & Norbert Schwarz, 'Questionable Research Practices Revisited', *Social Psychological and Personality Science* 7, no. 1 (Jan. 2016): pp. 45–52; https://doi.org/10.1177/1948550615612150

67 Min Qi Wang et al., 'Identifying Bioethical Issues in Biostatistical Consulting: Findings from a US National Pilot Survey of Biostatisticians', *BMJ Open* 7, no. 11 (Nov. 2017): e018491; https://doi.org/10.1136/bmjopen-2017-018491. I've added together the percentages for statisticians who reported having these requests made of them between one and nine times, and over ten times, in the past five years.

68 Sarah Necker, 'Scientific Misbehavior in Economics', *Research Policy* 43, no. 10 (Dec. 2014): pp. 1747–59; https://doi.org/10.1016/j.respol.2014.05.002. In that same survey, an eye-opening 2 per cent of economists also admitted to 'accept[ing] or offer[ing] sex in exchange for (co-)authorship, access to data, or promotion of particular persons.'

69 E. J. Masicampo & Daniel R. Lalande, 'A Peculiar Prevalence of p Values Just Below .05', *Quarterly Journal of Experimental Psychology* 65, no. 11 (Nov. 2012): pp. 2271–79; https://doi.org/10.1080/17470218.2012.711335. See also Adrian Gerard Barnett and Jonathan D. Wren, 'Examination of CIs in Health and Medical Journals from 1976 to 2019: An Observational Study', *BMJ Open* 9, no. 11 (Nov. 2019): e032506; https://doi.org/10.1136/bmjopen-2019-032506. This looked at the same issue using confidence intervals, another statistical method that essentially gives the same information as a *p*-value. Incidentally, this same

kind of pattern is found in graphs of school exam results, where there's a sudden bump in grades just above the arbitrary pass-mark: well-meaning teachers have revised failing children's grades upwards just enough so they pass the test. See 'Another Case of Teacher Cheating, or Is It Just Altruism?', https://freakonomics.com/2011/07/07/another-case-of-teacher-cheating-or-is-it-just-altruism/ (7 July 2011).

70 R. Silberzahn et al., 'Many Analysts, One Data Set: Making Transparent How Variations in Analytic Choices Affect Results', *Advances in Methods and Practices in Psychological Science* 1, no. 3 (Sept. 2018): pp. 337–56; https://doi.org/10.1177/2515245917747646; Justin F. Landy et al., 'Crowdsourcing Hypothesis Tests: Making Transparent How Design Choices Shape Research Results', *Psychological Bulletin* (16 Jan. 2020); https://doi.org/10.1037/bul0000220

71 Tal Yarkoni & Jacob Westfall, 'Choosing Prediction Over Explanation in Psychology: Lessons from Machine Learning', *Perspectives on Psychological Science* 12, no. 6 (Nov. 2017): pp. 1100–1122, p. 1104; https://doi.org/10.1177/1745691617693393

72 Andrew Gelman & Eric Loken, 'The Garden of Forking Paths: Why Multiple Comparisons can be a Problem, Even When There is no "Fishing Expedition" or "*p*-Hacking" and the Research Hypothesis was Posited Ahead of Time', unpublished, 4 Nov. 2013; http://www.stat.columbia.edu/~gelman/research/unpublished/p_hacking.pdf. And Jorge Luis Borges, 'The Garden of Forking Paths', *Labyrinths*, tr. Donald A. Yates (New York: New Directions, 1962, 1964).

73 This framing of the *p*-hacking problem is due to Yarkoni & Westfall, who call *p*-hacking 'procedural overfitting': Yarkoni & Westfall, 'Choosing Prediction', p. 1103.

74 Roger Giner-Sorolla, 'Science or Art? How Aesthetic Standards Grease the Way Through the Publication Bottleneck but Undermine Science', *Perspectives on Psychological Science* 7, no. 6 (Nov. 2012): pp. 567-571; https://doi.org/10.1177/1745691612457576

75 Ernest Hugh O'Boyle et al., 'The Chrysalis Effect: How Ugly Initial Results Metamorphosize Into Beautiful Articles', *Journal of Management* 43, no. 2 (Feb. 2017): pp. 376–99; https://doi.org/10.1177/0149206314527133. The chrysalis effect has been replicated more recently when looking at psychology research: Athena H. Cairo et al., 'Gray (Literature) Matters: Evidence of Selective Hypothesis Reporting in Social Psychological Research', *Personality and Social Psychology Bulletin*, 24 Feb. 2020; https://doi.org/10.1177/0146167220903896

76 One of the classic examples of this bad advice is a chapter in an edited book by none other than Daryl Bem (of the psychic study that I failed to replicate, discussed in the Preface). He implores young academics to 'go on a fishing expedition for something – anything – interesting' in their data. Sure, he argues, this might in some cases lead to false positives, but we should still 'err on the side of discovery'. Daryl J. Bem, 'Writing the Empirical Journal Article', in *The Compleat Academic: A Career Guide*, eds. John M. Darley, Mark P. Zanna, and Henry L. Roediger III, 2nd ed., pp. 171–201 (Washington, DC: American Psychological Association, 2003): p.172.

77 Sabine Hossenfelder, *Lost in Math: How Beauty Leads Physics Astray* (New York: Basic Books, 2018). Hossenfelder also argues that tricky philosophical and theoretical problems in physics are being '[swept] under the rug,' as physicists favour 'questions that are more likely to produce publishable results in a short period of time' (p.194). This is an issue we'll encounter more broadly later in the book. For further discussion of problems in physics, see Lee Smolin, *The Trouble with Physics: The Rise of String Theory, the Fall of a Science and What*

Comes Next (London: Allen Lane, 2007) and Peter Woit, *Not Even Wrong: The Failure of String Theory and the Continuing Challenge to Unify the Laws of Physics* (London: Vintage Books, 2007).

78 Catherine De Angelis et al., 'Clinical Trial Registration: A Statement from the International Committee of Medical Journal Editors', *New England Journal of Medicine* 351, no. 12 (16 Sept. 2004): pp. 1250–51; https://doi.org/10.1056/NEJMe048225

79 We'll discuss more about pre-registering studies in Chapter 8.

80 http://compare-trials.org

81 Goldacre's team also tried to get letters published in the journals that showed that the trials hadn't accurately reported their findings. Most editors weren't interested. Ben Goldacre, 'Make Journals Report Clinical Trials Properly', *Nature* 530, no. 7588 (Feb. 2016): p. 7; https://doi.org/10.1038/530007a

82 Philip M. Jones et al., 'Comparison of Registered and Reported Outcomes in Randomized Clinical Trials Published in Anesthesiology Journals', *Anesthesia & Analgesia* 125, no. 4 (Oct. 2017): pp. 1292–1300; https://doi.org/10.1213/ANE.0000000000002272; see also Douglas G. Altman et al., 'Harms of Outcome Switching in Reports of Randomised Trials: CONSORT Perspective', *BMJ* (14 Feb. 2017): j396; https://doi.org/10.1136/bmj.j396

83 For book-length treatments of the problems with medical trials, see Ben Goldacre, *Bad Pharma: How Drug Companies Mislead Doctors and Harm Patients* (London: Fourth Estate, 2012) and Richard F. Harris, *Rigor Mortis: How Sloppy Science Creates Worthless Cures, Crushes Hope, and Wastes Billions* (New York: Basic Books, 2017).

84 This has been called the 'Garbage In, Garbage Out' principle in meta-analysis. Morton Hunt, *How Science Takes Stock: The Story of Meta-Analysis* (New York: Russell Sage Foundation, 1998).

85 The absolute number of industry-funded trials has increased over time, although their proportion of all trials has gone down in size. Stephan Ehrhardt et al., 'Trends in National Institutes of Health Funding for Clinical Trials Registered in ClinicalTrials.Gov', *JAMA* 314, no. 23 (15 Dec. 2015): pp. 2566–67; https://doi.org/10.1001/jama.2015.12206

86 A potential retort is that pharmaceutical companies have enough money and other resources to pay for better quality trials, so we'd expect the results to be different. But if the trials were higher quality, it would mean they wouldn't have as much bias and would thus be *less* likely to have positive results, since we know that the main biases in such studies tend to lead to more false positives. (If a treatment works, an unbiased trial would show just that, but a biased study would both exaggerate the size of the effect and potentially find some false-positive results as well.) Also, the recent review study controlled for sample size – one of the things that industry trials, given their greater levels of funding, can often do better than those funded by other sources – and still found that industry-funded trials were more likely to get positive results. Stig Waldorff, 'Results of Clinical Trials Sponsored by For-Profit vs Nonprofit Entities', *JAMA* 290, no. 23 (17 Dec. 2003): p. 3071; https://doi.org/10.1001/jama.290.23.3071-a

87 D. N. Lathyris et al., 'Industry Sponsorship and Selection of Comparators in Randomized Clinical Trials', *European Journal of Clinical Investigation* 40, no. 2 (Feb. 2010): pp. 172–82; https://doi.org/10.1111/j.1365-2362.2009.02240.x. And Candice Estellat, 'Lack of Head-to-Head Trials and Fair Control Arms: Randomized Controlled Trials of Biologic Treatment for Rheumatoid Arthritis', *Archives of Internal Medicine* 172, no. 3 (13 Feb. 2012): pp. 237–44; https://doi.org/10.1001/archinternmed.2011.1209

88 C. W. Jones, L. Handler et al., 'Non-Publication of Large Randomized Clinical Trials: Cross Sectional Analysis', *BMJ* 347, no. oct28 9 (29 Oct. 2013): f6104; https://doi.org/10.1136/bmj.f6104. It should be said, though, that outcome switching doesn't seem to be systematically worse in trials paid for by drug companies (Christopher W. Jones et al., 'Primary Outcome Switching among Drug Trials with and without Principal Investigator Financial Ties to Industry: A Cross-Sectional Study', *BMJ Open* 8, no. 2 (Feb. 2018): e019831; https://doi.org/10.1136/bmjopen-2017-019831). One review even found outcome-switching to be worse in non-profit-funded research: Alberto Falk Delgado & Anna Falk Delgado, 'Outcome Switching in Randomized Controlled Oncology Trials Reporting on Surrogate Endpoints: A Cross-Sectional Analysis', *Scientific Reports* 7, no. 1 (Dec. 2017): 9206; https://doi.org/10.1038/s41598-017-09553-y.

89 At a lecture series I sometimes attend at my university, there's a lunch provided and it's always clearly marked as a 'non-pharma lunch'. This is how scared medical researchers are of inadvertently accepting any gifts from pharmaceutical companies – gifts that they'd have to declare in every single 'Conflict of Interest' section from that point on.

90 See Tom Chivers, 'Does Psychology Have a Conflict-of-Interest Problem?', *Nature* 571, no. 7763 (July 2019): pp. 20–23; https://doi.org/10.1038/d41586-019-02041-5

91 See Lisa A. Bero & Quinn Grundy, 'Why Having a (Nonfinancial) Interest is Not a Conflict of Interest', *PLOS Biology* 14, no. 12 (21 Dec. 2016): e2001221; https://doi.org/10.1371/journal.pbio.2001221. A counter-argument to mine, Bero and Grundy argue that financial conflicts of interest are a different kind of thing from intellectual ones and that it 'muddies the waters' to conflate the two. Either way, next time I write a scientific paper that has anything to do with the replication crisis, should I have to note something to the effect of 'I wrote a book about the replication crisis and it would be a bit awkward for me if it turned out that science was actually in completely fine shape'? There's genuinely a strong argument for doing just that.

92 Sharon Begley, 'The Maddening Saga of How an Alzheimer's "Cabal" Thwarted Progress toward a Cure for Decades', *STAT News*, 25 June 2019; https://www.statnews.com/2019/06/25/alzheimers-cabal-thwarted-progress-toward-cure/

93 Yan-Mei Huang et al., 'Major Clinical Trials Failed the Amyloid Hypothesis of Alzheimer's Disease', *Journal of the American Geriatrics Society* 67, no. 4 (April 2019): pp. 841–44; https://doi.org/10.1111/jgs.15830; and Francesco Panza et al., 'A Critical Appraisal of Amyloid-β-Targeting Therapies for Alzheimer Disease', *Nature Reviews Neurology* 15, no. 2 (Feb. 2019): pp. 73–88; https://doi.org/10.1038/s41582-018-0116-6

94 Karl Herrup, 'The Case for Rejecting the Amyloid Cascade Hypothesis', *Nature Neuroscience* 18, no. 6 (June 2015): pp. 794–99; https://doi.org/10.1038/nn.4017

95 Judith R. Harrison & Michael J. Owen, 'Alzheimer's disease: The amyloid hypothesis on trial', *British Journal of Psychiatry* 208, no. 1 (Jan. 2016): pp. 1–3; http://doi.org/10.1192/bjp.bp.115.167569

96 G. McCartney et al., 'Why the Scots Die Younger: Synthesizing the Evidence', *Public Health* 126, no. 6 (June 2012): pp. 459-470, p. 467; https://doi.org/10.1016/j.puhe.2012.03.007. There's discussion of this unusual conflict of interest statement here: G. L. McCartney et al., 'When Do Your Politics Become a Competing Interest?', *BMJ* 342 (25 Jan. 2011): d269; https://doi.org/10.1136/bmj.d269

97 If not necessarily for his political beliefs. Since we're in the mood for full disclosure, my own politics are: very high social liberalism, moderately high economic liberalism. If it helps, I'm usually somewhere on the left-hand side of the bottom-right quadrant on www.politicalcompass.org

98   José L. Duarte et al., 'Political Diversity Will Improve Social Psychological
      Science', *Behavioral and Brain Sciences* 38 (2015): e130; https://doi.org/10.1017/
      S0140525X14000430

99   In one survey study, there was a correlation between how liberal psychologists
      were and how much they were willing to discriminate against perceived con-
      servatives when reviewing papers or making decisions on who to hire. See Yoel
      Inbar & Joris Lammers, 'Political Diversity in Social and Personality Psychology',
      Perspectives on Psychological Science 7, no. 5 (Sept. 2012): pp. 496–503; https://
      doi.org/10.1177/1745691612448792. It's interesting to note, though, that a 2019
      study found no difference in the replicability of studies that were rated as 'liberal'
      or 'conservative' (in terms of who would like the conclusions best). Diego A.
      Reinero et al., 'Is the Political Slant of Psychology Research Related to Scientific
      Replicability?', preprint, *PsyArXiv* (7 Feb. 2019); https://doi.org/10.31234/osf.
      io/6k3j5

100   Lee Jussim, 'Is Stereotype Threat Overcooked, Overstated, and Oversold?',
      Rabble Rouser, 30 Dec. 2015; https://www.psychologytoday.com/gb/blog/
      rabble-rouser/201512/is-stereotype-threat-overcooked-overstated-and-oversold

101   Paulette C. Flore & Jelte M. Wicherts, 'Does Stereotype Threat Influence
      Performance of Girls in Stereotyped Domains? A Meta-Analysis', *Journal of
      School Psychology* 53, no. 1 (Feb. 2015): pp. 25–44; https://doi.org/10.1016/j.
      jsp.2014.10.002 and Paulette C. Flore et al., 'The Influence of Gender
      Stereotype Threat on Mathematics Test Scores of Dutch High School Students:
      A Registered Report', *Comprehensive Results in Social Psychology* 3, no. 2
      (4 May 2018): pp. 140–74; https://doi.org/10.1080/23743603.2018.15596
      47. For more evidence on publication bias in stereotype threat studies, see
      Oren R. Shewach et al., 'Stereotype Threat Effects in Settings with Features
      Likely versus Unlikely in Operational Test Settings: A Meta-Analysis', *Journal
      of Applied Psychology* 104, no. 12 (Dec. 2019): pp. 1514–34; https://doi.
      org/10.1037/apl0000420

102   Subsequently, the same authors ran a fully preregistered, large-scale experiment
      on stereotype threat and the sex difference in mathematics performance. It
      found no effect of stereotype threat. Flore et al., 'Gender Stereotype Threat'.

103   Corinne A. Moss-Racusin et al., 'Gender Bias Produces Gender Gaps in
      STEM Engagement', *Sex Roles* 79, no. 11–12 (Dec. 2018): pp. 651–70;
      https://doi.org/10.1007/s11199-018-0902-z. For a range of others, see *The
      Underrepresentation of Women in Science: International and Cross-Disciplinary
      Evidence and Debate*, eds., Stephen J. Ceci et al., (Frontiers Research Topics:
      Frontiers Media SA, 2018); https://doi.org/10.3389/978-2-88945-434-1

104   Jill B. Becker et al, 'Female Rats Are Not More Variable than Male Rats: A
      Meta-Analysis of Neuroscience Studies', *Biology of Sex Differences* 7, no. 1
      (Dec. 2016): 34; https://doi.org/10.1186/s13293-016-0087-5

105   International Mouse Phenotyping Consortium, Natasha A. Karp, et al., 'Prevalence
      of Sexual Dimorphism in Mammalian Phenotypic Traits', *Nature Communi-
      cations* 8, no. 1 (Aug. 2017): 15475; https://doi.org/10.1038/ncomms15475

106   Rebecca M. Shansky, 'Are Hormones a "emale Problem" for Animal Research?',
      Science 364, no. 6443 (31 May 2019): pp. 825–6; https://doi.org/10.1126/sci-
      ence.aaw7570, p. 826. As Shansky explains, many funders and journals now
      mandate researchers to include both males and females in their experiments.
      See also Janine Austin Clayton, 'Applying the New SABV (Sex as a Biological
      Variable) Policy to Research and Clinical Care', *Physiology & Behavior* 187
      (April 2018): pp. 2–5; https://doi.org/10.1016/j.physbeh.2017.08.012

107   Cordelia Fine, *Testosterone Rex: Unmaking the Myths of Sex of Our Gendered
      Minds* (London: Icon Books, 2017).

108   Cordelia Fine, 'Feminist Science: Who Needs It?', *Lancet* 392, no. 10155 (Oct. 2018): pp. 1302–3; https://doi.org/10.1016/S0140-6736(18)32400-0

109   Ibid. p. 1303. This is a version of 'Standpoint Theory', a philosophical position that has its roots in Karl Marx's writings, and emphasises that a person's own identity or experiences (for Marx, it was the identity and experience of the working class) are what shapes their view on reality, and that we should listen particularly to the perspectives of marginalised people, whose views we might otherwise miss. See the section on Standpoint Theory in Elizabeth Anderson, 'Feminist Epistemology and Philosophy of Science', *The Stanford Encyclopedia of Philosophy*, ed. Edward N. Zalta (Spring 2020 Edition); https://plato.stanford.edu/archives/spr2020/entries/feminism-epistemology

110   Jason E. Lewis et al., 'The Mismeasure of Science: Stephen Jay Gould versus Samuel George Morton on Skulls and Bias', *PLOS Biology* 9, no. 6 (7 June 2011): e1001071; https://doi.org/10.1371/journal.pbio.1001071. The new measurements weren't made with seeds or shot, but with tiny acrylic balls. This incidentally shows that Morton's nineteenth century methods weren't too far off the twenty-first century gold standard. There had been one previous remeasuring of the skulls in 1988 – an analysis that drew conclusions broadly sympathetic to Morton's measurements, if not his racial conclusions. John S. Michael, 'A New Look at Morton's Craniological Research', *Current Anthropology* 29, no. 2 (April 1988): pp. 349–54; https://doi.org/10.1086/203646

111   'My original reasons for writing *The Mismeasure of Man* mixed the personal with the professional. I confess, first of all, to strong feelings on this particular issue. I grew up in a family with a tradition of participation in campaigns for social justice'. Gould, *The Mismeasure of Man*, p. 36.

112   Lewis et al., 'The Mismeasure of Science', p. 6.

113   Michael Weisberg, 'Remeasuring Man', *Evolution & Development* 16, no. 3 (May 2014): pp. 166–78; https://doi.org/10.1111/ede.12077. See also Michael Weisberg & Diane B. Paul, 'Morton, Gould, and Bias: A Comment on "The Mismeasure of Science"', ed. David Penny, *PLOS Biology* 14, no. 4 (19 April 2016): e1002444; https://doi.org/10.1371/journal.pbio.1002444

114   Mitchell, 'The Fault in his Seeds'.

115   Jonathan Michael Kaplan et al., 'Gould on Morton, Redux: What Can the Debate Reveal about the Limits of Data?', *Studies in History and Philosophy of Science Part C: Studies in History and Philosophy of Biological and Biomedical Sciences* 52 (Aug. 2015): pp. 22–31; https://doi.org/10.1016/j.shpsc.2015.01.001. The point is also made by Joseph L. Graves, 'Great Is Their Sin: Biological Determinism in the Age of Genomics', *Annals of the American Academy of Political and Social Science* 661, no. 1 (Sept. 2015): pp. 24–50; https://doi.org/10.1177/0002716215586558

116   Of course, racism is by far the most pernicious of these in terms of societal effects. My point here is that they can all bias our view of scientific results.

5: *Negligence*

Epigraph: Charles Caleb Cotton, *Lacon, or Many Things in Few Words* (London, 1820).

1   Daniel Hirschman, 'Stylized Facts in the Social Sciences', *Sociological Science* 3 (2016): pp. 604–26; https://doi.org/10.15195/v3.a26

2   The study was an online 'working paper' for a while (as is normal in econom-
     ics, as we'll see in the final chapter), but it was eventually officially published
     as Carmen M. Reinhart and Kenneth S. Rogoff, 'Growth in a Time of Debt',
     American Economic Review 100, no. 2 (May 2010): pp. 573–78; https://doi.
     org/10.1257/aer.100.2.573

3   Osborne: George Osborne, 'Mais Lecture – A New Economic Model', 24
     Feb. 2010; https://conservative-speeches.sayit.mysociety.org/speech/601526;
     Republican members: United States Senate Committee on the Budget, 'Sessions,
     Ryan Issue Joint Statement On Jobs Report, Call For Senate Action On Budget',
     8 July 2011; https://www.budget.senate.gov/chairman/newsroom/press/sessions-
     ryan-issue-joint-statement-on-jobs-report-call-for-senate-action-on-budget

4   Paul Krugman, 'How the Case for Austerity Has Crumbled', *New York
     Review of Books*, 6 June 2013; https://www.nybooks.com/articles/2013/06/06/
     how-case-austerity-has-crumbled/

5   Thomas Herndon et al., 'Does High Public Debt Consistently Stifle Economic
     Growth? A Critique of Reinhart and Rogoff', *Cambridge Journal of Economics*
     38, no. 2 (April 2013): pp. 257–79; https://doi.org/10.1093/cje/bet075

6   Reinhart and Rogoff admitted the Excel error, though they didn't agree with the
     critics on many of their other points: Carmen M. Reinhart & Kenneth S. Rogoff,
     'Reinhart-Rogoff Response to Critique', *Wall Street Journal*, 16 April 2013;
     https://blogs.wsj.com/economics/2013/04/16/reinhart-rogoff-response-to-
     critique/

7   Herndon et al., 'High Public Debt', p. 14.

8   Betsey Stevenson & Justin Wolfers, 'Refereeing Reinhart-Rogoff Debate',
     Bloomberg Opinion, 28 April 2013; https://www.bloomberg.com/opinion/
     articles/2013-04-28/refereeing-the-reinhart-rogoff-debate

9   Michèle B. Nuijten, 'statcheck – a Spellchecker for Statistics', *LSE Impact of Social
     Sciences*, 28 Feb. 2018; https://blogs.lse.ac.uk/impactofsocialsciences/2018/02/28/statch-
     eck-a-spellchecker-for-statistics/. You can find the statcheck app at: http://statcheck.io/

10   Michèle B. Nuijten et al., 'The Prevalence of Statistical Reporting Errors in
     Psychology (1985–2013)', *Behavior Research Methods* 48, no. 4 (Dec. 2016):
     pp. 1205–26; https://doi.org/10.3758/s13428-015-0664-2. It should be noted
     that statcheck has its critics – see Thomas Schmidt, 'Statcheck Does Not Work:
     All the Numbers. Reply to Nuijten et al. (2017)', *PsyArXiv* (preprint), 22 Nov.
     2017; https://doi.org/10.31234/osf.io/hr6qy

11   Nicholas J. L. Brown & James A. J. Heathers, 'The GRIM Test: A Simple Technique
     Detects Numerous Anomalies in the Reporting of Results in Psychology', *Social
     Psychological and Personality Science* 8, no. 4 (May 2017): pp. 363–69; https://
     doi.org/10.1177/1948550616673876

12   You can try it with a calculator, or use the app at: http://nickbrown.fr/GRIM

13   Leon Festinger & James M. Carlsmith, 'Cognitive Consequences of Forced
     Compliance', *Journal of Abnormal and Social Psychology* 58, no. 2 (1959):
     pp. 203–10; https://doi.org/10.1037/h0041593

14   There was actually a third group, who were paid $20. They reported finding the
     task as tedious as did those who weren't paid anything – supposedly because
     they had reduced their cognitive dissonance by thinking of all that nice money,
     rather than by altering their beliefs.

15   Matti Heino, 'The Legacy of Social Psychology', *Data Punk*, 13 Nov. 2016;
     https://mattiheino.com/2016/11/13/legacy-of-psychology/

16   As of January 2020, the paper has over 4,200 citations, according to Google
     Scholar.

17 Carlisle (2012), *Anaesthesia*. See also this profile of Carlisle: David Adam, 'How a Data Detective Exposed Suspicious Medical Trials', *Nature* 571, no. 7766 (July 2019): pp. 462–64; https://doi.org/10.1038/d41586-019-02241-z

18 See J. M. Kendall, 'Designing a Research Project: Randomised Controlled Trials and Their Principles', *Emergency Medicine Journal* 20, no. 2 (1 March 2003): pp. 164–68; https://doi.org/10.1136/emj.20.2.164

19 J. B. Carlisle (2012), 'The Analysis of 168 Randomised Controlled Trials to Test Data Integrity: Analysis of 168 Randomised Controlled Trials to Test Data Integrity', *Anaesthesia* 67, no. 5 (May 2012): pp. 521–37; https://doi.org/10.1111/j.1365-2044.2012.07128.x

20 J. B. Carlisle, 'Data Fabrication and Other Reasons for Non-Random Sampling in 5087 Randomised, Controlled Trials in Anaesthetic and General Medical Journals', *Anaesthesia* 72, no. 8 (Aug. 2017): pp. 944–52; https://doi.org/10.1111/anae.13938. One of Carlisle's main objectives was to test whether dodgy-looking trials in anaesthesia were any worse than those in other medical fields. He concluded that the mistakes were just as bad in anaesthesia as in non-anaesthesia research.

21 Not everyone was impressed by Carlisle's method: the editors of *Anesthesiology* wrote a bitingly critical piece pointing out some flaws in Carlisle's statistics, and admonishing him for implying that fraud, rather than error, was the chief cause of the randomisation failures (Evan D. Kharasch & Timothy T. Houle, 'Errors and Integrity in Seeking and Reporting Apparent Research Misconduct', *Anesthesiology* 127, no. 5 (Nov. 2017): pp. 733–37; https://doi.org/10.1097/ALN.0000000000001875). Carlisle has responded, I think fairly convincingly (J. B. Carlisle, 2018, 'Seeking and Reporting Apparent Research Misconduct: Errors and Integrity – a Reply', *Anaesthesia* 73, no. 1 (Jan. 2018): pp. 126–28; https://doi.org/10.1111/anae.14148), but it's another interesting instance of the watchers themselves needing to be watched. In any case, in the next chapter we'll see that the method really did result in randomisation failures being discovered in an extremely important trial from nutrition research, so the Carlisle method can't have been *completely* off-base.

22 Jelte M. Wicherts et al., 'The Poor Availability of Psychological Research Data for Reanalysis', *American Psychologist* 61, no. 7 (2006): pp. 726–28; https://doi.org/10.1037/0003-066X.61.7.726. See also Caroline J. Savage & Andrew J. Vickers, 'Empirical Study of Data Sharing by Authors Publishing in PLoS Journals', *PLOS ONE* 4, no. 9 (18 Sept. 2009): e7078; https://doi.org/10.1371/journal.pone.0007078. And Carol Tenopir et al., 'Data Sharing by Scientists: Practices and Perceptions', *PLOS ONE* 6, no. 6 (29 June 2011): e21101; https://doi.org/10.1371/journal.pone.0021101. And Garret Christensen & Edward Miguel, 'Transparency, Reproducibility, and the Credibility of Economics Research' (Cambridge, MA: National Bureau of Economic Research, Dec. 2016); https://doi.org/10.3386/w22989. For the point about data becoming less available with increasing age, see Timothy H. Vines et al., 'The Availability of Research Data Declines Rapidly with Article Age', *Current Biology* 24, no. 1 (Jan. 2014): pp. 94–97; https://doi.org/10.1016/j.cub.2013.11.014

23 American Type Culture Collection Standards Development Organization Workgroup ASN-0002, 'Cell Line Misidentification: The Beginning of the End', *Nature Reviews Cancer* 10, no. 6 (June 2010): pp. 441-48; https://doi.org/10.1038/nrc2852, see the timeline on p. 444.

24 Colon cancer: 'Retraction: Critical Role of Notch Signaling in Osteosarcoma Invasion and Metastasis', *Clinical Cancer Research* 19, no. 18 (15 Sept. 2013);

pp. 5256–57; https://doi.org/10.1158/1078-0432.CCR-13-1914; pigs: E. Milanesi et al., 'Molecular Detection of Cell Line Cross-Contaminations Using Amplified Fragment Length Polymorphism DNA Fingerprinting Technology', *In Vitro Cellular & Developmental Biology – Animal* 39, no. 3–4 (March 2003): pp. 124–30; https://doi.org/10.1007/s11626-003-0006-z; rats: Janyaporn Phuchareon et al., 'Genetic Profiling Reveals Cross-Contamination and Misidentification of 6 Adenoid Cystic Carcinoma Cell Lines: ACC2, ACC3, ACCM, ACCNS, ACCS and CAC2', *PLOS ONE* 4, no. 6 (25 June 2009): e6040; https://doi.org/10.1371/journal.pone.0006040

25 American Type Culture Collection Standards Development Organization Workgroup ASN-0002, 'Cell Line Misidentification'.

26 Serge P. J. M. Horbach & Willem Halffman, 'The Ghosts of HeLa: How Cell Line Misidentification Contaminates the Scientific Literature', *PLOS ONE* 12, no. 10 (12 Oct. 2017): e0186281; https://doi.org/10.1371/journal.pone.0186281. Numbers translate to around 0.8 per cent of all papers on cells using misidentified lines, but 10 per cent of papers that contain a reference to one of the problematic studies (excluding self-citations).

27 Yaqing Huang et al., 'Investigation of Cross-Contamination and Misidentification of 278 Widely Used Tumor Cell Lines', *PLOS ONE* 12, no. 1 (20 Jan. 2017): e0170384; https://doi.org/10.1371/journal.pone.0170384

28 85 per cent contaminated: Fang Ye et al., 'Genetic Profiling Reveals an Alarming Rate of Cross-Contamination among Human Cell Lines Used in China', *The FASEB Journal* 29, no. 10 (Oct. 2015): pp. 4268–72; https://doi.org/10.1096/fj.14-266718; see also Xiaocui Bian et al., 'A Combination of Species Identification and STR Profiling Identifies Cross-Contaminated Cells from 482 Human Tumor Cell Lines', *Scientific Reports* 7, no. 1 (Dec. 2017): 9774; https://doi.org/10.1038/s41598-017-09660-w. You can find a depressing register of all the known misidentified cell lines – 529 cell lines at the time of writing – at the following link: https://iclac.org/databases/cross-contaminations/. Incidentally, the 'STAP' stem cells we encountered in our discussion of Haruko Obokata in Chapter 3 were – aside from the faked pictures and other issues – also misidentified, coming from a different kind of mouse than that described in the article. Haruko Obokata et al., 'Retraction Note: Bidirectional Developmental Potential in Reprogrammed Cells with Acquired Pluripotency', *Nature* 511, no. 7507 (July 2014): p. 112; https://doi.org/10.1038/nature13599

29 Horbach & Halffmann, 'The Ghosts of HeLa'.

30 Editorial, 'Towards What Shining City, Which Hill?', *Nature* 289, no. 5795 (Jan. 1981): p. 212; https://doi.org/10.1038/289211a0

31 Christopher Korch & Marileila Varella-Garcia, 'Tackling the Human Cell Line and Tissue Misidentification Problem Is Needed for Reproducible Biomedical Research', *Advances in Molecular Pathology* 1, no. 1 (Nov. 2018): pp. 209–228, e36; https://doi.org/10.1016/j.yamp.2018.07.003

32 2010: American Type Culture Collection Standards Development Organization Workgroup ASN-0002, 'Cell Line Misidentification'; 2012: John R. Masters, 'End the Scandal of False Cell Lines', *Nature* 492, no. 7428 (Dec. 2012): p. 186; https://doi.org/10.1038/492186a; 2015: 'Announcement: Time to Tackle Cells' Mistaken Identity', *Nature* 520, no. 7547 (April 2015): p. 264; https://doi.org/10.1038/520264a; 2017: Norbert E. Fusenig, 'The Need for a Worldwide Consensus for Cell Line Authentication: Experience Implementing a Mandatory Requirement at the International Journal of Cancer', *PLOS Biology* 15, no. 4 (17 April 2017): e2001438; https://doi.org/10.1371/journal.pbio.2001438; 2018:

Jaimee C. Eckers et al., 'Identity Crisis – Rigor and Reproducibility in Human Cell Lines', *Radiation Research* 189, no. 6 (June 2018): pp. 551–52; https://doi.org/10.1667/RR15086.1

33 Korch & Varella-Garcia, 'Tackling the Human Cell Line'.

34 Of course, many people also believe that it's immoral to experiment on animals in general. When there's no choice but to do *some* research on animals, scientists subscribe to a set of principles to make their research as ethical as possible. These are known as the 'Three Rs' – Replacement (trying to use something other than animals in research – for example, humans, who can normally consent), Reduction (trying to get as much useful information from as few animals as possible) and Refinement (making sure that the animals' welfare is as high as possible while they're being used in research). The term was first coined here: William M. S. Russell & Rex L. Burch, *The Principles of Humane Experimental Technique*, Special ed. (Potters Bar: UFAW, 1992). Much more information about these principles and how scientists are trying to meet them can be found on the website of the UK's National Centre for the Three 3 Rs, at https://www.nc3rs.org.uk/

35 Malcolm R. Macleod et al., 'Risk of Bias in Reports of In Vivo Research: A Focus for Improvement', *PLOS Biology* 13, no. 10 (13 Oct. 2015): e1002273; https://doi.org/10.1371/journal.pbio.1002273

36 See Jennifer A. Hirst et al., 'The Need for Randomization in Animal Trials: An Overview of Systematic Reviews', *PLOS ONE* 9, no. 6 (6 June 2014): e98856; https://doi.org/10.1371/journal.pone.0098856

37 Of course, blinding is also particularly important in human medical trials, where the *participants* should be blinded to which treatment they're getting, lest their expectations interfere with the results of the study. In this case, where both the researcher and the participant lack information that could bias the results, the experiment is referred to as a 'double-blind' study. When the subjects of the study aren't humans, ensuring that they're blinded as to whether they're getting the real treatment or a placebo is obviously less of a priority (though it could still be a consideration in some cases).

38 Macleod and colleagues also looked at the inclusion of conflict-of-interest statements, which we covered in Chapter 4.

39 Malcolm R. Macleod et al., 'Evidence for the Efficacy of NXY-059 in Experimental Focal Cerebral Ischaemia Is Confounded by Study Quality', *Stroke* 39, no. 10 (Oct. 2008): pp. 2824–29; https://doi.org/10.1161/STROKEAHA.108.515957

40 The classic real-life case to illustrate this point, used in many statistics textbooks, is that of the 1936 US presidential election. The *Literary Digest* magazine ran a massive poll, sampling 2 million people, but failed to collect a *random* sample because they contacted their participants via telephone. At the time only the well-to-do had telephones at home, so their sample was biased and they got the election completely wrong, predicting that the Republican candidate Alf Landon would crush Franklin D. Roosevelt. FDR got 61 per cent of the vote, and the *Literary Digest* folded soon after. See Sharon L. Lohr and J. Michael Brick, 'Roosevelt Predicted to Win: Revisiting the 1936 Literary Digest Poll', *Statistics, Politics and Policy* 8, no. 1, 26 Jan. 2017; https://doi.org/10.1515/spp-2016-0006

41 Joseph P. Simmons et al., 'Life after P-Hacking: Meeting of the Society for Personality and Social Psychology', *SSRN*, (New Orleans, LA, 17–19 Jan. 2013); https://doi.org/10.2139/ssrn.2205186

42 When they talked about reliably detecting an effect, the authors were referring to a standard much used in the literature. Your statistical power is normally described as acceptable if you have an 80 per cent or higher chance of finding

an effect with a statistical test (that is, getting a p-value below 0.05), if it really exists. Obviously, higher power is even better, and with big enough samples (or big enough effects), your power can be substantially above this minimal criterion. For 80 per cent statistical power, the 20 per cent chance that you miss the effect when it's really there is the chance of a false negative.

43 Katherine S. Button et al., 'Power Failure: Why Small Sample Size Undermines the Reliability of Neuroscience', *Nature Reviews Neuroscience* 14, no. 5 (May 2013): pp. 365–76; https://doi.org/10.1038/nrn3475. See in particular Table 2.

44 There is, though, quite a lot of variation across different neuroscientific subfields. Camilla L. Nord et al., 'Power-up: A Reanalysis of "Power Failure" in Neuroscience Using Mixture Modeling', *Journal of Neuroscience* 37, no. 34 (23 Aug. 2017): pp. 8051–61; https://doi.org/10.1523/JNEUROSCI.3592-16.2017

45 Medical trials: Herm J. Lamberink et al., 'Statistical Power of Clinical Trials Increased While Effect Size Remained Stable: An Empirical Analysis of 136,212 Clinical Trials between 1975 and 2014', *Journal of Clinical Epidemiology* 102 (Oct. 2018): pp. 123–28; https://doi.org/10.1016/j.jclinepi.2018.06.014. Biomedical research: Estelle Dumas-Mallet et al., 'Low Statistical Power in Biomedical Science: A Review of Three Human Research Domains', *Royal Society Open Science* 4, no. 2 (Feb. 2017): 160254; https://doi.org/10.1098/rsos.160254. Economics: John P. A. Ioannidis et al., 'The Power of Bias in Economics Research', *Economic Journal* 127, no. 605 (1 Oct. 2017): F236–65; https://doi.org/10.1111/ecoj.12461. Brain imaging: Henk R. Cremers et al., 'The Relation between Statistical Power and Inference in FMRI', ed. Eric-Jan Wagenmakers, *PLOS ONE* 12, no. 11 (20 Nov. 2017): e0184923; https://doi.org/10.1371/journal.pone.0184923. Nursing research: Cadeyrn J. Gaskin & Brenda Happell, 'Power, Effects, Confidence, and Significance: An Investigation of Statistical Practices in Nursing Research', *International Journal of Nursing Studies* 51, no. 5 (May 2014): 795–806; https://doi.org/10.1016/j.ijnurstu.2013.09.014. Behavioural ecology: M. D. Jennions & Anders Pape Møller, 'A Survey of the Statistical Power of Research in Behavioral Ecology and Animal Behavior', *Behavioral Ecology* 14, no. 3 (1 May 2003): pp. 438–45; https://doi.org/10.1093/beheco/14.3.438. Psychology: Denes Szucs & John P. A. Ioannidis, 'Empirical Assessment of Published Effect Sizes and Power in the Recent Cognitive Neuroscience and Psychology Literature', ed. Eric-Jan Wagenmakers, *PLOS Biology* 15, no. 3 (2 Mar. 2017): e2000797; https://doi.org/10.1371/journal.pbio.2000797

46 Leif D. Nelson et al., 'Psychology's Renaissance', *Annual Review of Psychology* 69, no. 1 (4 Jan. 2018): pp. 511–34; https://doi.org/10.1146/annurev-psych-122216-011836

47 This is a version of the 'winner's curse' that's sometimes discussed in auctions, where the person who puts in the winning bid overvalues whatever's being auctioned. In science it's also been named the 'Proteus phenomenon', after the shape-shifting character from Greek mythology. The basic idea is that in the early stages of the discovery of an effect, its size will often change dramatically between different studies, partly due to the issues that we've been discussing of statistical power and the inability of some studies to see small effects. See John P. A. Ioannidis & Thomas A. Trikalinos, 'Early Extreme Contradictory Estimates May Appear in Published Research: The Proteus Phenomenon in Molecular Genetics Research and Randomized Trials', *Journal of Clinical Epidemiology* 58, no. 6 (June 2005): pp. 543–49; https://doi.org/10.1016/j.jclinepi.2004.10.019. Nathan P. Lemoine et al., 'Underappreciated Problems of Low Replication in Ecological Field Studies', *Ecology* 97, no. 10 (Oct. 2016): pp. 2554–61; https://doi.org/10.1002/ecy.1506; and Button et al., 'Power Failure'.

48 A similar issue affects the studies that I cited above, which surveyed the power in specific fields. They estimate the power of those studies post hoc: that is, they ask 'how much power did they have to detect the effect they found?' But if the studies overestimated the size of the true effect, this post hoc method will thus overestimate their power. So, testing for post hoc power might lull you into thinking that the power in your study was absolutely fine when it wasn't. A better idea is to take an ideal effect size – one that, based on the practical meaning of your effect, you'd consider to be small, medium or large (perhaps using a meaningful difference in a more solid indicator, like movement on a pain scale, income in dollars, temperature, or speed) – and power your study (that is, include enough participants or observations) to detect it reliably. See Andrew Gelman, 'Don't Calculate Post-Hoc Power Using Observed Estimate of Effect Size' (2018); http://www.stat.columbia.edu/~gelman/research/unpublished/power_surgery.pdf

49 Herm J. Lamberink et al., 'Statistical Power of Clinical Trials Increased While Effect Size Remained Stable: An Empirical Analysis of 136,212 Clinical Trials between 1975 and 2014', Journal of Clinical Epidemiology 102 (Oct. 2018): pp. 123–28; https://doi.org/10.1016/j.jclinepi.2018.06.014. The effect size I'm talking about here is a Cohen's d-value of 0.21. The interpretation I gave, about the number of people who'd benefit from the treatment, is from the very useful calculator at the following website, created by Kristoffer Magnusson: https://rpsychologist.com/d3/cohend/

50 Stefan Leucht et al., 'How Effective Are Common Medications: A Perspective Based on Meta-Analyses of Major Drugs', BMC Medicine 13, no. 1 (Dec. 2015): 253; https://doi.org/10.1186/s12916-015-0494-1. This study, which looked at the effect sizes of well-used medical treatments, did highlight some common treatments with large effects (like proton-pump inhibitors such as omeprazole, which have huge effects on gastric acid levels), but also some where the effects were surprisingly small (like aspirin for prevention of cardiovascular disease). Of course, even drugs with small effects can be highly beneficial on a societal level – saving millions in medical costs – if they're prescribed to millions of people who need them. Nevertheless, the authors of the study recommended that 'we need to be more realistic about drug efficacy' (p. 4). The effect sizes for the three treatments mentioned in the text are glossed as having a Cohen's d-value of 0.55. See also Tiago V. Pereira et al., 'Empirical Evaluation of Very Large Treatment Effects of Medical Interventions', JAMA 308, no. 16 (24 Oct. 2012): 1676–84; https://doi.org/10.1001/jama.2012.13444

51 E.g. Gilles E. Gignac & Eva T. Szodorai, 'Effect Size Guidelines for Individual Differences Researchers', Personality and Individual Differences 102 (Nov. 2016): pp. 74–78; https://doi.org/10.1016/j.paid.2016.06.069

52 Thankfully there have been a great many reliable twin studies since the time of Cyril Burt. For a review, see Tinca J. C. Polderman et al., 'Meta-Analysis of the Heritability of Human Traits Based on Fifty Years of Twin Studies', Nature Genetics 47, no. 7 (July 2015): 702–9; https://doi.org/10.1038/ng.3285

53 For a review of the candidate-genetic links to cognitive abilities, see Antony Payton, 'The Impact of Genetic Research on Our Understanding of Normal Cognitive Ageing: 1995 to 2009', Neuropsychology Review 19, no. 4 (Dec. 2009): pp. 451–77; https://doi.org/10.1007/s11065-009-9116-z

54 Dominique J-F de Quervain et al., 'A Functional Genetic Variation of the 5-HT2a Receptor Affects Human Memory', Nature Neuroscience 6, no. 11 (Nov. 2003): pp. 1141–42; https://doi.org/10.1038/nn1146

55 Marcus R. Munafò et al., 'Serotonin Transporter (5-HTTLPR) Genotype and Amygdala Activation: A Meta-Analysis', *Biological Psychiatry* 63, no. 9 (May 2008): pp. 852–57; https://doi.org/10.1016/j.biopsych.2007.08.016

56 Nowadays you can send away a saliva sample to a direct-to-consumer genotyping company for somewhere around £100 and learn which genetic variants you carry within a couple of weeks.

57 You might wonder whether, in testing the link of the trait in question to so many thousands of genetic variations, GWAS falls prey to the problem of multiple comparisons that we saw in the previous chapter, where the risk of a false-positive result ramps up along with the number of *p*-values that are calculated. GWAS researchers are well aware of this issue and have lowered their *p*-value criterion accordingly – and dramatically. Instead of using the 0.05 cut-off, they only accept as statistically significant *p*-values that are lower than 5×10^{-08} (or 0.00000005).

58 Laramie E. Duncan et al., 'How Genome-Wide Association Studies (GWAS) Made Traditional Candidate Gene Studies Obsolete', *Neuropsychopharmacology* 44, no. 9 (Aug. 2019): pp. 1518–23; https://doi.org/10.1038/s41386-019-0389-5

59 'Rare' being the operative word here. For instance, we're aware of many rare mutations that are linked to learning disabilities and some forms of Autism Spectrum Disorder. For example, Mari E. K. Niemi et al., 'Common Genetic Variants Contribute to Risk of Rare Severe Neurodevelopmental Disorders', *Nature* 562, no. 7726 (Oct. 2018): pp. 268–71; https://doi.org/10.1038/s41586-018-0566-4. To my knowledge, the only more common 'candidate gene' that survived the onslaught of the large genome-wide association studies is a variation in the *APOE* gene, which does seem reliably to be related to risk for Alzheimer's disease. See Riccardo E. Marioni et al., 'GWAS on Family History of Alzheimer's Disease', *Translational Psychiatry* 8, no. 1 (Dec. 2018): 99; https://doi.org/10.1038/s41398-018-0150-6

60 IQ: Christopher F. Chabris et al., 'Most Reported Genetic Associations with General Intelligence are Probably False Positives', *Psychological Science* 23, no. 11 (Nov. 2012): pp. 1314–23; https://doi.org/10.1177/0956797611435528. Depression: Richard Border et al., 'No Support for Historical Candidate Gene or Candidate Gene-by-Interaction Hypotheses for Major Depression Across Multiple Large Samples', *American Journal of Psychiatry* 176, no. 5 (May 2019): pp. 376–87; https://doi.org/10.1176/appi.ajp.2018.18070881. Schizophrenia: M. S. Farrell et al., 'Evaluating Historical Candidate Genes for Schizophrenia', *Molecular Psychiatry* 20, no. 5 (May 2015): pp. 555–62; https://doi.org/10.1038/mp.2015.16

61 Scott Alexander, '5-HTTLPR: A Pointed Review', *Slate Star Codex*, 7 May 2019; https://slatestarcodex.com/2019/05/07/5-httlpr-a-pointed-review/

62 For example, low power is mentioned in the Abstract of: H. Clarke et al., 'Association of the *5-HTTLPR* Genotype and Unipolar Depression: A Meta-Analysis', *Psychological Medicine* 40, no. 11 (Nov. 2010): pp. 1767–78; https://doi.org/10.1017/S0033291710000516. Incidentally, you can bet that there was a great deal of publication bias in the candidate gene literature. For some evidence of this in studies of how candidate genes interact with the environment, see Laramie E. Duncan & Matthew C. Keller, 'A Critical Review of the First 10 Years of Candidate Gene-by-Environment Interaction Research in Psychiatry', *American Journal of Psychiatry* 168, no. 10 (Oct. 2011): pp. 1041–49; https://doi.org/10.1176/appi.ajp.2011.11020191

63 R. A. Fisher, 'XV. – The Correlation between Relatives on the Supposition of Mendelian Inheritance', *Transactions of the Royal Society of Edinburgh* 52,

no. 2 (1919): pp. 399–433; https://doi.org/10.1017/S0080456800012163. See the historical discussion in Peter M. Visscher et al., '10 Years of GWAS Discovery: Biology, Function, and Translation', *American Journal of Human Genetics* 101, no. 1 (July 2017): pp. 5–22; https://doi.org/10.1016/j.ajhg.2017.06.005

64 Which isn't to say geneticists are anywhere near fully understanding the genetics of complex traits. They've simply cast off this faulty old method and replaced it with better ones. There's still an enormous amount of work to do on which genes are involved, how we find them across different groups of people, how our analyses might be frustrated by complexities of social and demographic differences, specifically how the genes have their effects, what we *do* with our knowledge about genetics in terms of helping people with medical conditions, and so on. For a useful review, see Vivian Tam et. al, 'Benefits and Limitations of Genome-Wide Association Studies', *Nature Reviews Genetics* 20, no. 8 (Aug. 2019): pp. 467–84; https://doi.org/10.1038/s41576-019-0127-1

65 There's an interesting paradox of statistical power in animal research. Perhaps counter-intuitively, we might be better off using *more* animals in our research, at least in the short term. Increasing sample size, and thus statistical power, means we get more reliable results – ones that will stand the tests of time and replication, avoiding long cycles of inconclusive, uninformative experiments, and probably avoiding more animal death in the long run.

66 For example, see Christine R. Critchley, 'Public Opinion and Trust in Scientists: The Role of the Research Context, and the Perceived Motivation of Stem Cell Researchers', *Public Understanding of Science* 17, no. 3 (July 2008): pp. 309–27; https://doi.org/10.1177/0963662506070162

6: *Hype*

Epigraph: Reid Harrison, 'The Springfield Files', *The Simpsons*, Steven Dean Moore, dir. (Season 8, Episode 10, 12 Jan. 1997).

1 F. Wolfe-Simon et al., 'A Bacterium That Can Grow by Using Arsenic Instead of Phosphorus', *Science* 332, no. 6034 (3 June 2011): pp. 1163–66; https://doi.org/10.1126/science.1197258

2 Technically they're not stalagmites but *tufa*, which look similar but have slightly different properties, such as a spongier interior; https://itotd.com/articles/2773/tufa/

3 Wolfe-Simon's study belongs to the sub-field of astrobiology. In the absence of actual alien life from other planets to study in their labs, astrobiologists study what that life *might* look like. One way to do so is to look at so-called 'extremophiles', such as the bacteria who live in harsh environments like Mono Lake.

4 Paul Davies, 'The "Give Me a Job" Microbe', *Wall Street Journal*, 4 Dec. 2010; https://on.wsj.com/2PAX4ut

5 Quoted in Tom Clynes, 'Scientist in a Strange Land', *Popular Science*, 26 Sept. 2011; https://www.popsci.com/science/article/2011-09/scientist-strange-land/. Wolfe-Simon was also interviewed for a profile in *Glamour* magazine: Anne Gowen, 'This Rising Star's Four Rules for *You*', *Glamour*, June 2011; https://bit.ly/2wbLLCb

6 An article by the science journalist Carl Zimmer contains many quotations from the sceptics: Carl Zimmer, '"This Paper Should Not Have Been Published": Scientists See Fatal Flaws in the NASA Study of Arsenic-Based Life', *Slate*, 7

Dec. 2010; https://slate.com/technology/2010/12/the-nasa-study-of-arsenic-based-life-was-fatally-flawed-say-scientists.html

7   The many posts can be found if you search for the #arseniclife hashtag on Redfield's blog: http://rrresearch.fieldofscience.com/

8   Editorial, 'Response Required', *Nature* 468, no. 7326 (Dec. 2010): p. 867; https://doi.org/10.1038/468867a

9   The press release can be found at the following link: https://www.nasa.gov/home/hqnews/2010/nov/HQ_M10-167_Astrobiology.html

10 Jason Kottke, 'Has NASA Discovered Extraterrestrial Life?', *Kottke*, 29 Nov. 2010; https://kottke.org/10/11/has-nasa-discovered-extraterrestrial-life. See also the following article, which includes a still from the movie *E.T. The Extra-Terrestrial*: 'NASA to Unveil Details of Quest for Alien Life', *Fox News*, 2 Dec. 2010; https://www.foxnews.com/science/nasa-to-unveil-details-of-quest-for-alien-life

11 Quoted in Tony Phillips, ed. 'Discovery of "Arsenic-Bug" Expands Definition of Life', 2 Dec. 2010; https://science.nasa.gov/science-news/science-at-nasa/2010/02dec_monolake

12 All are referred to in a note by the then-Editor-in-Chief of *Science*, Bruce Alberts. B. Alberts, 'Editor's Note', *Science* 332, no. 6034 (3 June 2011): p. 1149; https://doi.org/10.1126/science.1208877

13 M. L. Reaves et al., 'Absence of Detectable Arsenate in DNA from Arsenate-Grown GFAJ-1 Cells', *Science* 337, no. 6093 (27 July 2012): pp. 470–73; https://doi.org/10.1126/science.1219861

14 Erb et al., 'GFAJ-1 Is an Arsenate-Resistant, Phosphate-Dependent Organism', *Science* 337, no. 6093 (27 July 2012): pp. 467–70; https://doi.org/10.1126/science.1218455

15 Clynes, 'Scientist in a Strange Land'.

16 This isn't to say it never occurs. An occupational hazard of publishing science is that your findings will be misrepresented, misunderstood or mangled in the media. The mistakes run from the trivial to the serious and damaging. Of the latter, inaccurate media fearmongering in 2011 about the side-effects of statins – well-evidenced and safe drugs that reduce the risk of heart disease – appears to have influenced people to stop taking them in the subsequent years. Anthony Matthews et al., 'Impact of Statin Related Media Coverage on Use of Statins: Interrupted Time Series Analysis with UK Primary Care Data', *BMJ* (28 June 2016): i3283; https://doi.org/10.1136/bmj.i3283

17 P. S. Sumner et al., 'The Association between Exaggeration in Health-Related Science News and Academic Press Releases: Retrospective Observational Study', *BMJ* 349, (9 Dec. 2014): g7015; https://doi.org/10.1136/bmj.g7015

18 A nice potted history of animal models is: Aaron C. Ericsson et al., 'A Brief History of Animal Modeling', *Missouri Medicine* 110, no. 3 (June 2013): pp. 201–5; https://www.ncbi.nlm.nih.gov/pubmed/23829102

19 D. G. Contopoulos-Ioannidis et al., 'Life Cycle of Translational Research for Medical Interventions', *Science* 321, no. 5894 (5 Sept. 2008): pp. 1298–99, https://doi.org/10.1126/science.1160622

20 J. P. Garner, 'The Significance of Meaning: Why Do Over 90% of Behavioral Neuroscience Results Fail to Translate to Humans and What Can We Do to Fix It?', *ILAR Journal* 55, no. 3 (20 Dec. 2014): pp. 438–56; https://doi.org/10.1093/ilar/ilu047

21 https://twitter.com/justsayinmice. Incidentally, we shouldn't forget what we saw in the previous chapter, about the low quality of much of the research on animals like mice.

22   You often see a slightly different version: 'correlation does not *imply* causation'. There's an ambiguity here due to the multiple meanings of the word 'imply'. In its strong definition (thing A logically involves thing B, in the same way that the existence of, say, a dance *implies* that there's a dancer), it's certainly true. But in its weak version (thing A suggests thing B without explicitly saying thing B, in the same way that a receiving a slightly terse email from your boss might *imply* that they're unhappy with you), then it's not correct. In that weak sense, a correlation does sometimes imply causation, even if there's no causation there at all. Put it this way: if correlation didn't imply causation in this latter sense, there wouldn't be so much confusion between the two.

23   Janie Corley et al., 'Caffeine Consumption and Cognitive Function at Age 70: The Lothian Birth Cohort 1936 Study', *Psychosomatic Medicine* 72, no. 2 (Feb. 2010): pp. 206–14; https://doi.org/10.1097/PSY.0b013e3181c92a9c

24   There's also another, less well-known reason why there might be a correlation between two variables: 'collider bias'. The following excellent blog post describes the issue: Julia Rohrer, 'That One Weird Third Variable Problem Nobody Ever Mentions: Conditioning on a Collider', *The 100% CI*, March 14, 2017; http://www.the100.ci/2017/03/14/that-one-weird-third-variable-problem-nobody-ever-mentions-conditioning-on-a-collider/. The example given there is that if we only look in a sample of college students, we might, to our surprise, find a *negative* correlation between IQ and conscientiousness even if in the whole population there's no correlation at all. This is because IQ and conscientiousness both increase the chances of attending college in the first place, selecting out people who are low on both those traits. The fact that low-IQ, low-conscientiousness people are missing from the college sample produces a spurious correlation between those two variables. It's a wicked problem, and more pervasive in studies than we'd like to think. See also Marcus R. Munafò et al., 'Collider Scope: When Selection Bias Can Substantially Influence Observed Associations', *International Journal of Epidemiology* 47, no. 1 (27 Sept. 2017): pp. 226–35; https://doi.org/10.1093/ije/dyx206

25   If you really want to freak yourself out, read about the philosopher David Hume's 'Problem of Induction', which essentially states that correlation isn't even *correlation* – there's no rational basis for arguing that things that have happened before will happen again. (The classic challenge is to prove that the sun will rise tomorrow morning – 'it's always done so before' isn't truly a logical basis for believing it.) This problem has been debated in philosophy for hundreds of years and while many extremely smart thinkers have tried to crack it, others think it's flatly unsolvable. An excellent discussion is provided in Leah Henderson, 'The Problem of Induction', *Stanford Encyclopedia of Philosophy*, ed. Edward N. Zalta, Winter 2019; https://plato.stanford.edu/archives/win2019/entries/induction-problem

26   Rachel C. Adams et al., 'Claims of Causality in Health News: A Randomised Trial', *BMC Medicine* 17, no. 1 (Dec. 2019): 91; https://doi.org/10.1186/s12916-019-1324-7

27   Isabelle Boutron et al., 'Three Randomized Controlled Trials Evaluating the Impact of "Spin" in Health News Stories Reporting Studies of Pharmacologic Treatments on Patients'/Caregivers' Interpretation of Treatment Benefit', *BMC Medicine* 17, no. 1 (Dec. 2019): 105; https://doi.org/10.1186/s12916-019-1330-9

28   Nick Davies, *Flat Earth News: An Award-Winning Reporter Exposes Falsehood, Distortion and Propaganda in the Global Media* (London: Vintage Books, 2009). Daniel Jackson and Kevin Moloney, 'Inside Churnalism: PR, Journalism and

Power Relationships in Flux', *Journalism Studies* 17, no. 6 (17 Aug. 2016): pp. 763–80; https://doi.org/10.1080/1461670X.2015.1017597

29 Estelle Dumas-Mallet et al., 'Poor Replication Validity of Biomedical Association Studies Reported by Newspapers', *PLOS ONE* 12, no. 2 (21 Feb. 2017): e0172650; https://doi.org/10.1371/journal.pone.0172650

30 This isn't to let popular science books by non-scientists off the hook: they're also prone to major problems. In his review of Malcolm Gladwell's book of collected essays, *What the Dog Saw*, Steven Pinker coined the term 'Igon Values Problem' to describe a case where Gladwell had butchered the word 'eigenvalues' (a mathematical concept that's important in many statistical analyses) – he'd presumably heard one of his interviewees say it and then never bothered to look it up. Igon Values are all too common in popular science writing, highlighting gaps in understanding that can occur when the writer isn't an expert in the subject at hand. But as we're about to see, scientists *themselves*, even writing about their own topics of expertise, can produce books with problems just as bad as the Igon Values. Steven Pinker, 'Malcolm Gladwell, Eclectic Detective', *New York Times*, 7 Nov. 2009; https://www.nytimes.com/2009/11/15/books/review/Pinker-t.html

31 Carol S. Dweck, *Mindset: The New Psychology of Success* (New York: Ballantine Books, 2008): pp. 6, 15. Dweck also presented a very popular TEDx talk, with a total of 13.5 million views at this writing – 10.2m at the TED website and 3.3 on YouTube – in which she states that it is a 'basic human right of children, all children, to live in places that promote growth'. Carol Dweck, 'The Power of Believing That You Can Improve', presented at *TEDxNorrkoping*, Nov. 2014; https://www.ted.com/talks/carol_dweck_the_power_of_believing_that_you_can_improve

32 Ibid. ix.

33 Holly Yettick et al., 'Mindset in the Classroom: A National Study of K-12 Teachers', Editorial Projects in Education, Bethesda, MD: Education Week Research Center, 2016; https://www.edweek.org/media/ewrc_mindsetintheclassroom_sept2016.pdf. Note that this wasn't a representative sample, so *caveat emptor*. In February 2020, a Google search for 'mindset' on only '.sch.uk' domains – that is, just on the websites of UK schools – gives 43,200 results, giving some indication of the popularity of the idea. There are currently just over 32,000 schools in the UK: 'Key UK Education Statistics', British Educational Suppliers Association, 28 Oct. 2019; https://www.besa.org.uk/key-uk-education-statistics/

34 Victoria F. Sisk et al., 'To What Extent and Under Which Circumstances Are Growth Mind-Sets Important to Academic Achievement? Two Meta-Analyses', *Psychological Science* 29, no. 4 (April 2018): pp. 549–71; https://doi.org/10.1177/0956797617739704. For further research from the same mindset-sceptical researchers, see Alexander P. Burgoyne et al., 'How Firm Are the Foundations of Mind-Set Theory? The Claims Appear Stronger Than the Evidence', *Psychological Science*, 3 Feb. 2020; https://doi.org/10.1177/0956797619897588

35 For stats fans, the correlational and experimental effects correspond to a Pearson's r-value of 0.10 and a Cohen's d-value of 0.08, respectively. Another way of thinking about the d-value of 0.08, aside from seeing it as 96.8 per cent overlap in the distributions, is to think about selecting a random person from the mindset group and seeing whether their grades are above or below the average of the controls. If there was no effect, the probability of this would be 50 per cent (the averages are identical). In the meta-analysis, the mindset effect meant that those who'd been trained to have a growth mindset had a 52.3 per cent chance of being higher than the average of the control group. Calculated using https://rpsychologist.com/d3/cohend/

36 In the meta-analysis, there was some evidence that particularly at-risk children (those from poorer backgrounds, for example) might benefit more from mindset interventions. This was also the case in a recent large-scale study by proponents of growth mindsets, which found similar results to the meta-analysis in general. David S. Yeager et al., 'A National Experiment Reveals Where a Growth Mindset Improves Achievement', *Nature* 573, no. 7774 (Sept. 2019): pp. 364–69; https://doi.org/10.1038/s41586-019-1466-y

37 If you want a perfect illustration of how hype can induce a sense of scientific 'mission creep', witness the paper published by Dweck and colleagues in *Science* in 2011 that suggested – on the basis of some rather weak-looking evidence – that growth mindsets could be used to promote peace in the Middle East. E. Halperin et al., 'Promoting the Middle East Peace Process by Changing Beliefs About Group Malleability', *Science* 333, no. 6050 (23 Sept. 2011): pp. 1767–69; https://doi.org/10.1126/science.1202925

38 Another example of educational hype is in some ways the second coming of the growth mindset concept: 'grit'. This is the idea, promoted by the psychologist Angela Duckworth, that the ability to stick to a task you're passionate about, and not give up even when life puts obstacles in your path, is key to life success, and far more important than innate talent. The appetite for her message was immense: at the time of this writing, her TED talk on the subject has received 25.5 million views (19.5m on the TED website and a further 6m on YouTube; Angela Lee Duckworth, 'Grit: The Power of Passion and Perseverance', presented at *TED Talks Education*, April 2013; https://www.ted.com/talks/angela_lee_duckworth_grit_the_power_of_passion_and_perseverance), and her subsequent book, *Grit: The Power of Passion and Perseverance*, became a *New York Times* bestseller and continues to sell steadily. Like mindset, grit has become part of the philosophy of many schools, including KIPP (Knowledge is Power Program) schools, the biggest charter school group in the US, which teaches almost 90,000 students (https://www.kipp.org/approach/character/). To her credit, Duckworth has been concerned about how overhyped her results have become. She told an *NPR* interviewer in 2015 that 'the enthusiasm is getting ahead of the science' (Anya Kamenetz, 'A Key Researcher Says "Grit" Isn't Ready For High-Stakes Measures', *NPR*, 13 May 2015; https://www.npr.org/sections/ed/2015/05/13/405891613/a-key-researcher-says-grit-isnt-ready-for-high-stakes-measures). A wise statement, given that the meta-analytic evidence for the impact of grit (or interventions trying to teach it) is extremely weak. See Credé et al., 'Much Ado about Grit: A Meta-Analytic Synthesis of the Grit Literature', *Journal of Personality and Social Psychology* 113, no. 3 (Sept. 2017): pp. 492–511; https://doi.org/10.1037/pspp0000102. And Marcus Credé, 'What Shall We Do About Grit? A Critical Review of What We Know and What We Don't Know', *Educational Researcher* 47, no. 9 (Dec. 2018): pp. 606–11; https://doi.org/10.3102/0013189X18801322

39 Indeed, it was in response to a failure to replicate one of Bargh's results – the 'priming the idea of elderly people makes you walk more slowly' study that we encountered in Chapter 3 – that Daniel Kahneman wrote his open letter to social psychologists, telling them that he saw a 'train wreck looming' and urging them to change the way they go about their research.

40 John Bargh, *Before You Know It: The Unconscious Reasons We Do What We Do* (London: Windmill Books, 2018).

41 The following blog post, by the psychologist Ulrich Schimmack, provides a detailed 'quantitative review' of Bargh's book, assessing how much we should trust each of the studies Bargh cites: Ulrich Schimmack, '"Before You Know It" by

John A. Bargh: A Quantitative Book Review', *Replication Index*, 28 Nov. 2017; https://replicationindex.com/2017/11/28/before-you-know-it-by-john-a-bargh-a-quantitative-book-review/. Incidentally, the 'quantitative book review' is a great idea that should be followed by lots of other scientists.

42 Bargh, *Before You Know It*, p. 16.

43 Serena Chen et al., 'Relationship Orientation as a Moderator of the Effects of Social Power', *Journal of Personality and Social Psychology* 80, no. 2 (2001): pp. 173–87; https://doi.org/10.1037/0022-3514.80.2.173

44 Christopher F. Chabris et al., 'No Evidence that Experiencing Physical Warmth Promotes Interpersonal Warmth: Two Failures to Replicate', *Social Psychology* 50, no. 2 (Mar. 2019): pp. 127–32; https://doi.org/10.1027/1864-9335/a000361. To be scrupulously fair to Bargh, this specific coffee-cup replication attempt was published after his book. But since his other, very similar studies on the concept of 'warmth' (which involved holding one of those therapeutic heat packs instead of a cup of coffee) had also failed to replicate, one would think he'd have been a little more careful in relying on those results. See Dermot Lynott et al., 'Replication of "Experiencing Physical Warmth Promotes Interpersonal Warmth" by Williams and Bargh (2008)', *Social Psychology* 45, no. 3 (May 2014): pp. 216–22; https://doi.org/10.1027/1864-9335/a000187

45 You might also remember the case of Amy Cuddy, whose bestselling book was based on a study that was later revealed to be a paradigm example of *p*-hacking. See Chapters 2 and 4.

46 Matthew Walker, *Why We Sleep: The New Science of Sleep and Dreams* (London: Allen Lane, 2017).

47 https://www.ted.com/talks/matt_walker_sleep_is_your_superpower. The viewing figures are a combination of the 6.7m on TED's site and a further 3.3m on YouTube, as of November 2019.

48 Richard Smith, 'Why We Sleep – One of Those Rare Books That Changes Your Worldview and Should Change Society and Medicine', *TheBMJOpinion*, 20 June 2018; https://blogs.bmj.com/bmj/2018/06/20/richard-smith-why-we-sleep-one-of-those-rare-books-that-changes-your-worldview-and-should-change-society-and-medicine/

49 Walker, *Why We Sleep*, pp. 3–4.

50 Alexey Guzey, 'Matthew Walker's "Why We Sleep" Is Riddled with Scientific and Factual Errors', 15 Nov. 2019; https://guzey.com/books/why-we-sleep/

51 Xiaoli Shen et al., 'Nighttime Sleep Duration, 24-Hour Sleep Duration and Risk of All-Cause Mortality among Adults: A Meta-Analysis of Prospective Cohort Studies', *Scientific Reports* 6, no. 1 (Feb. 2016): p. 21480; https://doi.org/10.1038/srep21480

52 Yuheng Chen et al. (2018), 'Sleep Duration and the Risk of Cancer: A Systematic Review and Meta-Analysis Including Dose–Response Relationship', *BMC Cancer* 18, no. 1 (Dec. 2018): p. 1149; https://doi.org/10.1186/s12885-018-5025-y

53 Andrew Gelman, '"Why We Sleep" Data Manipulation: A Smoking Gun?', *Statistical Modeling, Causal Inference, and Social Science*, 27 Dec. 2019; https://statmodeling.stat.columbia.edu/2019/12/27/why-we-sleep-data-manipulation-a-smoking-gun/. Note that there is a response, seemingly from Walker, to some criticisms of *Why We Sleep* here: SleepDiplomat, 'Why We Sleep: Responses to Questions from Readers', 19 Dec. 2019; https://sleepdiplomat.wordpress.com/2019/12/19/why-we-sleep-responses-to-questions-from-readers/

54 For some reflections on popular-science writing, see Christopher F. Chabris, 'What Has Been Forgotten About Jonah Lehrer', 12 Feb. 2013; http://blog.chabris.com/2013/02/what-has-been-forgotten-about-jonah.html

55 Christiaan H. Vinkers et al., 'Use of Positive and Negative Words in Scientific PubMed Abstracts between 1974 and 2014: Retrospective Analysis', *BMJ* 351 (14 Dec. 2015): h6467; https://doi.org/10.1136/bmj.h6467. Note that the analysis adjusted for the increasing number of papers published each year. There are specific phrases that could've been analysed, too: I've lost track of the number of times I've read (and sometimes written!) 'this paper examines, for the first time …'

56 A later study looked specifically at the word 'unprecedented' in cancer research. It found that in about a third of cases, the word was factually incorrect: even though the authors claimed the result was 'unprecedented', there had been a previous paper of the same treatment that had reported an even bigger effect. See Kristy Tayapongsak Duggan et al., 'Use of Word "Unprecedented" in the Media Coverage of Cancer Drugs: Do "Unprecedented" Drugs Live up to the Hype?', *Journal of Cancer Policy* 14 (Dec. 2017): pp. 16–20; https://doi.org/10.1016/j.jcpo.2017.09.010

57 Vinkers et al., 'Use of Positive and Negative Words', p. 2. Oddly, the number of negative words in abstracts also increased in frequency, though only by a tiny touch, so perhaps we should say that abstracts have become more *extreme*. Neutral and randomly selected words didn't increase in frequency at all.

58 In fact, there's some evidence that scientific progress is slowing down over time. Tyler Cowen and Ben Southwood, 'Is the Rate of Scientific Progress Slowing Down?', 5 Aug. 2019; https://bit.ly/3ahf7om

59 *Nature*: https://www.nature.com/authors/author_resources/about_npg.html; *Science*: https://www.sciencemag.org/about/mission-and-scope; *Cell*: https://www.cell.com/cell/aims; *Proceedings of the National Academy of Sciences*: http://www.pnas.org/page/authors/purpose-scope

60 *New England Journal of Medicine*: https://www.nejm.org/about-nejm/about-nejm

61 Isabelle Boutron, 'Reporting and Interpretation of Randomized Controlled Trials with Statistically Nonsignificant Results for Primary Outcomes', *JAMA* 303, no. 20 (26 May 2010): pp. 2058–64; https://doi.org/10.1001/jama.2010.651. See also Isabelle Boutron & Philippe Ravaud, 'Misrepresentation and Distortion of Research in Biomedical Literature', *Proceedings of the National Academy of Sciences* 115, no. 11 (13 Mar. 2018): pp. 2613–19; https://doi.org/10.1073/pnas.1710755115

62 Matthew Hankins, 'Still Not Significant', *Probable Error*, 21 April 2013; https://mchankins.wordpress.com/2013/04/21/still-not-significant-2/. For an analysis of the prevalence of such statements in the oncology literature, see Kevin T. Nead, Mackenzie R. Wehner, & Nandita Mitra, 'The Use of "Trend" Statements to Describe Statistically Nonsignificant Results in the Oncology Literature', *JAMA Oncology* 4, no. 12 (1 Dec. 2018): pp. 1778–79; https://doi.org/10.1001/jamaoncol.2018.4524. It's been pointed out that these statements are written as if the result is always moving *towards* significance – but if you're going to think of your results in this weird teleological way, disregarding how *p*-values work, how do you know the numbers aren't moving *away* from it? For that matter, how do you know that those of your results just on the favourable side of the significance threshold aren't doing everything they can to get back above 0.05? For mysterious reasons, scientists never seem to feel the need to describe *p*-values that are just *under* 0.05 as 'trending away from significance' (see 'Dredging for P' at the following link: http://www.senns.demon.co.uk/wprose.html). Furthermore, it might appear as if there's a contradiction between my description of the *p*-value threshold as arbitrary in Chapter 4, and my criticism here of scientists who get a result slightly above the threshold but still interpret it as significant (or in

some sense approaching significance). The point is that if you're going to play the *p*-value game, you should stick to the rules. If you've declared at the start that you'll only accept results with $p < 0.05$ as significant, then you can't just move the goalposts once you've seen the results. Otherwise, the threshold loses the one useful function it has: to control the risk of a false-positive result when your hypothesis is wrong.

63 Mark Turrentine, 'It's All How You "Spin" It: Interpretive Bias in Research Findings in the Obstetrics and Gynecology Literature', *Obstetrics & Gynecology* 129, no. 2 (Feb. 2017): pp. 239–42; https://doi.org/10.1097/AOG.0000000000001818

64 Emmanuelle Kempf et al., 'Overinterpretation and Misreporting of Prognostic Factor Studies in Oncology: A Systematic Review', *British Journal of Cancer* 119, no. 10 (Nov. 2018): pp. 1288–96; https://doi.org/10.1038/s41416-018-0305-5. One example of this kind of spin, which occurred in thirty-one different studies in the review, was where scientists hid their non-significant *p*-values in a note below their results table but prominently reported the significant *p*-values in the table itself.

65 J. Austin et al., 'Evaluation of Spin within Abstracts in Obesity Randomized Clinical Trials: A Cross-Sectional Review: Spin in Obesity Clinical Trials', *Clinical Obesity* 9, no. 2 (April 2019): e12292; https://doi.org/10.1111/cob.12292

66 Lian Beijers et al., 'Spin in RCTs of Anxiety Medication with a Positive Primary Outcome: A Comparison of Concerns Expressed by the US FDA and in the Published Literature', *BMJ Open* 7, no. 3 (Mar. 2017): e012886; https://doi.org/10.1136/bmjopen-2016-012886

67 David Marc Anton Mehler & Konrad Paul Kording, 'The Lure of Causal Statements: Rampant Mis-Inference of Causality in Estimated Connectivity', *ArXiv:1812.03363 [q-Bio]*, 8 Dec. 2018; http://arxiv.org/abs/1812.03363. This paper notes the rise of the term 'Granger Causality'. It was introduced by the Nobel Prize-winning economist Clive Granger in the 1960s. The idea is that if data from one 'time series' – say, fluctuations in the stock market – predict later changes in another – say, changes in a country's other economic indicators – then that's a step beyond a basic correlation. In this case, some researchers would say that the stock market changes 'Granger caused' the economic changes. These correlations might be interesting, but they're still only correlations and the usual worries about third-variable confounders apply just as strongly (a third, parallel time trend might have caused the first market fluctuations *and* the later changes in the economic indicators). Using the word 'caused' when your study wasn't designed to uncover causal links (with an experiment, or another clever way to infer the causal structure of the data) is playing with fire.

68 Taixiang Wu et al., 'Randomized Trials Published in Some Chinese Journals: How Many Are Randomized?', *Trials* 10, no. 1 (Dec. 2009): p. 46; https://doi.org/10.1186/1745-6215-10-46

69 Trevor A. McGrath et al., 'Overinterpretation of Research Findings: Evidence of "Spin" in Systematic Reviews of Diagnostic Accuracy Studies', *Clinical Chemistry* 63, no. 8 (1 Aug. 2017): p. 1362; https://doi.org/10.1373/clinchem.2017.271544. See also Kellia Chiu et al., '"Spin" in Published Biomedical Literature: A Methodological Systematic Review', *PLOS Biology* 15, no. 9 (11 Sept. 2017): e2002173; https://doi.org/10.1371/journal.pbio.2002173

70 To my knowledge, there's been one randomised controlled trial of the effects of spin in scientific papers in a clinical context. The researchers took a selection of abstracts from studies in cancer research with null results, 'spun' to sound positive. They rewrote them to remove the hype, cutting out any unjustified

exaggeration and making sure all the results were reported honestly. They then showed these abstracts to a set of 300 clinicians – who are, after all, the target audience of abstracts, since they're regularly making decisions about drugs and other treatments. Sure enough, the clinicians rated the treatments with the hyped abstracts as being more beneficial. Importantly though, the effects in this study were relatively small, with p-values that were just below the 0.05 threshold. I'd want to see a replication before I put too much faith in the study – hence its relegation to just an endnote here. Isabelle Boutron et al., 'Impact of Spin in the Abstracts of Articles Reporting Results of Randomized Controlled Trials in the Field of Cancer: The SPIIN Randomized Controlled Trial', *Journal of Clinical Oncology* 32, no. 36 (20 Dec. 2014): pp. 4120–26; https://doi.org/10.1200/JCO.2014.56.7503

71 Ed Yong, *I Contain Multitudes: The Microbes within Us and a Grander View of Life* (New York: HarperCollins, 2016).

72 Timothy Caulfield, 'Microbiome Research Needs a Gut Check', *Globe and Mail*, 11 Oct. 2019; https://www.theglobeandmail.com/opinion/article-microbiome-research-needs-a-gut-check/

73 Andi L. Shane, 'The Problem of DIY Fecal Transplants', *Atlantic*, 16 July 2013; https://www.theatlantic.com/health/archive/2013/07/the-problem-of-diy-fecal-transplants/277813/

74 Dina Kao et al., 'Effect of Oral Capsule- vs Colonoscopy-Delivered Fecal Microbiota Transplantation on Recurrent *Clostridium Difficile* Infection: A Randomized Clinical Trial', *JAMA* 318, no. 20 (28 Nov. 2017): p. 1985; https://doi.org/10.1001/jama.2017.17077. The first recorded transplant of this nature was in 1958, but interest in the treatment didn't kick in for many decades. B. Eiseman et al., 'Fecal Enema as an Adjunct in the Treatment of Pseudomembranous Enterocolitis', *Surgery* 44, no. 5 (Nov. 1958): pp. 854–59; https://www.ncbi.nlm.nih.gov/pubmed/13592638

75 Wenjia Hui et al., 'Fecal Microbiota Transplantation for Treatment of Recurrent C. Difficile Infection: An Updated Randomized Controlled Trial Meta-Analysis', *PLOS ONE* 14, no. 1 (2019): e0210016; https://doi.org/10.1371/journal.pone.0210016; Theodore Rokkas et al., 'A Network Meta-Analysis of Randomized Controlled Trials Exploring the Role of Fecal Microbiota Transplantation in Recurrent Clostridium Difficile Infection', *United European Gastroenterology Journal* 7, no. 8 (Oct. 2019): pp. 1051–63; https://doi.org/10.1177/2050640619854587

76 Microbiome and depression, anxiety and schizophrenia: Jane A. Foster & Karen-Anne McVey Neufeld, 'Gut–Brain Axis: How the Microbiome Influences Anxiety and Depression', *Trends in Neurosciences* 36, no. 5 (May 2013): pp. 305–12; https://doi.org/10.1016/j.tins.2013.01.005; T. G. Dinan et al., 'Genomics of Schizophrenia: Time to Consider the Gut Microbiome?', *Molecular Psychiatry* 19, no. 12 (Dec. 2014): pp. 1252–57; https://doi.org/10.1038/mp.2014.93. Heart disease: Shadi Ahmadmehrabi Shadi & W. H. Wilson Tang, 'Gut Microbiome and Its Role in Cardiovascular Diseases', *Current Opinion in Cardiology* 32, no. 6 (Nov. 2017): pp. 761–66; https://doi.org/10.1097/HCO.0000000000000445. Obesity: Clarisse A. Marotz & Amir Zarrinpar, 'Treating Obesity and Metabolic Syndrome with Fecal Microbiota Transplantation', *Yale Journal of Biology and Medicine* 89, no. 3 (2016): pp. 383–88; https://www.ncbi.nlm.nih.gov/pmc/articles/PMC5045147/. Cancer: Chen et al., 'Fecal Microbiota Transplantation in Cancer Management: Current Status and Perspectives', *International Journal of Cancer* 145, no. 8 (15 Oct. 2019): pp. 2021–31; https://doi.org/10.1002/ijc.32003. Alzheimer's disease: Ana Sandoiu, 'Stool Transplants from "Super

Donors" Could Be a Cure-All', *Medical News Today*, 22 January 2019; https://www.medicalnewstoday.com/articles/324238. Parkinson's disease: T. Van Laar et al., 'Faecal Transplantation, Pro- and Prebiotics in Parkinson's Disease; Hope or Hype?', *Journal of Parkinson's Disease* 9, no. s2 (30 Oct. 2019): pp. S371–79; https://doi.org/10.3233/JPD-191802. Autism: Stefano Bibbò et al., 'Fecal Microbiota Transplantation: Past, Present and Future Perspectives', *Minerva Gastroenterologica e Dietologica*, no. 4 (Sept. 2017): pp. 420–30; https://doi.org/10.23736/S1121-421X.17.02374-1

77 I may have used the word 'cause' in that sentence, but the causal status of many claims about the microbiome is far from clear. See Kate E. Lynch et al., 'How Causal Are Microbiomes? A Comparison with the Helicobacter Pylori Explanation of Ulcers', *Biology & Philosophy* 34, no. 6 (Dec. 2019): 62; https://doi.org/10.1007/s10539-019-9702-2

78 It shouldn't escape our notice that the prospect of a link between gut health and autism was already raised by Andrew Wakefield, as discussed in Chapter 3. We're a million miles from being certain whether the microbiome differences cause autism or are caused by it – for example, by the fact that autistic people often have more restricted diets.

79 Though in some of their analyses, they only used samples from five autistic children and three controls.

80 Gil Sharon et al., 'Human Gut Microbiota from Autism Spectrum Disorder Promote Behavioral Symptoms in Mice', *Cell* 177, no. 6 (May 2019): 1600-1618. e17; https://doi.org/10.1016/j.cell.2019.05.004

81 Derek Lowe, 'Autism Mouse Models for the Microbiome?', *In the Pipeline*, 31 May 2019; https://blogs.sciencemag.org/pipeline/archives/2019/05/31/autism-mouse-models-for-the-microbiome

82 Sharon et al., 'Human Gut Microbiota', p.1162.

83 California Institute of Technology, 'Gut Bacteria Influence Autism-like Behaviors in Mice' (news release), 30 May 2019; https://www.eurekalert.org/pub_releases/2019-05/ciot-gbi052319.php

84 Jon Brock, 'Can Gut Bacteria Cause Autism (in Mice)?', *Medium*, 14 June 2019; https://medium.com/dr-jon-brock/can-gut-bacteria-cause-autism-in-mice-582306fd7235; see also Nicholette Zeliadt, 'Study of Microbiome's Importance in Autism Triggers Swift Backlash', *Spectrum News*, 27 June 2019, https://www.spectrumnews.org/news/study-microbiomes-importance-autism-triggers-swift-backlash/

85 Thomas Lumley, 'Analysing the Mouse Microbiome Autism Data', *Not Stats Chat*, 16 June 2019; https://notstatschat.rbind.io/2019/06/16/analysing-the-mouse-autism-data/; see also Jon Brock's own analysis, at the following page: https://rpubs.com/drbrocktagon/506022

86 Zheng et al., 'The Gut Microbiome from Patients with Schizophrenia Modulates the Glutamate-Glutamine-GABA Cycle and Schizophrenia-Relevant Behaviors in Mice', *Science Advances* 5, no. 2 (Feb. 2019): p. 8; https://doi.org/10.1126/sciadv.aau8317. A critique in the form of a Twitter thread can be found here: https://twitter.com/WiringTheBrain/status/1095012297200844800

87 There are additional reasons to be sceptical. A 2015 study of over 1,500 people who'd had colectomies – the surgical removal of the colon and thus all its associated microbiota – followed them up several years later to see whether they were less likely than control patients to have developed heart disease. If the microbiome was a big contributor to the disease – and, given their colectomies, these were patients with very unhealthy microbiomes – we might have expected to see a reduction in their risk of disease. There were no differences. Anders

Boeck Jensen et al., 'Long-Term Risk of Cardiovascular and Cerebrovascular Disease after Removal of the Colonic Microbiota by Colectomy: A Cohort Study Based on the Danish National Patient Register from 1996 to 2014', *BMJ Open* 5, no. 12 (Dec. 2015): e008702; https://doi.org/10.1136/bmjopen-2015-008702

88   William P. Hanage, 'Microbiology: Microbiome Science Needs a Healthy Dose of Scepticism', *Nature* 512, no. 7514 (Aug. 2014): pp. 247–48; https://doi.org/10.1038/512247a. Gwen Falony et al., 'The Human Microbiome in Health and Disease: Hype or Hope', *Acta Clinica Belgica* 74, no. 2 (4 Mar. 2019): pp. 53–64; https://doi.org/10.1080/17843286.2019.1583782; and J. Taylor, 'The Microbiome and Mental Health: Hope or Hype?', *Journal of Psychiatry and Neuroscience* 44, no. 4 (1 July 2019): pp. 219–22; https://doi.org/10.1503/jpn.190110

89   'boost your performance': Andrew Holtz, 'Harvard Researchers' Speculative, Poop-Based Sports Drink Company Raises Questions about Conflicts of Interest', *Health News Review*, 19 Oct. 2017; https://www.healthnewsreview.org/2017/10/harvard-researchers-speculative-poop-based-sports-drink-company-raises-questions-about-conflicts-of-interest/; see also the *Lancet Gastroenterology & Hepatology* (Editorial), 'Probiotics: Elixir or Empty Promise?', *Lancet Gastroenterology & Hepatology* 4, no. 2 (Feb. 2019): p. 81; https://doi.org/10.1016/S2468-1253(18)30415-1; 'rectal perforation': Shapiro, Nina, 'There Are Trillions Of Reasons Not To Cleanse Your Colon', *Forbes*, 19 Sept. 2019; https://www.forbes.com/sites/nina-shapiro/2019/09/19/there-are-trillions-of-reasons-not-to-cleanse-your-colon/; for the dangers, see also Doug V. Handley et al., 'Rectal Perforation from Colonic Irrigation Administered by Alternative Practitioners', *Medical Journal of Australia* 181, no. 10 (15 Nov. 2004): pp. 575–76; https://doi.org/10.5694/j.1326-5377.2004.tb06454.x. The whole endeavour is pointless, in any case: two weeks after all the microbes are flushed out by a colonic irrigation, everything is back to the way it was pre-irrigation. See Naoyoshi Nagata et al., 'Effects of Bowel Preparation on the Human Gut Microbiome and Metabolome', Scientific Reports 9, no. 1 (Dec. 2019): p. 4042; https://doi.org/10.1038/s41598-019-40182-9. The nationality of your microbiome: https://atlasbiomed.com/uk/microbiome/results. See also Kavin Senapathy, 'Keep Calm And Avoid Microbiome Mayhem', *Forbes*, 7 March 2016; https://www.forbes.com/sites/kavinsenapathy/2016/03/07/keep-calm-and-avoid-microbiome-mayhem/

90   Milk: Josh Harkinson, 'The Scary New Science That Shows Milk Is Bad For You', *Mother Jones*, Dec. 2015; https://www.motherjones.com/environment/2015/11/dairy-industry-milk-federal-dietary-guidelines/; bacon: 'Killer Full English: Bacon Ups Cancer Risk', *LBC News*, 17 April 2019; https://www.lbc.co.uk/news/killer-full-english-bacon-ups-cancer-risk/; eggs: Physicians' Committee for Responsible Medicine, 'New Study Finds Eggs Will Break Your Heart', 16 March 2016; https://www.pcrm.org/news/blog/new-study-finds-eggs-will-break-your-heart. This was followed up by the sub-heading: 'Americans are eating 279 eggs per person a year, and a new study finds it's **killing** them' [bold in the original]. The original study is Victor Zhong et al., 'Associations of Dietary Cholesterol or Egg Consumption with Incident Cardiovascular Disease and Mortality', *JAMA* 321, no. 11 (19 Mar. 2019): p. 1081; https://doi.org/10.1001/jama.2019.1572. For a detailed critique, see Zad Rafi, 'Revisiting Eggs and Dietary Cholesterol', *Less Likely*, 22 March 2019; https://lesslikely.com/nutrition/eggs-cholesterol/. Relatedly, my favourite parody of this media tendency is from the website Clickhole: 'Nutritional Shake-Up: The FDA Now Recommends That Americans Eat A Bowl Of 200 Eggs On Their 30th Birthday And Then Never Eat Any Eggs Again', *Clickhole*, 24 Oct. 2017; https://news.clickhole.com/nutritional-shake-up-the-fda-now-recommends-that-ameri-1825121901

91   After years of exaggerated findings, the public now lacks confidence and is sceptical of the field's research. This point has been made previously by the dietetics researcher Kevin Klatt: https://twitter.com/kcklatt/status/902558341414694912. There's some evidence from a survey by the British Nutrition Foundation that people find the 'mixed messages' about nutritional research to be highly confusing, though it's not clear how representative their survey sample was; https://www.nutrition.org.uk/press-office/pressreleases/1156-mixedmessages.html

92   A detailed summary of the case of Das, who died the year following his dismissal, can be found here: Geoffrey P. Webb, 'Dipak Kumar Das (1946–2013) Who Faked Data about Resveratrol – the Magic Red Wine Ingredient That Cures Everything?', Dr Geoff Nutrition, 10 Nov. 2017; https://drgeoffnutrition.wordpress.com/2017/11/10/dipak-kumar-das-1946-2013-who-faked-data-about-resveratrol-the-magic-red-wine-ingredient-that-cures-everything/

93   A recent controversy arose over the quality of studies on red meat, where the major players on one side had links to the meat industry, and those on the other side had links to companies selling vegetarian products that stood to benefit if people cut down on their red meat consumption. Rita Rubin, 'Backlash Over Meat Dietary Recommendations Raises Questions About Corporate Ties to Nutrition Scientists', JAMA 323, no. 5 (4 Feb. 2020): 401; https://doi.org/10.1001/jama.2019.21441

94   John P. A. Ioannidis and John F. Trepanowski, 'Disclosures in Nutrition Research: Why it is Different', JAMA 319, no. 6 (13 Feb. 2018): p. 547; https://doi.org/10.1001/jama.2017.18571

95   For example, see this webpage from the UK's National Health Service: https://www.nhs.uk/live-well/eat-well/different-fats-nutrition/, and this from the Mayo Clinic: Mayo Clinic Staff, 'Dietary Fats: Know Which Types to Choose', 1 Feb. 2019; https://www.mayoclinic.org/healthy-lifestyle/nutrition-and-healthy-eating/in-depth/fat/art-20045550

96   Steven Hamley, 'The Effect of Replacing Saturated Fat with Mostly N-6 Polyunsaturated Fat on Coronary Heart Disease: A Meta-Analysis of Randomised Controlled Trials', Nutrition Journal 16, no. 1 (Dec. 2017): p. 30; https://doi.org/10.1186/s12937-017-0254-5

97   Also, some trials showing no difference between the fatty acid types took an abnormally long time to be published, hinting that their authors – or their peer-reviewers – were reluctant to see them in print.

98   For a full explanation, see Matti Miettinen et al., 'Effect of Cholesterol-lowering Diet on Mortality from Coronary Heart-Disease and Other Causes', Lancet 300, no. 7782 (Oct. 1972): pp. 835–38; https://doi.org/10.1016/S0140-6736(72)92208-8

99   In any experiment, the only thing that should be changing is the variable of interest – in this case, whether saturated or unsaturated fats were eaten. But the groups in some trials differed in other ways: being given different dietary advice, for example, and in one case being administered different drugs as part of a hospital study.

100   Jonathan D. Schoenfeld & John P. A. Ioannidis, 'Is Everything We Eat Associated with Cancer? A Systematic Cookbook Review', American Journal of Clinical Nutrition 97, no. 1 (1 Jan. 2013): pp. 127–34; https://doi.org/10.3945/ajcn.112.047142

101   Indeed, as we've seen in other areas, the meta-analyses that followed up the original studies finding those large effects on the risks of cancer tended to find much smaller effects.

102   For a list of nutritional correlations that didn't hold up in randomised trials, see S. Stanley Young & Alan Karr, 'Deming, Data and Observational Studies: A Process out of Control and Needing Fixing', *Significance* 8, no. 3 (Sept. 2011): pp. 116–20; https://doi.org/10.1111/j.1740-9713.2011.00506.x. However, for some push-back against this idea see 'Myth 4' in Ambika Satija et al., 'Perspective: Are Large, Simple Trials the Solution for Nutrition Research?', *Advances in Nutrition* 9, no. 4 (1 July 2018): p. 381; https://doi.org/10.1093/advances/nmy030. A big problem is that randomised controlled trials, because they're so expensive to run, tend to be far smaller than the observational studies, making it an apples-to-oranges comparison (and I don't mean literally, though I'm sure that study does exist somewhere in the annals of nutritional epidemiology).

103   Jakob Westfall & Tai Yarkoni, 'Statistically Controlling for Confounding Constructs Is Harder than You Think', *PLOS ONE* 11, no. 3 (31 Mar. 2016): e0152719; https://doi.org/10.1371/journal.pone.0152719

104   Edward Archer et al., 'Controversy and Debate: Memory-Based Methods Paper 1: The Fatal Flaws of Food Frequency Questionnaires and Other Memory-Based Dietary Assessment Methods', *Journal of Clinical Epidemiology* 104 (Dec. 2018): p. 113; https://doi.org/10.1016/j.jclinepi.2018.08.003

105   For a strong attack on the validity of Food Frequency Questionnaires, see Edward Archer et al., 'Validity of U.S. Nutritional Surveillance: National Health and Nutrition Examination Survey Caloric Energy Intake Data, 1971–2010', *PLOS ONE* 8, no. 10 (9 Oct. 2013): e76632; https://doi.org/10.1371/journal.pone.0076632. Others think that the criticisms are themselves hyped and overblown. For a qualified defence, see James R. Hébert et al., 'Considering the Value of Dietary Assessment Data in Informing Nutrition-Related Health Policy', *Advances in Nutrition* 5, no. 4 (1 July 2014): pp. 447–55; https://doi.org/10.3945/an.114.006189. For additional context about the debate, see Alex Berezow, 'Is Nutrition Science Mostly Junk?', *American Council on Science and Health*, 20 Nov. 2018; https://www.acsh.org/news/2018/11/19/nutrition-science-mostly-junk-13611; David Nosowitz, 'The Bizarre Quest to Discredit America's Most Important Nutrition Survey', *TakePart*, 29 July 2015; http://www.takepart.com/article/2015/06/29/america-dietary-guidelines-self-reporting

106   Trepanowski & Ioannidis, 'Disclosures in Nutrition'.

107   Satija et al., 'Perspective'. This article also defends nutritional research against many of the criticisms I've made here. Similarly, see Edward Giovannucci, 'Nutritional Epidemiology: Forest, Trees and Leaves', *European Journal of Epidemiology* 34, no. 4 (April 2019): pp. 319–25; https://doi.org/10.1007/s10654-019-00488-4. Ioannidis's response: John P. A. Ioannidis, 'Unreformed Nutritional Epidemiology: A Lamp Post in the Dark Forest', *European Journal of Epidemiology* 34, no. 4 (April 2019): pp. 327–31; https://doi.org/10.1007/s10654-019-00487-5

108   Ramón Estruch et al., 'Primary Prevention of Cardiovascular Disease with a Mediterranean Diet', *New England Journal of Medicine* 368, no. 14 (4 April 2013): pp. 1279–90; https://doi.org/10.1056/NEJMoa1200303

109   Gina Kolata, 'Mediterranean Diet Shown to Ward Off Heart Attack and Stroke', *New York Times*, 25 Feb. 2013; https://www.nytimes.com/2013/02/26/health/mediterranean-diet-can-cut-heart-disease-study-finds.html

110   David Brown, 'Mediterranean Diet Reduces Cardiovascular Risk', *Washington Post*, 25 Feb. 2013; https://www.washingtonpost.com/national/health-science/mediterranean-diet-reduces-cardiovascular-risk/2013/02/25/20396e16-7f87-11e2-a350-49866afab584_story.html

111 California Walnut Commission, 'Landmark Clinical Study Reports Mediterranean Diet Supplemented with Walnuts Significantly Reduces Risk of Stroke and Cardiovascular Diseases' (news release), 25 Feb. 2013; https://www.prnewswire.com/news-releases/landmark-clinical-study-reports-mediterranean-diet-supplemented-with-walnuts-significantly-reduces-risk-of-stroke-and-cardiovascular-diseases-192989571.html

112 'proved': Universitat de Barcelona, 'Mediterranean Diet Helps Cut Risk of Heart Attack, Stroke: Results of PREDIMED Study Presented' (news release), 25 Feb. 2013; https://www.sciencedaily.com/releases/2013/02/130225181536.htm; 'strong evidence': M. Guasch-Ferré et al., 'The PREDIMED Trial, Mediterranean Diet and Health Outcomes: How Strong Is the Evidence?', *Nutrition, Metabolism and Cardiovascular Diseases* 27, no. 7 (July 2017): p. 6; https://doi.org/10.1016/j.numecd.2017.05.004

113 J. B. Carlisle, 'Data Fabrication and Other Reasons for Non-Random Sampling in 5087 Randomised, Controlled Trials in Anaesthetic and General Medical Journals', *Anaesthesia* 72, no. 8 (Aug. 2017): pp. 944–52; https://doi.org/10.1111/anae.13938

114 Ramón Estruch et al., 'Primary Prevention of Cardiovascular Disease with a Mediterranean Diet Supplemented with Extra-Virgin Olive Oil or Nuts', *New England Journal of Medicine* 378, no. 25 (21 June 2018): e34 (34); https://doi.org/10.1056/NEJMoa1800389

115 Citations calculated using Google Scholar; 'corrected version': Estruch et al., 'Primary Prevention … Olive Oil or Nuts'.

116 Julia Belluz, 'This Mediterranean Diet Study Was Hugely Impactful. The Science Has Fallen Apart', *Vox*, 13 Feb. 2019; https://www.vox.com/science-and-health/2018/6/20/17464906/mediterranean-diet-science-health-predimed

117 Also, the study was stopped early, so the effect size might be inflated. See Dirk Bassler et al., 'Early Stopping of Randomized Clinical Trials for Overt Efficacy Is Problematic', *Journal of Clinical Epidemiology* 61, no. 3 (Mar. 2008): pp. 241–46; https://doi.org/10.1016/j.jclinepi.2007.07.016

118 Arnav Agarwal & John P. A. Ioannidis, 'PREDIMED Trial of Mediterranean Diet: Retracted, Republished, Still Trusted?', *BMJ* (7 Feb. 2019): p. l341; https://doi.org/10.1136/bmj.l341. The fact that the diet only seemed to affect stroke, not heart attacks or mortality, was obscured in the study by the use of so-called 'composite endpoints'. These amalgams of several different outcomes are commonly found in clinical trials, because they increase statistical power. The downside is that it's difficult to interpret the specific effects of your intervention. See Christopher McCoy, 'Understanding the Use of Composite Endpoints in Clinical Trials', *Western Journal of Emergency Medicine* 19, no. 4 (29 June 2018): pp. 631–34; https://doi.org/10.5811/westjem.2018.4.38383. See also Eric Lim et al., 'Composite Outcomes in Cardiovascular Research: A Survey of Randomized Trials', *Annals of Internal Medicine* 149, no. 9 (4 Nov. 2008): pp. 612–17; https://doi.org/10.7326/0003-4819-149-9-200811040-00004. Finally, the authors tested a lot of different outcomes, and didn't make any correction for the substantial number of *p*-values this generated, increasing the false-positive risk.

119 OPERA stands for Oscillation Project with Emulsion-tRacking Apparatus. The broad idea was to see how neutrinos changed their properties ('oscillated') as they moved between the emitter in Switzerland and the detector in Italy. Details can be found at http://operaweb.lngs.infn.it/

120 Ransom Stephens, 'The Data That Threatened to Break Physics', *Nautilus*, 28 Dec. 2017; http://nautil.us/issue/55/trust/the-data-that-threatened-to-break-physics-rp

121   T. Adam et al., 'Measurement of the Neutrino Velocity with the OPERA Detector in the CNGS Beam', *Journal of High Energy Physics* 2012, no. 10 (Oct. 2012): 93; https://doi.org/10.1007/JHEP10(2012)093

122   CERN, 'OPERA Experiment Reports Anomaly in Flight Time of Neutrinos from CERN to Gran Sasso', 23 Sept. 2011; https://home.cern/news/press-release/cern/opera-experiment-reports-anomaly-flight-time-neutrinos-cern-gran-sasso

123   'CERN Scientists "Break the Speed of Light"', *Daily Telegraph*, 22 Sept. 2011; https://www.telegraph.co.uk/news/science/8782895/CERN-scientists-break-the-speed-of-light.html and 'The Speed of Light: Not So Fast?', ABC News, 24 Sept. 2011; https://www.youtube.com/watch?v=zgmL47lD7RA

124   In truth the underestimate was 73 nanoseconds; the researchers then found a second issue with a timing circuit that produced a slight lag. Both problems combined led to the 60-nanosecond underestimate. Stephens, 'The Data That Threatened'.

125   Lisa Grossman, 'Faster-than-Light Neutrino Result to Get Extra Checks', *New Scientist*, 25 Oct. 2011; https://www.newscientist.com/article/dn21093-faster-than-light-neutrino-result-to-get-extra-checks/

126   Antonio Ereditato, 'OPERA: Ereditato's Point of View', *Le Scienze*, 30 March 2012; http://www.lescienze.it/news/2012/03/30/news/opera_ereditatos_point_of_view-938232/

127   Jason Palmer, 'Faster-than-Light Neutrinos Could Be down to Bad Wiring', *BBC News*, 23 Feb. 2012; https://www.bbc.co.uk/news/science-environment-17139635; Lisa Grossman & Celeste Biever, 'Was Speeding Neutrino Claim a Human Error?', *New Scientist*, 23 Feb. 2012; https://www.newscientist.com/article/dn21510-was-speeding-neutrino-claim-a-human-error/

7: *Perverse Incentives*

Epigraph: Cormac McCarthy, *No Country for Old Men* (London: Picador, 2005).

1   Sukey Lewis, 'Cleaning Up: Inside the Wildfire Debris Removal Job That Cost Taxpayers $1.3 Billion', *KQED*, 19 July 2018; https://www.kqed.org/news/11681280/cleaning-up-inside-the-wildfire-debris-removal-job-that-cost-taxpayers-1-3-billion

2   A related perverse incentive is that of the US Endangered Species Act, which incentivised landowners to destroy perfectly good habitat for rare animals, since doing so avoided their land being regulated. Jacob P. Byl, 'Accurate Economics to Protect Endangered Species and Their Critical Habitats', *SSRN* preprint (2018); https://doi.org/10.2139/ssrn.3143841

3   Cary Funk & Meg Hefferon, 'As the Need for Highly Trained Scientists Grows, a Look at Why People Choose These Careers', *Fact Tank*, 24 Oct. 2016; https://www.pewresearch.org/fact-tank/2016/10/24/as-the-need-for-highly-trained-scientists-grows-a-look-at-why-people-choose-these-careers/

4   Melissa S. Anderson et al., 'Extending the Mertonian Norms: Scientists' Subscription to Norms of Research', *Journal of Higher Education* 81, no. 3 (May 2010): pp. 366–93; https://doi.org/10.1080/00221546.2010.11779057. They don't, however, necessarily agree that their colleagues follow those norms – see Melissa Anderson et al., 'Normative Dissonance in Science: Results from a National Survey of U.S. Scientists', *Journal of Empirical Research on Human Research Ethics* 2, no. 4 (Dec. 2007): pp. 3–14; https://doi.org/10.1525/jer.2007.2.4.3

5   Darwin Correspondence Project, 'Letter no. 5986' (6 March 1868); https://www. darwinproject.ac.uk/letter/DCP-LETT-5986.xml

6   '400,000 studies': Steven Kelly, 'The Continuing Evolution of Publishing in the Biological Sciences', *Biology Open* 7, no. 8 (15 Aug. 2018): bio037325; https:// doi.org/10.1242/bio.037325; '2.4 million papers': Andrew Plume & Daphne van Weijen, 'Publish or Perish? The Rise of the Fractional Author …', *Research Trends*, Sept. 2014; https://www.researchtrends.com/issue-38-september-2014/ publish-or-perish-the-rise-of-the-fractional-author/. According to a report from the US National Science Foundation, in 2016 China became the largest single-country producer of scientific papers, just above the US. Jeff Tollefson, 'China Declared World's Largest Producer of Scientific Articles', *Nature* 553, no. 7689 (18 Jan. 2018): p. 390; https://doi.org/10.1038/d41586-018-00927-4

7   Lutz Bornmann & Rüdiger Mutz, 'Growth Rates of Modern Science: A Biblio-metric Analysis Based on the Number of Publications and Cited References: Growth Rates of Modern Science: A Bibliometric Analysis Based on the Number of Publications and Cited References', *Journal of the Association for Information Science and Technology* 66, no. 11 (Nov. 2015): pp. 2215–22; https://doi.org/ 10.1002/asi.23329

8   One investigation found rewards of US$165,000, which can be around 20 times the annual salary. The average reward amounts are around US$44,000. All of this, incidentally, highlights the low salaries of Chinese academics, which according to Quan et al. average around $8,600 per annum. In some cases, it appears that only a fraction of the awarded money is an individual prize, and the rest gets invested in the scientists' future research. The real figures and the ways in which they're spent are often mysterious. Wei Quan et al., 'Publish or Impoverish: An Investigation of the Monetary Reward System of Science in China (1999-2016)', *Aslib Journal of Information Management* 69, no. 5 (18 Sept. 2017): pp. 486–502; https://doi.org/10.1108/AJIM-01-2017-0014

9   Alison Abritis, 'Cash Bonuses for Peer-Reviewed Papers Go Global', *Science*, 10 Aug. 2017; https://doi.org/10.1126/science.aan7214. See also Editorial, 'Don't Pay Prizes for Published Science', *Nature* 547, no. 7662 (July 2017): p. 137; https://doi.org/10.1038/547137a

10 The policy does appear to work, at least for the narrow purpose of encouraging attempts at high-impact publication. One analysis showed that, after countries introduced cash incentives, submissions to the journal *Science* from those countries increased by 46 per cent (a larger increase than with other incentives) – though they were negatively associated with the actual acceptance rate of those papers. That is, the scientists started taking more pot-shots at a high-end journal, without much success. See C. Franzoni et al., 'Changing Incentives to Publish', *Science* 333, no. 6043 (5 Aug. 2011): pp. 702–3; https://doi.org/10.1126/science.1197286. Perhaps ironically, one of the authors of that study, which was itself published in *Science*, received a cash bonus of $3,500 for the publication – though in this case, he donated it to charity. See Alison Abritis, 'Cash Bonuses for Peer-Reviewed Papers Go Global', *Science*, 10 Aug. 2017; https://doi.org/10.1126/ science.aan7214

11 https://www.ref.ac.uk. Other countries have debated and ultimately decided not to use similar processes: Gunnar Sivertsen, 'Why Has No Other European Country Adopted the Research Excellence Framework?', *LSE Impact of Social Sciences*, 18 Jan. 2018; https://blogs.lse.ac.uk/impactofsocialsciences/2018/01/16/ why-has-no-other-european-country-adopted-the-research-excellence-framework/

12   For an (inconclusive) attempt to track down the origins of the phrase 'publish or perish', see Eugene Garfield, 'What is the Primordial Reference for the Phrase "Publish or Perish"?', *Scientist* 10, no. 2 (10 June 1996): p. 11.

13   Albert N. Link et al., 'A Time Allocation Study of University Faculty', *Economics of Education Review* 27, no. 4 (Aug. 2008): pp. 363–74; https://doi.org/10.1016/j.econedurev.2007.04.002

14   My quotation is from the King James version. It was first highlighted in the context of science by Robert Merton, he of the Mertonian norms. R. K. Merton, 'The Matthew Effect in Science: The Reward and Communication Systems of Science Are Considered', *Science* 159, no. 3810 (5 Jan. 1968): pp. 56–63; https://doi.org/10.1126/science.159.3810.56

15   Thijs Bol et al., 'The Matthew Effect in Science Funding', *Proceedings of the National Academy of Sciences* 115, no. 19 (8 May 2018): pp. 4887–90; https://doi.org/10.1073/pnas.1719557115

16   This is borne out by two main lines of evidence. First, the vast majority of those who get a PhD in science don't end up staying scientists permanently. In a 2010 analysis for the Royal Society, it was found that 53 per cent of PhDs immediately left science, then a further 26.5 per cent joined them, leaving science after the early part of their career. 17 per cent left to do research outside academia – for instance, for industry or government. Overall only 3.5 per cent remained in science permanently (and 0.45 per cent made it to full professorship). Royal Society, *The Scientific Century: Securing Our Future Prosperity* (London: Royal Society, 2010); https://royalsociety.org/-/media/Royal_Society_Content/policy/publications/2010/4294970126.pdf. The second line of evidence comes from surveys of scientists. One such survey by the Wellcome Trust in early 2020 found that no less than 78 per cent agreed that the high levels of competition in science 'have created unkind and aggressive conditions'. Wellcome Trust, *What Researchers Think about the Culture They Work In* (London: Wellcome Trust, 2020); https://wellcome.ac.uk/reports/what-researchers-think-about-research-culture

17   François Brischoux & Frédéric Angelier, 'Academia's Never-Ending Selection for Productivity', *Scientometrics* 103, no. 1 (April 2015): pp. 333–36; https://doi.org/10.1007/s11192-015-1534-5. Other studies show similar trends. For parallel evidence from the Canadian psychology job market, see Gordon Pennycook & Valerie A. Thompson, 'An Analysis of the Canadian Cognitive Psychology Job Market (2006–2016)', *Canadian Journal of Experimental Psychology/Revue Canadienne de Psychologie Expérimentale* 72, no. 2 (June 2018): pp. 71–80; https://doi.org/10.1037/cep0000149

18   David Cyranoski et al., 'Education: The PhD Factory', *Nature* 472, no. 7343 (April 2011): pp. 276–79; https://doi.org/10.1038/472276a. You may be wondering why, if there are all these excess PhDs hanging around without research jobs, one often hears of a crisis where not enough people are going into science, technology, engineering and mathematics jobs to keep up with the demands of the modern industrialized economy. One review of the literature found that, in fact, both are true at the same time: there are too many PhDs looking for jobs in the university sector, but not enough for the needs of government and industry. Yi Xue & Richard Larson, 'STEM Crisis or STEM Surplus? Yes and Yes', *Monthly Labor Review* (26 May 2015); https://doi.org/10.21916/mlr.2015.14

19   Richard P. Heitz, 'The Speed-Accuracy Tradeoff: History, Physiology, Methodology, and Behavior', *Frontiers in Neuroscience* 8 (11 June 2014): p. 150; https://doi.org/10.3389/fnins.2014.00150

20 This is set out mathematically by Remco Heesen, 'Why the Reward Structure of Science Makes Reproducibility Problems Inevitable', *Journal of Philosophy* 115, no. 12 (2018): pp. 661–74; https://doi.org/10.5840/jphil20181151239. See also Daniel Sarewitz, 'The Pressure to Publish Pushes down Quality', *Nature* 533, no. 7602 (May 2016): p. 147; https://doi.org/10.1038/533147a

21 All that said, I should acknowledge a study that appears to go against my argument: Daniele Fanelli et al., 'Misconduct Policies, Academic Culture and Career Stage, Not Gender or Pressures to Publish, Affect Scientific Integrity', *PLOS ONE* 10, no. 6 (17 June 2015): e0127556; https://doi.org/10.1371/journal.pone.0127556. In 2015, researchers counted up all the public corrections to, and retractions of, scientific articles that had occurred in the years 2010 and 2011 – essentially, all the times that fixes had to be made to the scientific record in these years. They correlated the number of times each researcher had to make these fixes with their other characteristics, like their overall number of publications and the country they worked in. The overall conclusion was that countries that pay their academics by the paper, and those without good policies on dealing with scientific misconduct, had higher rates of retractions: so far, so good for my thesis. But they also found that researchers who published more papers per year had slightly *fewer* retractions overall – a finding they interpreted as going against the idea that the publish-or-perish culture causes more research misconduct. Retractions are quite rare and quite extreme, though: retraction involves removing a paper from the literature and is often due to major infractions like fraud. In my view, therefore, the 2015 paper can't be used as a defence of publish-or-perish, since it didn't measure anything about the quality of the articles. The authors also found that researchers who publish more regularly tend to publish more corrections. They argue that this is a good thing, since corrections 'carry no stigma'. I find this argument odd: not only do corrections, in my experience, definitely carry stigma, but by definition they represent errors: they're correcting mistakes that shouldn't have been made in the first place.

22 Quan et al., 'Publish or Impoverish'.

23 The first published instance I could find of the term 'salami-slicing' being used in the context of science is John Maddox, 'Is the Salami Sliced Too Thinly?', *Nature* 342, no. 6251 (Dec. 1989): p. 733; https://doi.org/10.1038/342733a0. This source mentions that the metaphor had been around for a while prior to that. The term had also been used previously in other contexts, such as in cases where employees steal a large amount from their workplace by repeatedly taking tiny quantities over a long space of time.

24 I refuse to cite all the papers, as that would be letting those authors win. But one is Xing Chen et al., 'A Novel Relationship for Schizophrenia, Bipolar and Major Depressive Disorder Part 5: A Hint from Chromosome 5 High Density Association Screen', *American Journal of Translational Research* 9, no. 5 (2017): pp. 2473–91; https://www.ncbi.nlm.nih.gov/pubmed/28559998

25 By the way, neither paper found any difference – so at least these were publications of null results. Glen I. Spielmans et al., 'A Case Study of Salami Slicing: Pooled Analyses of Duloxetine for Depression', *Psychotherapy and Psychosomatics* 79, no. 2 (2010): pp. 97–106; https://doi.org/10.1159/000270917

26 For a further example, this time relating to antipsychotic medication, see Glen. I. Spielmans et al., '"Salami Slicing" in Pooled Analyses of Second-Generation Antipsychotics for the Treatment of Depression', *Psychotherapy and Psychosomatics* 86, no. 3 (2017): pp. 171–72; https://doi.org/10.1159/000464251

27 Of course, an important plank of this book's argument is that the usual peer-review and editorial standards at scientific journals are very much lacking – they must be, given the amount of bad research that's accepted and published. But predatory journals don't even *try*.

28 There's a similar scam industry of fake academic *conferences*, to which researchers are invited in regular junk emails. Good overviews are given by James McCrostie, '"Predatory Conferences" Stalk Japan's Groves of Academia', *Japan Times*, 11 May 2016; https://www.japantimes.co.jp/community/2016/05/11/issues/predatory-conferences-stalk-japans-groves-academia/ and Emma Stoye, 'Predatory Conference Scammers Are Getting Smarter', *Chemistry World*, 6 Aug. 2018; https://www.chemistryworld.com/news/predatory-conference-scammers-are-getting-smarter/3009263.article

29 The University of Colorado Denver librarian Jeffrey Beall waged a one-man campaign against predatory journals. See Jeffrey Beall, 'What I Learned from Predatory Publishers', *Biochemia Medica* 27, no. 2 (15 June 2017): pp. 273–78; https://doi.org/10.11613/BM.2017.029. His list of dodgy-looking outlets eventually disappeared from the internet (https://retractionwatch.com/2017/01/17/bealls-list-potential-predatory-publishers-go-dark/), but a newer version, with a very long list of 'possibly predatory journals' can be found at https://predatory-journals.com/journals/. See also Pravin Bolshete, 'Analysis of Thirteen Predatory Publishers: A Trap for Eager-to-Publish Researchers', *Current Medical Research and Opinion* 34, no. 1 (2 Jan. 2018): pp. 157–62; https://doi.org/10.1080/03007995.2017.1358160 and Agnes Grudniewicz et al., 'Predatory Journals: No Definition, No Defence', *Nature* 576, no. 7786 (Dec. 2019): pp. 210–12; https://doi.org/10.1038/d41586-019-03759-y

30 Alas, the paper was never officially published, because Vamplew refused to pay the journal's fee of $150 for publication (more details in Joseph Stromberg, '"Get Me Off Your Fucking Mailing List" Is an Actual Science Paper Accepted by a Journal', *Vox*, 21 Nov. 2014; https://www.vox.com/2014/11/21/7259207/scientific-paper-scam). The manuscript, originally written several years earlier for a similar purpose by the computer scientists David Mazières and Eddie Kohler, can be read in its glorious entirety at the following link: http://www.scs.stanford.edu/~dm/home/papers/remove.pdf

31 Quoted in Ivan Oransky, 'South Korean Plant Compound Researcher Faked Email Addresses so He Could Review His Own Studies', *Retraction Watch*, 24 Aug. 2012; https://retractionwatch.com/2012/08/24/korean-plant-compound-researcher-faked-email-addresses-so-he-could-review-his-own-studies/

32 For a detailed discussion of one example, and a *mea culpa* from editors who let it through, see Adam Cohen et al., 'Organised Crime against the Academic Peer Review System: Organised Crime against the Academic Peer Review System', *British Journal of Clinical Pharmacology* 81, no. 6 (June 2016): pp. 1012–17; https://doi.org/10.1111/bcp.12992

33 Alison McCook, 'A New Record: Major Publisher Retracting More than 100 Studies from Cancer Journal over Fake Peer Reviews', *Retraction Watch*, 20 April 2017; https://retractionwatch.com/2017/04/20/new-record-major-publisher-retracting-100-studies-cancer-journal-fake-peer-reviews/

34 Vincent Larivière et al., 'The Decline in the Concentration of Citations, 1900-2007', *Journal of the American Society for Information Science and Technology* 60, no. 4 (April 2009): pp. 858–62; https://doi.org/10.1002/asi.21011.9

35 Ibid. At least things aren't as bad in science as in the humanities: fewer than 20 per cent of humanities papers are cited within five years of publication. Of course,

there are different citation practices between science and the humanities, with the latter emphasising books more than papers. Even so, one wonders if that 80 per cent of non-cited papers have really made much of a contribution to knowledge.

36 J. E. Hirsch, 'An Index to Quantify an Individual's Scientific Research Output', *Proceedings of the National Academy of Sciences* 102, no. 46 (15 Nov. 2005): pp. 16569–72; https://doi.org/10.1073/pnas.0507655102

37 I'd be remiss if I didn't say that Google Scholar, if anything, overestimates the *h*-index, because it's very liberal about what it includes as a citation. There are other ways of calculating *h*-indices, such as the site Web of Science, that are more conservative and thus come up with lower numbers.

38 Bram Duyx et al., 'Scientific Citations Favor Positive Results: A Systematic Review and Meta-Analysis', *Journal of Clinical Epidemiology* 88 (Aug. 2017): pp. 92–101; https://doi.org/10.1016/j.jclinepi.2017.06.002; see also R. Leimu and J. Koricheva, 'What Determines the Citation Frequency of Ecological Papers?', *Trends in Ecology & Evolution* 20, no. 1 (Jan. 2005): pp. 28–32; https://doi.org/10.1016/j.tree.2004.10.010

39 'a third of all citations': Dag W. Aksnes, 'A Macro Study of Self-Citation', *Scientometrics* 56, no. 2 (2003): pp. 235–46; https://doi.org/10.1023/A:1021919228368. It also works alongside salami-slicing: if your salami-sliced papers all cite one another, you'll get far more citations than you would otherwise, even if nobody except you has actually read or cited your work. I'm tempted to suggest that this is the salami turning back on itself and becoming one of those ring sausages, but that may be taking the metaphor too far.

40 For some extreme examples, see John P. Ioannidis (2015), 'A Generalized View of Self-Citation: Direct, Co-Author, Collaborative, and Coercive Induced Self-Citation', *Journal of Psychosomatic Research* 78, no. 1 (Jan. 2015): pp. 7–11; https://doi.org/10.1016/j.jpsychores.2014.11.008

41 Colleen Flaherty, 'Revolt Over an Editor', *Inside Higher Ed*, 30 April 2018; https://www.insidehighered.com/news/2018/04/30/prominent-psychologist-resigns-journal-editor-over-allegations-over-self-citation. For another, similar story, this time from the world of research on Autism Spectrum Disorder, see Pete Etchells & Chris Chambers, 'The Games We Play: A Troubling Dark Side in Academic Publishing', *Guardian*, 12 March 2015; https://www.theguardian.com/science/head-quarters/2015/mar/12/games-we-play-troubling-dark-side-academic-publishing-matson-sigafoos-lancioni

42 Eiko Fried, '7 Sternberg Papers: 351 References, 161 Self-Citations', *Eiko Fried*, 29 March 2018; https://eiko-fried.com/sternberg-selfcitations/

43 Brett D. Thombs et al., 'Potentially Coercive Self-Citation by Peer Reviewers: A Cross-Sectional Study', *Journal of Psychosomatic Research* 78, no. 1 (Jan. 2015): pp. 1–6; https://doi.org/10.1016/j.jpsychores.2014.09.015

44 That Sternberg tried to use this specific method to game the system is quite ironic, since in 2017 he'd written the following in his book of advice for academics: 'Self-plagiarism occurs when you fail adequately to cite your own work ... In extreme cases, someone might attempt to publish exactly the same paper twice without noting that the paper has been previously published.' Robert J. Sternberg, & Karin Sternberg, *The Psychologist's Companion: A Guide to Professional Success for Students, Teachers, and Researchers* (Cambridge: CUP, 2016): p.141.

45 The original paper (where the text was originally published) is Robert J. Sternberg, 'WICS: A New Model for Cognitive Education', *Journal of Cognitive Education and Psychology* 9, no. 1 (Feb. 2010): pp. 36–47; https://doi.org/10.1891/1945-8959.9.1.36. The retraction notice 'for reasons of redundant publication' is

Editorial, 'Retraction Notice for "WICS: A New Model for School Psychology" by Robert J. Sternberg', *School Psychology International* 39, no. 3 (June 2018): p. 329; https://doi.org/10.1177/0143034318782213. See also Nicholas J. L. Brown, 'Some Instances of Apparent Duplicate Publication by Dr. Robert J. Sternberg', *Nick Brown's Blog*, 25 April 2018; https://steamtraen.blogspot.com/2018/04/some-instances-of-apparent-duplicate.html

46 Tracey Bretag & Saadia Carapiet, 'A Preliminary Study to Identify the Extent of Self-Plagiarism in Australian Academic Research', *Plagiary: Cross-Disciplinary Studies in Plagiarism, Fabrication, and Falsification* 2, no. 5 (2007): pp. 1–12.

47 Éric Archambault & Vincent Larivière, 'History of the Journal Impact Factor: Contingencies and Consequences', *Scientometrics* 79, no. 3 (June 2009): pp. 635–49; https://doi.org/10.1007/s11192-007-2036-x

48 Technically, and somewhat confusingly, the impact factor is the average number of citations in a given year to papers published by that journal in the two years before that. Of course, a whole year has to elapse for this to be calculated, so it's always slightly out of date. For instance, in 2020, a journal's impact factor will be the average number of citations that were made in 2019 to articles published by that journal in 2017 and 2018.

49 The *American Journal of Potato Research*, in case you're wondering, currently has an impact factor of 1.095.

50 Vincent Larivière et al., 'A Simple Proposal for the Publication of Journal Citation Distributions', *bioRxiv*, 5 July 2016; https://doi.org/10.1101/062109; see also Vincent Larivière & Cassidy R. Sugimoto, 'The Journal Impact Factor: A Brief History, Critique, and Discussion of Adverse Effects', in *Springer Handbook of Science and Technology Indicators*, eds. Wolfgang Glänzel, Henk F. Moed, Ulrich Schmoch, & Mike Thelwall, pp. 3–24 (Cham: Springer International Publishing, 2019); https://doi.org/10.1007/978-3-030-02511-3_1. For the other side of the argument, see Lutz Bornmann & Alexander I. Pudovkin, 'The Journal Impact Factor Should Not Be Discarded', *Journal of Korean Medical Science* 32, no. 2 (2017): p. 180–82; https://doi.org/10.3346/jkms.2017.32.2.180

51 Richard Monastersky, 'The Number That's Devouring Science', *Chronicle of Higher Education*, 14 Oct. 2005; https://www.chronicle.com/article/the-number-thats-devouring/26481

52 A. W. Wilhite & E. A. Fong, 'Coercive Citation in Academic Publishing', *Science* 335, no. 6068 (2 Feb. 2012): pp. 542–43; https://doi.org/10.1126/science.1212540

53 Phil Davis, 'The Emergence of a Citation Cartel', *Scholarly Kitchen*, 10 April 2012; https://scholarlykitchen.sspnet.org/2012/04/10/emergence-of-a-citation-cartel/

54 Paul Jump, 'Journal Citation Cartels on the Rise', *Times Higher Education*, 21 June 2013; https://www.timeshighereducation.com/news/journal-citation-cartels-on-the-rise/2005009.article. The cartels may also have found their Eliot Ness: an algorithm was developed in 2017 which, when fed data on citations, flags up groups of authors who appear to be citing each other disproportionately. Iztok Fister Jr. et al., 'Toward the Discovery of Citation Cartels in Citation Networks', *Frontiers in Physics* 4:49 (15 Dec. 2016); https://doi.org/10.3389/fphy.2016.00049

55 Charles Goodhart, 'Monetary Relationships: A View from Threadneedle Street', *Papers in Monetary Economics* I (Reserve Bank of Australia, 1975). The specific phrasing is due to Marilyn Strathearn, 'Improving Ratings: Audit in the British University System', *European Review* 5, no. 3 (July 1997): pp. 305–21; https://doi.org/10.002/(SICI)1234-981X(199707)5:3<305::AID-EURO184>3.0.CO;2-4

56 Paul E. Smaldino & Richard McElreath, 'The Natural Selection of Bad Science', *Royal Society Open Science* 3, no. 9 (Sept. 2016): 160384; https://doi.org/10.1098/rsos.160384

57 Andrew D. Higginson & Marcus R. Munafò, 'Current Incentives for Scientists Lead to Underpowered Studies with Erroneous Conclusions', *PLOS Biology* 14, no. 11 (10 Nov. 2016): p.6; https://doi.org/10.1371/journal.pbio.2000995

58 David Robert Grimes et al., 'Modelling Science Trustworthiness under Publish or Perish Pressure', *Royal Society Open Science* 5, no. 1 (Jan. 2018); https://doi.org/10.1098/rsos.171511

59 There's another version by the same artist, with a different alchemist, in the National Gallery of Art in Washington DC.

60 See Anton Howes, 'Age of Invention: When Alchemy Works', *Age on Invention*, 6 Oct. 2019; https://antonhowes.substack.com/p/age-of-invention-when-alchemy-works and Richard Conniff, 'Alchemy May Not Have Been the Pseudoscience We All Thought It Was', *Smithsonian Magazine*, Feb. 2014; https://www.smithsonianmag.com/history/alchemy-may-not-been-pseudoscience-we-thought-it-was-180949430/

61 The big difference between the alchemists and modern scientists is that for modern scientists, the gold is real. As we saw in Chapter 6, there are enormous monetary rewards for scientists who are willing to hype up and dumb down their findings into best-selling books and high-profile lecture-circuit tours.

62 Marc A. Edwards & Siddhartha Roy, 'Academic Research in the 21st Century: Maintaining Scientific Integrity in a Climate of Perverse Incentives and Hypercompetition', *Environmental Engineering Science* 34, no. 1 (Jan. 2017): pp. 51–61; https://doi.org/10.1089/ees.2016.0223

63 Tal Yarkoni, 'No, It's Not The Incentives – It's You', *[Citation Needed]*, 2 Oct. 2018; https://www.talyarkoni.org/blog/2018/10/02/no-its-not-the-incentives-its-you/

64 See Edwards & Roy, 'Academic Research', Fig. 1.

8: *Fixing Science*

Epigraph: Michael Nielsen http://michaelnielsen.org/blog/the-future-of-science-2/

1 Y. A. de Vries et al., 'The Cumulative Effect of Reporting and Citation Biases on the Apparent Efficacy of Treatments: The Case of Depression', *Psychological Medicine* 48, no. 15 (Nov. 2018): pp. 2453–55; https://doi.org/10.1017/S0033291718001873

2 Ibid. p. 2453.

3 Specifically, they found examples of spin and citation bias; it was more difficult to assess the other problems since psychotherapy trials aren't required to be pre-registered like those of new drugs.

4 In some countries this is due partly to privacy laws. Charles Seife, 'Research Misconduct Identified by the US Food and Drug Administration: Out of Sight, Out of Mind, Out of the Peer-Reviewed Literature', *JAMA Internal Medicine* 175, no. 4 (1 April 2015): pp. 567–577; https://doi.org/10.1001/jamainternmed.2014.7774. See also Michael Robinson, 'Canadian Researchers Who Commit Scientific Fraud Are Protected by Privacy Laws', *The Star*, 12 July 2016; https://www.thestar.com/news/canada/2016/07/12/canadian-researchers-who-commit-scientific-fraud-are-protected-by-privacy-laws.html

5 Ivan Oransky & Adam Marcus, 'Governments Routinely Cover up Scientific Misdeeds. Let's End That', *STAT News*, 15 Dec. 2015; https://www.statnews.com/2015/12/15/governments-scientific-misdeeds/

6 Chia-Yi Hou, 'Sweden Passes Law For National Research Misconduct Agency', *Scientist*, 10 July 2019; https://www.the-scientist.com/news-opinion/sweden-passes-law-for-national-research-misconduct-agency-66129

7   Morten P. Oksvold, 'Incidence of Data Duplications in a Randomly Selected Pool of Life Science Publications', *Science and Engineering Ethics* 22, no. 2 (April 2016): pp. 487–96; https://doi.org/10.1007/s11948-015-9668-7. M. Enrico Bucci et al., 'Automatic Detection of Image Manipulations in the Biomedical Literature', *Cell Death & Disease* 9, no. 3 (Mar. 2018): p. 400; https://doi.org/10.1038/s41419-018-0430-3. Incidentally, the latter paper notes that the AI algorithm picked up on an 'alarming' amount of image duplication in the cell biology literature.

8   Technology can also help scientists see the scientific literature more clearly. In Chapter 3, we discussed the problem of zombie papers that keep on being cited after their retraction. There's an automated solution: Zotero, the freely downloadable reference-managing software (which many scientists use to store and organise the bibliography for each paper, which is already a major time-saver and error-preventer – I used it for this book) recently announced a partnership with Retraction Watch, whereby it'll flag up any withdrawn publications that you're about to cite, informing you that 'this work has been retracted'. Dan Stillman, 'Retracted Item Notifications with Retraction Watch Integration', *Zotero*, 14 June 2019; https://www.zotero.org/blog/retracted-item-notifications/

9   There are even suggestions that sophisticated algorithms could scan through tens of thousands of papers and find all the features, perhaps not obvious to human readers, that predict whether their findings will replicate. Adam Rogers, 'Darpa Wants to Solve Science's Reproducibility Crisis With AI', *Wired*, 15 Feb. 2019; https://www.wired.com/story/darpa-wants-to-solve-sciences-replication-crisis-with-robots/

10   One such program is *RMarkdown*: https://rmarkdown.rstudio.com/

11   An even more radical proposal along these lines is the psychologist Jeff Rouder's 'born-open data', where the data from any new participant tested in an experiment is automatically uploaded into an online dataset at the end of each test day. Jeffrey N. Rouder, 'The What, Why, and How of Born-Open Data', *Behavior Research Methods* 48, no. 3 (Sept. 2016): pp. 1062–69; https://doi.org/10.3758/s13428-015-0630-z

12   Mark Ziemann et al., 'Gene Name Errors are Widespread in the Scientific Literature', *Genome Biology* 17, no. 1 (Dec. 2016): 177; https://doi.org/10.1186/s13059-016-1044-7

13   'Ottoline Leyser on How Plants Decide What to Do', *The Life Scientific*, BBC Radio 4, 16 May 2017.

14   Brian A. Nosek et al., 'Scientific Utopia: II. Restructuring Incentives and Practices to Promote Truth Over Publishability', *Perspectives on Psychological Science* 7, no. 6 (Nov. 2012): pp. 615–631; https://doi.org/10.1177/1745691612459058, p.619.

15   You can find an archive of papers from the *Journal of Negative Results in Biomedicine* here: https://jnrbm.biomedcentral.com/articles. There's also the *Journal of Negative Results: Ecology and Evolutionary Biology*, which is a less professional-looking outlet that had hardly any submissions. In 2014 and 2015 it didn't publish a single paper, then published two papers in 2016, one more in 2018, and hasn't been heard from since. There couldn't be a better illustration of how unpopular it is to publish your paper in a journal that only accepts null studies; http://www.jnr-eeb.org/index.php/jnr

16   The first word stands for 'Public Library of Science'. *PLOS ONE*'s strategy of publishing anything that passes peer review regardless of 'interest' has been enormously successful, turning it into the world's largest journal in terms of papers published – until 2017, when it was overtaken by *Scientific Reports*, a megajournal with the same publishing strategy: Phil Davis, 'Scientific Reports Overtakes PLoS ONE As Largest Megajournal', 6 April 2017; https://scholarlykitchen.sspnet.org/2017/04/06/scientific-reports-overtakes-plos-one-as-largest-megajournal/

Scientific Reports is published by the Nature Publishing Group and thus gets some of the halo effect of the world's top journal, *Nature*, which might partly explain its popularity. Other examples of journals with the 'accept anything that's high quality, regardless of perceived impact' model include *PeerJ* (https://peerj.com/) and *Royal Society Open Science* (https://royalsocietypublishing.org/journal/rsos).

17 https://www.apa.org/pubs/journals/psp/?tab=4

18 Sanjay Srivastava, 'A Pottery Barn Rule for Scientific Journals', *The Hardest Science*, 27 Sept. 2012; https://thehardestscience.com/2012/09/27/a-pottery-barn-rule-for-scientific-journals/ – this is known as the 'Pottery Barn rule', after the specific US retail chain that supposedly uses it. But this is something of an urban legend, for there's no such rule at Pottery Barn. See Daniel Grant, 'You Break It, You Buy It? Not According to the Law', *Crafts Report*, April 2005; https://web.archive.org/web/20061207233337/http:/www.craftsreport.com/april05/break_not_buy.html

19 B. A. Nosek et al., 'Promoting an Open Research Culture', *Science* 348, no. 6242 (26 June 2015): 1422–25; https://doi.org/10.1126/science.aab2374. See also https://cos.io/top/

20 Jop de Vrieze, '"Replication Grants" Will Allow Researchers to Repeat Nine Influential Studies That Still Raise Questions', *Science*, 11 July 2017; https://doi.org/10.1126/science.aan7085

21 To take just two well-known examples from the mid- and late-twentieth century: David Bakan, 'The Test of Significance in Psychological Research', *Psychological Bulletin* 66, no. 6 (1966): pp. 423–37; https://doi.org/10.1037/h0020412. And: Jacob Cohen, 'The Earth Is Round (p < .05)', *American Psychologist* 49, no. 12 (1994): pp. 997–1003; https://doi.org/10.1037/0003-066X.49.12.997

22 Stephen Thomas Ziliak & Deirdre N. McCloskey, *The Cult of Statistical Significance: How the Standard Error Costs Us Jobs, Justice, and Lives, Economics, Cognition, and Society* (Ann Arbor: University of Michigan Press, 2008): p. 33.

23 Valentin Amrhein et al., 'Scientists Rise up against Statistical Significance', *Nature* 567, no. 7748 (March 2019): pp. 305–7; https://doi.org/10.1038/d41586-019-00857-9

24 For more from this perspective see Geoff Cumming, 'The New Statistics: Why and How', *Psychological Science* 25, no. 1 (Jan. 2014): pp. 7–29; https://doi.org/10.1177/0956797613504966; and Lewis G. Halsey, 'The Reign of the *p*-Value Is Over: What Alternative Analyses Could We Employ to Fill the Power Vacuum?', *Biology Letters* 15, no. 5 (31 May 2019): 20190174; https://doi.org/10.1098/rsbl.2019.0174

25 As long as you know the effect size and the *p*-value, you can derive the confidence intervals and vice versa. See D. G. Altman & J. M. Bland, 'How to Obtain the Confidence Interval from a P Value', *BMJ* 343 (16 July 2011): d2090; https://doi.org/10.1136/bmj.d2090 and D. G. Altman & J. M. Bland, 'How to Obtain the P Value from a Confidence Interval', *BMJ* 343 (8 Aug. 2011): d2304; https://doi.org/10.1136/bmj.d2304

26 John P. A. Ioannidis, 'The Importance of Predefined Rules and Prespecified Statistical Analyses: Do Not Abandon Significance', *JAMA* 321, no. 21 (4 June 2019): p. 2067; https://doi.org/10.1001/jama.2019.4582

27 Quoted in Andrew Gelman, '"Retire Statistical Significance": The Discussion', *Statistical Modeling, Causal Inference, and Social Science*, 20 March 2019; https://statmodeling.stat.columbia.edu/2019/03/20/retire-statistical-significance-the-discussion/

28 In this sense, a Bayesian prior is a mathematical way of invoking Carl Sagan's famous dictum that 'extraordinary claims require extraordinary evidence'.

29 The broader statistical tradition where p-values sit, incidentally, is called *frequentist* statistics. That's because, fundamentally, users of p-values are interested in *frequencies* – most notably the frequency with which you'll find results with p-values below 0.05 if you run your study an infinite number of times and the hypothesis you're testing isn't true.

30 A useful annotated reading list that serves as an introduction to Bayesian statistics is given by Etz et al., 'How to Become a Bayesian in Eight Easy Steps: An Annotated Reading List', *Psychonomic Bulletin & Review* 25, no. 1 (Feb. 2018): 219–34; https://doi.org/10.3758/s13423-017-1317-5. See also Richard McElreath, *Statistical Rethinking: A Bayesian Course with Examples in R and Stan*, Chapman & Hall/CRC Texts in Statistical Science Series 122 (Boca Raton: CRC Press/Taylor & Francis Group, 2016).

31 For (often very qualified) defences of the beleaguered p-value, see Victoria Savalei & Elizabeth Dunn, 'Is the Call to Abandon P-Values the Red Herring of the Replicability Crisis?', *Frontiers in Psychology* 6:245 (6 March 2015); https://doi.org/10.3389/fpsyg.2015.00245; Paul A. Murtaugh, 'In Defense of P Values', *Ecology* 95, no. 3 (March 2014): pp. 611–17; https://doi.org/10.1890/13-0590.1; and S. Senn, 'Two Cheers for P-Values?', *Journal of Epidemiology and Biostatistics* 6, no. 2 (1 March 2001): 193–204; https://doi.org/10.1080/135952201753172953

32 Daniel J. Benjamin et al., 'Redefine Statistical Significance', *Nature Human Behaviour* 2, no. 1 (Jan. 2018): pp. 6–10; https://doi.org/10.1038/s41562-017-0189-z. Though see the response from Daniël Lakens et al., 'Justify Your Alpha', *Nature Human Behaviour* 2, no. 3 (March 2018): pp. 168–71; https://doi.org/10.1038/s41562-018-0311-x. There are also new statistical methods that still use p-values yet move away from the naïve scenario where they're always comparing results to zero (or zero difference). See Daniël Lakens et al., 'Equivalence Testing for Psychological Research: A Tutorial', *Advances in Methods and Practices in Psychological Science* 1, no. 2 (June 2018): pp. 259–69; https://doi.org/10.1177/2515245918770963

33 Alternatively, the researchers could run the analysis but allow independent statisticians to interpret it. Isabelle Boutron & Philippe Ravaud, 'Misrepresentation and Distortion of Research in Biomedical Literature', *Proceedings of the National Academy of Sciences* 115, no. 11 (13 March 2018): pp. 2613–19; https://doi.org/10.1073/pnas.1710755115

34 Or when the statistician thinks that the study design wasn't up to scratch in the first place. Ronald Fisher famously said in 1938 that 'to consult the statistician after an experiment is finished is often merely to ask him to conduct a post mortem examination. He can perhaps say what the experiment died of.' https://www.gwern.net/docs/statistics/decision/1938-fisher.pdf

35 'specification-curve analysis': Uri Simonsohn et al., 'Specification Curve: Descriptive and Inferential Statistics on All Reasonable Specifications', *SSRN Electronic Journal* (2015); https://doi.org/10.2139/ssrn.2694998. 'Vibration-of-effects': Chirag J. Patel et al., 'Assessment of Vibration of Effects Due to Model Specification Can Demonstrate the Instability of Observational Associations', *Journal of Clinical Epidemiology* 68, no. 9 (Sept. 2015): pp. 1046–58; https://doi.org/10.1016/j.jclinepi.2015.05.029. 'Multiverse analysis': Sara Steegen et al., 'Increasing Transparency Through a Multiverse Analysis', *Perspectives on Psychological Science* 11, no. 5 (Sept. 2016): pp. 702–12; https://doi.org/10.1177/1745691616658637

36 Amy Orben & Andrew K. Przybylski, 'The Association between Adolescent Well-Being and Digital Technology Use', *Nature Human Behaviour* 3, no. 2 (Feb. 2019): pp. 173–82; https://doi.org/10.1038/s41562-018-0506-1. Full disclosure: Orben and Przybylski are friends and colleagues of mine.

37 Sam Blanchard, 'Smartphones and Tablets Are Causing Mental Health Problems in Children as Young as TWO by Crushing Their Curiosity and Making Them Anxious', *MailOnline*, 2 Nov. 2018; https://www.dailymail.co.uk/health/article-6346349/Smartphones-tablets-causing-mental-health-problems-children-young-two.html

38 For example, see Jean M. Twenge, 'Have Smartphones Destroyed a Generation?', *Atlantic*, Sept. 2017; https://www.theatlantic.com/magazine/archive/2017/09/has-the-smartphone-destroyed-a-generation/534198/; and Jean M. Twenge, *IGEN: Why Today's Super-Connected Kids Are Growing up Less Rebellious, More Tolerant, Less Happy – and Completely Unprepared for Adulthood (and What This Means for the Rest of Us)* (New York: Atria Books, 2017).

39 'Video gaming disorder': https://www.who.int/features/qa/gaming-disorder/en/; Antonius J. van Rooij et al., 'A Weak Scientific Basis for Gaming Disorder: Let Us Err on the Side of Caution', *Journal of Behavioral Addictions* 7, no. 1 (March 2018): pp. 1–9; https://doi.org/10.1556/2006.7.2018.19; 'online porn disorder': Rubén de Alarcón et al., 'Online Porn Addiction: What We Know and What We Don't – A Systematic Review', *Journal of Clinical Medicine* 8, no. 1 (15 Jan. 2019): p. 91; https://doi.org/10.3390/jcm8010091; 'iPhone addiction': André Spicer, 'The iPhone Is the Crack Cocaine of Technology. Don't Celebrate Its Birthday', *Guardian*, 29 June 2017; https://www.theguardian.com/commentisfree/2017/jun/29/apple-iphone-ten-years-old-crippling-addiction; 'the list goes on': Christopher Snowdon, 'Evidence-Based Puritanism', *Velvet Glove, Iron Fist*, 10 Jan. 2019; https://velvet-gloveironfist.blogspot.com/2019/01/evidence-based-puritanism.html

40 Those same researchers have also produced other studies using multiverse analyses on different datasets that broadly come to the same conclusion: the 'screen time' panic is overblown. See e.g. Amy Orben & Andrew K. Przybylski, 'Screens, Teens, and Psychological Well-Being: Evidence from Three Time-Use-Diary Studies', *Psychological Science* 30, no. 5 (May 2019): pp. 682–96; https://doi.org/10.1177/0956797619830329. Amy Orben et al., 'Social Media's Enduring Effect on Adolescent Life Satisfaction', *Proceedings of the National Academy of Sciences* 116, no. 21 (21 May 2019): 10226–28; https://doi.org/10.1073/pnas.1902058116. This isn't to say that all online activities are harmless, or that screen time isn't problematic for some children. But on average, the effects are a lot smaller than the media panic would lead you to believe.

41 A timeline is provided on the ClinicalTrials.gov website here: https://clinicaltrials.gov/ct2/about-site/history. See also Jamie L. Todd et al., 'Using ClinicalTrials.Gov to Understand the State of Clinical Research in Pulmonary, Critical Care, and Sleep Medicine', *Annals of the American Thoracic Society* 10, no. 5 (Oct. 2013): pp. 411–17; https://doi.org/10.1513/AnnalsATS.201305-111OC

42 Sophie Scott, 'Pre-Registration Would Put Science in Chains', *Times Higher Education*, 25 July 2013; https://www.timeshighereducation.com/comment/opinion/pre-registration-would-put-science-in-chains/2005954.article

43 Eric-Jan Wagenmakers et al., 'An Agenda for Purely Confirmatory Research', *Perspectives on Psychological Science* 7, no. 6 (Nov. 2012): pp. 632–38; https://doi.org/10.1177/1745691612463078

44 Robert M. Kaplan & Veronica L. Irvin, 'Likelihood of Null Effects of Large NHLBI Clinical Trials has Increased Over Time', *PLOS ONE* 10, no. 8 (5 Aug. 2015): e0132382; https://doi.org/10.1371/journal.pone.0132382

45 What we don't yet have, it should be said, is an answer to the ultimate question: are pre-registered results more likely to *replicate* in future studies? Only time, and more meta-scientific data, will tell.

46 Kent Anderson, 'Why Is ClinicalTrials.Gov Still Struggling?', *Scholarly Kitchen*, 15 March 2016; https://scholarlykitchen.sspnet.org/2016/03/15/why-is-clinicaltrials-gov-still-struggling/; Monique Anderson, 'Compliance with Results Reporting at ClinicalTrials.Gov', *New England Journal of Medicine* 372, no. 11 (12 Mar. 2015): pp. 1031–39; https://doi.org/10.1056/NEJMsa1409364; Ruijan Chen et al., 'Publication and Reporting of Clinical Trial Results: Cross Sectional Analysis across Academic Medical Centers', *BMJ* (17 Feb. 2016): i637; https://doi.org/10.1136/bmj.i637; Ben Goldacre et al., 'Compliance with Requirement to Report Results on the EU Clinical Trials Register: Cohort Study and Web Resource', *BMJ* (12 Sept. 2018): k3218; https://doi.org/10.1136/bmj.k3218. There's also some initial evidence from psychology that the dream of the pre-registered analysis doesn't always match the reality: Aline Claesen et al., 'Preregistration: Comparing Dream to Reality', preprint, *PsyArXiv*, 9 May 2019; https://doi.org/10.31234/osf.io/d8wex

47 'one investigation by the journal *Science*': Charles Piller, 'FDA and NIH Let Clinical Trial Sponsors Keep Results Secret and Break the Law', *Science*, 13 Jan. 2020; https://doi.org/10.1126/science.aba8123; 'scofflaws': Charles Piller, 'Clinical Scofflaws?' *Science*, 13 Jan. 2020; https://doi.org/10.1126/science.aba8575

48 S. D. Turner et al., 'Publication Rate for Funded Studies from a Major UK Health Research Funder: A Cohort Study', *BMJ Open* 3, no. 5 (2013): e002521; https://doi.org/10.1136/bmjopen-2012-002521; Fay Chinnery et al., 'Time to Publication for NIHR HTA Programme-Funded Research: A Cohort Study', *BMJ Open* 3, no. 11 (Nov. 2013): e004121; https://doi.org/10.1136/bmjopen-2013-004121. See also Paul Glasziou & Iain Chalmers, 'Funders and Regulators Are More Important than Journals in Fixing the Waste in Research', *TheBMJOpinion*, 6 Sept. 2017; https://blogs.bmj.com/bmj/2017/09/06/paul-glasziou-and-iain-chalmers-funders-and-regulators-are-more-important-than-journals-in-fixing-the-waste-in-research/

49 See Christopher Chambers, 'Registered Reports: A New Publishing Initiative at Cortex', *Cortex* 49, no. 3 (March 2013): pp. 609–10; https://doi.org/10.1016/j.cortex.2012.12.016. At the time of writing, 225 journals offer the option of such Registered Reports – a tiny minority, but a growing one (https://cos.io/rr/). See also Chris Chambers, *The Seven Deadly Sins of Psychology: A Manifesto for Reforming the Culture of Scientific Practice* (Princeton: Princeton University Press, 2017).

50 Initial work on Registered Reports finds similarly attention-grabbing results to those in Figure 4: a 61 per cent rate of null results in Registered Reports for psychological and biomedical studies compared to an estimate of only 10 per cent in the regular, non-registered research. Christopher Allen & David M. A. Mehler, 'Open Science Challenges, Benefits and Tips in Early Career and Beyond', *PLOS Biology* 17, no. 5 (1 May 2019): e3000246; https://doi.org/10.1371/journal.pbio.3000246 (see their Figure 1); see also Anne M. Scheel et al., 'An Excess of Positive Results: Comparing the Standard Psychology Literature with Registered Reports', *PsyArXiv*, preprint, 5 Feb. 2020; https://doi.org/10.31234/osf.io/p6e9c, who found nearly identical results for psychology. Again, there are alternative explanations that might play a role in explaining this difference. If the kinds of studies where scientists tend to run Registered Reports are different from the ones where Registered Reports are less common (for example, scientists might

be more likely to set up a Registered Report for findings about which they're very sceptical, and thus that are probably less likely to be true), you'd already expect to see differences in the rates of positive and negative findings that have nothing to do with the Registered Report *per se*.

51 Indeed, you can even think of the invention of the scientific journal by the Royal Society in the seventeenth century as one of the first steps in opening up science and making it more than just the private pastime of a few individuals. See Paul A. David, 'The Historical Origins of "Open Science": An Essay on Patronage, Reputation and Common Agency Contracting in the Scientific Revolution', *Capitalism and Society* 3, no. 2 , 24 Jan. 2008; https://doi.org/10.2202/1932-0213.1040

52 A useful summary can be found in: Marcus R. Munafò et al., 'A Manifesto for Reproducible Science', *Nature Human Behaviour* 1, no. 1 (Jan. 2017): 0021; https://doi.org/10.1038/s41562-016-0021

53 The Open Science Framework (https://osf.io/) is one online repository where data can be stored, along with time-stamped pre-registrations, working papers, and much more.

54 One publication that does this is the general biology journal *eLife*: if you look up any of its articles, you'll find the previous versions and the reviews linked from its page; https://elifesciences.org

55 But it does happen. Indeed, as I write this, just such a case might well be playing out in the field of behavioural ecology. See Giuliana Viglione, '"Avalanche" of Spider-Paper Retractions Shakes Behavioural-Ecology Community', *Nature* 578, no. 7794 (Feb. 2020): 199–200; https://doi.org/10.1038/d41586-020-00287-y

56 It's not quite as simple as posting data online somewhere. It makes sense for there to be a record of who's accessed data, in the form of some kind of gatekeeper to whom researchers can make data requests. It's the same principle as a registry: it acts as a barrier against researchers who might be tempted to download a freely available dataset, *p*-hack their way to significant results, then publish a paper pretending they hypothesized those results right from the start. Gatekeepers, along with repositories where data are stored, require funding, though. This is one reason why scientists haven't done enough of this in the past, and one of the ways funders could do better.

57 S. Herfst et al., 'Airborne Transmission of Influenza A/H5N1 Virus Between Ferrets', *Science* 336, no. 6088 (22 June 2012): pp. 1534–41; https://doi.org/10.1126/science.1213362

58 National Research Council et al., *Perspectives on Research with H5N1 Avian Influenza: Scientific Inquiry, Communication, Controversy: Summary of a Workshop* (Washington, D.C.: National Academies Press, 2013); https://doi.org/10.17226/18255. Appendix B: Official Statements.

59 Daniel Stokols et al., (2008), 'The Science of Team Science', *American Journal of Preventive Medicine* 35, no. 2 (Aug. 2008): S77–89; https://doi.org/10.1016/j.amepre.2008.05.002

60 One such team is the OPERA group, the discoverers of the faster-than-light neutrino that wasn't.

61 For example, see the Psychiatric Genomics Consortium: https://www.med.unc.edu/pgc/

62 For discussion see Peter M. Visscher et al., '10 Years of GWAS Discovery: Biology, Function, and Translation', *The American Journal of Human Genetics* 101, no. 1 (July 2017): pp. 5–22; https://doi.org/10.1016/j.ajhg.2017.06.005. Of course, genetics research has also been subject to huge and unjustified hype – see Timothy

Caulfield, 'Spinning the Genome: Why Science Hype Matters', *Perspectives in Biology and Medicine* 61, no. 4 (2018): pp. 560–71; https://doi.org/10.1353/pbm.2018.0065

63   'neuroscience': http://enigma.ini.usc.edu/; 'cancer epidemiology': https://epi.grants.cancer.gov/InterLymph/; 'psychology': https://psysciacc.org/; 'translational medical research': http://www.dcn.ed.ac.uk/camarades/default.htm. Some researchers have also begun 'adversarial collaborations', which are the organised-scepticism part of science made manifest: scientists attempt to counter their biases by deliberately working with researchers who take the opposite view on some scientific debate. The idea is that if both 'sides' can agree on a fair test of the theory, and work together to perform that test, the outcome will be more convincing to everyone involved. My favourite example of an adversarial collaboration is one where proponents and sceptics of the existence of psychic powers collaborated to run a set of experiments testing whether people could psychically tell when another person was staring at the back of their head. I won't spoil the outcome, because it's well worth a read. Marilyn Schlitz et al., 'Of Two Minds: Sceptic-Proponent Collaboration within Parapsychology', *British Journal of Psychology* 97, no. 3 (Aug. 2006): pp. 313–22; https://doi.org/10.1348/000712605X80704

64   https://www.usa.gov/government-works

65   https://www.coalition-s.org/

66   Holly Else, 'Radical Open-Access Plan Could Spell End to Journal Subscriptions', *Nature* 561, no. 7721 (Sept. 2018): pp. 17–18; https://doi.org/10.1038/d41586-018-06178-7. I should note that not everyone is instinctually a fan of Open Access. Probably the best and most detailed discussion of the pros and cons of Open Access is the one by the artificial intelligence researcher Daniel Allington: Daniel Allington, 'On Open Access, and Why It's Not the Answer', *Daniel Allington*, 15 Oct. 2013; http://www.danielallington.net/2013/10/open-access-why-not-answer/

67   Randy Schekman, 'Scientific Research Shouldn't Sit behind a Paywall', *Scientific American*, 20 June 2019; https://blogs.scientificamerican.com/observations/scientific-research-shouldnt-sit-behind-a-paywall/. I chose to mention the University of California and Elsevier because, at the time of this writing, they're embroiled in a massive dispute where the UC system tried to get Elsevier to reduce its costs, Elsevier refused, and then UC played hardball by simply cancelling its subscription to Elsevier's journals. Good on them. University of California Office of Scholarly Communication, 'UC and Elsevier', 20 March 2019; https://osc.universityofcalifornia.edu/uc-publisher-relationships/uc-and-elsevier/

68   According to one report, in at least some years the scientific publisher Elsevier has made a bigger percentage profit margin than Apple, Google, or Amazon: Stephen Buranyi, 'Is the Staggeringly Profitable Business of Scientific Publishing Bad for Science?', *Guardian*, 27 June 2017; https://www.theguardian.com/science/2017/jun/27/profitable-business-scientific-publishing-bad-for-science. The true ridiculousness of Elsevier's position on scientific publication was revealed in 2016, when they patented the idea of 'online peer review'. This was awarded Stupid Patent of the Month for August 2016 by the Electronic Frontier Foundation. Daniel Nazer & Elliot Harmon, 'Stupid Patent of the Month: Elsevier Patents Online Peer Review', *Electronic Freedom Foundation*, 31 Aug. 2016; https://www.eff.org/deeplinks/2016/08/stupid-patent-month-elsevier-patents-online-peer-review. For much more on the sins of Elsevier, see Tal Yarkoni, 'Why I Still Won't

Review for or Publish with Elsevier – and Think You Shouldn't Either', *[Citation Needed]*, 12 Dec. 2016; https://www.talyarkoni.org/blog/2016/12/12/why-i-still-wont-review-for-or-publish-with-elsevier-and-think-you-shouldnt-either/

69 Buranyi, 'Is the Staggeringly Profitable ... Bad for Science?'

70 At least the publisher Wiley, very much unlike Elsevier, has shown interest in negotiating new publishing models: Diana Kwon, 'As Elsevier Falters, Wiley Succeeds in Open-Access Deal Making', *Scientist*, 26 March 2019; https://www.the-scientist.com/news-opinion/as-elsevier-falters--wiley-succeeds-in-open-access-deal-making-65664

71 Preprints from physics can be found on *arXiv*, (https://arxiv.org/); it's pronounced 'archive', because the X is actually a Greek letter chi (χ). Preprints from economics, where they're usually called 'working papers', can be found, among other places, at the National Burea of Economic Research (https://www.nber.org/papers.html). The main biology preprint server is *bioRxiv* (https://www.biorxiv.org/); the medical one is *medRxiv* (https://www.medrxiv.org); for psychology, it's *PsyArXiv* (https://psyarxiv.com/). There's a list of servers for other subjects here: https://osf.io/preprints/. For a description of just how popular biology preprinting has become, and how quickly, see Richard J. Abdill & Jan Blekhman, 'Tracking the Popularity and Outcomes of All bioRxiv Preprints', *eLife* 8 (24 April 2019): e45133; https://doi.org/10.7554/eLife.45133. You might also be wondering how preprint servers are funded. In some cases, it's by scientific grants, but some archives have more sustainable models: for example, the original *arXiv* has agreements with various universities who each chip in funds that rise as a function of how heavily their scientists download and use its preprints (see https://arxiv.org/about/ourmembers).

72 As noted in footnote 54 above, some standard journals are also opening up their peer review process.

73 Versions of this idea have been discussed by Brian A. Nosek & Yoav Bar-Anan, 'Scientific Utopia: I. Opening Scientific Communication', *Psychological Inquiry* 23, no. 3 (July 2012): pp. 217–43; https://doi.org/10.1080/1047840X.2012.692215; and by Bodo M. Stern & Erin K. O'Shea, 'A Proposal for the Future of Scientific Publishing in the Life Sciences', *PLOS Biology* 17, no. 2 (12 Feb. 2019): e3000116; https://doi.org/10.1371/journal.pbio.3000116. See also Aliaksandr Birukou et al., 'Alternatives to Peer Review: Novel Approaches for Research Evaluation', *Frontiers in Computational Neuroscience* 5 (2011); https://doi.org/10.3389/fncom.2011.00056

74 A promising instantiation of this idea is 'Peer Community In' (PCI), which 'aims to create specific communities of researchers reviewing and recommending, for free, unpublished preprints in their field'. At the time of writing, PCI includes peer communities in evolutionary biology, ecology, paleontology, animal science, entomology, circuit neuroscience and genomics; https://peercommunityin.org/

75 Stern & O'Shea ('A Proposal for the Future') even suggest that journals could curate issues of 'papers that need replication' or 'papers that contain questionable claims'.

76 One can imagine combinations of this system with the 'Registered Reports' framework that we saw above, where study plans rather than finished papers are preprinted and evaluated. It could also work with the error-checking algorithms, which could just as easily be implemented by a preprint repository as an old-style journal.

77 Indeed, something is already being tested on a small scale, and it's described at the following link: https://asapbio.org/eisen-appraise

78 It should be said that most preprint archives do *screen* papers, to avoid the very worst nonsense being posted up by malicious actors or crackpot 'independent researchers'. But they don't (and can't, given the time it would take) do full reviews.

79 Roland Fryer (2016), 'An Empirical Analysis of Racial Differences in Police Use of Force [Working Paper]', (Cambridge, MA: National Bureau of Economic Research, July 2016); https://doi.org/10.3386/w22399

80 Ibid. p. 5.

81 Quoctrung Bui & Amanda Cox, 'Surprising New Evidence Shows Bias in Police Use of Force but Not in Shootings', *New York Times*, 11 July 2016; https://www.nytimes.com/2016/07/12/upshot/surprising-new-evidence-shows-bias-in-police-use-of-force-but-not-in-shootings.html

82 Larry Elder, 'Ignorance of Facts Fuels the Anti-Cop "Movement"', *RealClear Politics*, 14 July 2016; https://www.realclearpolitics.com/articles/2016/07/14/ignorance_of_facts_fuels_the_anti-cop_movement_131188.html

83 Uri Simonsohn, 'Teenagers in Bikinis: Interpreting Police-Shooting Data', *Data Colada*, 14 July 2016; http://datacolada.org/50

84 This fact was noted in the next line of the working paper after the 23.8 per cent figure was reported but didn't appear to make it into Fryer's interactions with the press. Further criticism of Fryer's study can be found at the following blogs by economists: Rajiv Sethi, 'Police Use of Force: Notes on a Study', 11 July 2016; https://rajivsethi.blogspot.com/2016/07/police-use-of-force-notes-on-study.html; and Justin Feldman, 'Roland Fryer is Wrong: There is Racial Bias in Shootings by Police', 12 July 2016; https://scholar.harvard.edu/jfeldman/blog/roland-fryer-wrong-there-racial-bias-shootings-police

85 Roland Fryer, 'An Empirical Analysis of Racial Differences in Police Use of Force', *Journal of Political Economy* 127, no. 3 (June 2019): pp. 1210–61; https://doi.org/10.1086/701423

86 In fact, Fryer's preprint had been looked over by several of his fellow economists – it just hadn't been formally reviewed and published. Daniel Engber, 'Was This Study Even Peer-Reviewed?', *Slate*, 25 July 2016; https://slate.com/technology/2016/07/roland-fryers-research-on-racial-bias-in-policing-wasnt-peer-reviewed-does-that-matter.html

87 https://www.biorxiv.org/content/early/recent (note that the warning is likely a temporary one; it may have changed or been removed by the time this book goes to press).

88 Kai Kupferschmidt, 'Preprints Bring "Firehose" of Outbreak Data', *Science* 367, no. 6481 (28 Feb. 2020): pp. 963–64; https://doi.org/10.1126/science.367.6481.963

89 Erin McKiernan et al., 'The "Impact" of the Journal Impact Factor in the Review, Tenure, and Promotion Process', *Impact of Social Sciences*, 26 April 2019; https://blogs.lse.ac.uk/impactofsocialsciences/2019/04/26/the-impact-of-the-journal-impact-factor-in-the-review-tenure-and-promotion-process/

90 Jeffrey S. Flier (2019), 'Credit and Priority in Scientific Discovery: A Scientist's Perspective', *Perspectives in Biology and Medicine* 62, no. 2 (2019): pp. 189–215, https://doi.org/10.1353/pbm.2019.0010

91 It's very different in some other fields, like mathematics and to a lesser extent physics, where alphabetical author lists are used (though this appears to be becoming less common over time). See Ludo Waltman, 'An Empirical Analysis of the Use of Alphabetical Authorship in Scientific Publishing', *Journal of Informetrics* 6, no. 4 (Oct. 2012): pp. 700–711; https://doi.org/10.1016/j.joi.2012.07.008

92 Smriti Mallapaty, 'Paper Authorship Goes Hyper', *Nature Index*, 30 Jan. 2018; https://www.natureindex.com/news-blog/paper-authorship-goes-hyper

93 Dan L. Longo & Jeffrey M. Drazen, 'Data Sharing', *New England Journal of Medicine* 374, no. 3 (21 Jan. 2016): p. 276–277; https://doi.org/10.1056/NEJMe1516564. The *NEJM* editors also fretted that the research parasites might 'even use the data to try to disprove what the original investigators had posited' (p. 276); a statement self-evidently silly. There is, though, a related fear that's worth discussing: if your data are open, and you post online your plans for what you want to do with them, you might get scooped: other scientists could grab your data, run your analysis and rush the study to publication before you. In an industry so focused on novelty, being scooped could be a disaster for a scientist's career. It's unclear how often this happens, though, and there are a couple of points to make about it. First, in sensitive situations you can embargo pre-registrations, so that they only become public after you've run the analysis (but there's a timestamp of when you registered the hypotheses). Second, although it's surely frustrating if you get scooped, having different people running the same analyses on the same data might actually be a good thing: comparing between the different versions could pick up on mistakes or oversights. For this reason, at least one journal (*PLOS Biology*) allows researchers who've been scooped to send in their article and it'll still get a fair chance at publication: The *PLOS Biology* Staff Editors, 'The Importance of Being Second', *PLOS Biology* 16, no. 1 (29 Jan. 2018): e2005203; https://doi.org/10.1371/journal.pbio.2005203

94 One group of researchers has proposed the T-index, which re-calculates an author's *h*-index based on how much they actually contributed to each paper they co-authored. If a scientist tends to take the lead on most of their projects, their T-index would get higher; if they tend just to support the work of others, they would end up with a lower score. This would be dependent, though, on everyone honestly admitting how much work they did for each paper in the first place, which might be too optimistic an assumption. And, of course, such a metric could still be gamed, so as with the *h*-index it should never be the sole focus of attention. Mohammad Tariqur Rahman et al., 'The Need to Quantify Authors' Relative Intellectual Contributions in a Multi-Author Paper', *Journal of Informetrics* 11, no. 1. (Feb. 2017): pp. 275–281; https://doi.org/10.1016/j.joi.2017.01.002. See also H. W. Shen & A. L. Barabási, 'Collective Credit Allocation in Science', *Proceedings of the National Academy of Sciences* 111, no. 34 (26 Aug. 2014): pp. 12325–30; https://doi.org/10.1073/pnas.1401992111

95 For one example, see David Moher et al., 'Assessing Scientists for Hiring, Promotion, and Tenure', *PLOS Biology* 16, no. 3 (29 Mar. 2018): e2004089; https://doi.org/10.1371/journal.pbio.2004089. The San Francisco Declaration on Research Assessment, which argues strongly against the use of game-able metrics in hiring scientists and evaluating research, can be found here: https://sfdora.org

96 Florian Naudet et al., 'Six Principles for Assessing Scientists for Hiring, Promotion, and Tenure', *Impact of Social Sciences*, 4 June 2018; https://blogs.lse.ac.uk/impactofsocialsciences/2018/06/04/six-principles-for-assessing-scientists-for-hiring-promotion-and-tenure/

97 Scott O. Lilienfeld, 'Psychology's Replication Crisis and the Grant Culture: Righting the Ship', *Perspectives on Psychological Science* 12, no. 4 (July 2017): pp. 661–64; https://doi.org/10.1177/1745691616687745

98 John P. A. Ioannidis, 'Fund People Not Projects', *Nature* 477, no. 7366 (Sept. 2011): pp. 529–31; https://doi.org/10.1038/477529a and Emma Wilkinson, 'Wellcome Trust to Fund People Not Projects', *Lancet* 375, no. 9710 (Jan. 2010): pp. 185–86; https://doi.org/10.1016/S0140-6736(10)60075-X

99   Ferris C. Fang & Arturo Casadevall, 'Research Funding: The Case for a Modified Lottery', *MBio* 7, no. 2 (4 May 2016): p. 5; https://doi.org/10.1128/mBio.00422-16

100   Ferris C. Fang et al., 'NIH Peer Review Percentile Scores Are Poorly Predictive of Grant Productivity', *eLife* 5 (16 Feb. 2016): e13323; https://doi.org/10.7554/eLife.13323

101   Kevin Gross & Carl T. Bergstrom, 'Contest Models Highlight Inherent Inefficiencies of Scientific Funding Competitions', ed. John P. A. Ioannidis, *PLOS Biology* 17, no. 1 (2 Jan. 2019): p.1, e3000065; https://doi.org/10.1371/journal.pbio.3000065

102   Dorothy Bishop, 'Luck of the Draw', *Nature Index*, 7 May 2018; https://www.natureindex.com/news-blog/luck-of-the-draw. See also Simine Vazire, 'Our Obsession with Eminence Warps Research', *Nature* 547, no. 7661 (July 2017): p. 7; https://doi.org/10.1038/547007a

103   See Paul Smaldino et al., 'Open Science and Modified Funding Lotteries Can Impede the Natural Selection of Bad Science', *Open Science Framework,* preprint (28 Jan. 2019); https://doi.org/10.31219/osf.io/zvkwq

104   The *American Journal of Political Science* not only requires researchers to share their data, but explicitly verifies that the findings are reproducible for every paper during the review process: https://ajps.org/ajps-verification-policy/

105   Nosek et al., 'Promoting an Open Science Culture'.

106   There are also sets of guidelines that many journals are signed up to, such as that from the Committee on Publication Ethics (COPE): https://publicationethics.org/guidance/Guidelines, which gives editors a set of best practice guidelines for dealing with misconduct. The problem is often in ensuring that the editors actually follow these guidelines.

107   There's evidence that including cautionary statements in press releases can improve media coverage of science. In the randomised controlled trial where the researchers manipulated the contents of press releases that we discussed in Chapter 6, the researchers found that when they inserted a statement to the effect that 'these claims are correlational, not causal', 20 per cent of news stories picked up on it, where almost none had previously included this kind of statement. And lest scientists worry that fewer journalists will be interested in their press releases if they include these kinds of caveats, the same study found no evidence that more carefully written press releases (ones with less exaggeration) were less likely to be turned into news articles, something which has been noted by several other press-release studies. Rachel Adams et al., 'Claims of Causality in Health News'. See also Petroc Sumner et al., 'Exaggerations and Caveats in Press Releases and Health-Related Science News', *PLOS ONE* 11, no. 12 (15 Dec. 2016): e0168217; https://doi.org/10.1371/journal.pone.0168217. And Lewis Bott et al., 'Caveats in Science-Based News Stories Communicate Caution without Lowering Interest', *Journal of Experimental Psychology: Applied* 25, no. 4 (Dec. 2019): pp. 517–42; https://doi.org/10.1037/xap0000232. We've seen how a cautiously worded press release in the case of the faster-than-light neutrino translated into more equivocally worded news stories.

108   Leonid Tiokhin et al., 'Honest Signaling in Academic Publishing', *Open Science Framework*, preprint (13 June 2019); https://doi.org/10.31219/osf.io/gyeh8

109   See, for example, the following editorial from the journal *Psychological Science*: Eric Eich, 'Business Not as Usual', *Psychological Science* 25, no. 1 (Jan. 2014): pp. 3–6; https://doi.org/10.1177/0956797613512465

110   Marcus Munafò, 'Raising Research Quality Will Require Collective Action', *Nature* 576, no. 7786 (10 Dec. 2019): p. 183 https://doi.org/10.1038/

d41586-019-03750-7. There are equivalents elsewhere, such as the Open Science Communities initiative in the Netherlands (https://osf.io/vz2sy/) and the Open Science Center at LMU Munich (https://www.osc.uni-muenchen. de/index.html). Another successful grassroots movement is ReproducibiliTea, which holds 'journal club' meetings for early-career researchers to discuss issues surrounding Open Science: https://reproducibilitea.org/

111   Tom E. Hardwicke et al., 'Calibrating the Scientific Ecosystem Through Meta-Research', *Annual Review of Statistics and Its Application* 7, no. 1 (7 March 2020); https://doi.org/10.1146/annurev-statistics-031219-041104

112   It can, of course, go too far. The case of Amy Cuddy, whom we've encountered previously, is one example of where, regardless of how bad the science was, the online discussion took on a disproportionate, bullying and almost gleeful air – see Susan Dominus, 'When the Revolution Came for Amy Cuddy', *New York Times*, 18 Oct. 2017; https://www.nytimes.com/2017/10/18/magazine/ when-the-revolution-came-for-amy-cuddy.html. Incidentally, the mention of bullying raises an issue that's beyond the scope of this book, but still deserves a mention. One major aspect of science that's also in great need of reform is its power culture. There exist far too many (though even *one* is too many) stories of academics abusing their position of seniority over students or other researchers to bully, harass, or even sexually assault them. Two recent stories from psychology illustrate this depressing phenomenon. The social neuroscientist Tania Singer resigned as Director of the Max Planck Institute for Human Cognitive and Brain Sciences in Leipzig in 2018 after allegations that she had viciously bullied her research team for years, for example reportedly screaming at a postdoctoral researcher who had become pregnant (because her maternity leave would interrupt Singer's research). The irony of the situation was that Singer's main research interest is human empathy (Kai Kupferschmidt, 'She's the World's Top Empathy Researcher. But Colleagues Say She Bullied and Intimidated Them', *Science*, 8 Aug. 2018; https://doi. org/10.1126/science.aavo199). And at Dartmouth University, three psychology professors – Todd Heatherton, William Kelley and Paul Whalen – lost their jobs and were banned from campus (or had their access restricted) after nine women came forward with accusations about their alleged harassment, sexual assault and even rape of their students over the course of sixteen years. The Dartmouth Senior Staff, 'New Allegations of Sexual Assault Made in Ongoing Lawsuit against Dartmouth', *The Dartmouth*, 2 May 2019; https://www. thedartmouth.com/article/2019/05/new-allegations-of-sexual-assault-made-in-ongoing-lawsuit-against-dartmouth

113   Brian Nosek, 'Strategy for Culture Change', *Center for Open Science*, 11 July 2019; https://cos.io/blog/strategy-culture-change/

114   Florian Markowetz, 'Five Selfish Reasons to Work Reproducibly', *Genome Biology* 16:274 (Dec. 2015); https://doi.org/10.1186/s13059-015-0850-7

115   William Robin, 'How a Somber Symphony Sold More Than a Million Records', *New York Times*, 9 June 2017; https://www.nytimes.com/2017/06/09/arts/music/ how-a-somber-symphony-sold-more-than-a-million-records.html. There's a particularly good 2019 recording by the Polish National Radio Symphony Orchestra, conducted by Krzysztof Penderecki and published by Domino Recording Co Ltd.: https://open.spotify.com/album/6r4bpBHOQzQ8oJoYmzmKZK

116   Luke B. Howard, 'Henry M. Górecki's Symphony No. 3 (1976) As A Symbol of Polish Political History', *Polish Review* 52, no. 2 (2007): pp. 215–22; https:// www.jstor.org/stable/25779666

117   This distinction brings to mind the old debate in evolutionary biology about whether evolution occurs steadily and incrementally ('gradualism'), or whether it's mainly static, punctuated by sudden large bursts in the appearance of new species ('punctuated equilibrium'). See Kim Sterelny, *Dawkins vs. Gould: Survival of the Fittest* (Thriplow: Icon Books, 2007).

118   Once we correct all the problems described in this book, it'll be time to move on to bigger concerns. Ultimately, we want to build our scientific findings into strong theories that explain the world and predict future observations. See Michael Muthukrishna & Joseph Henrich, 'A Problem in Theory', *Nature Human Behaviour* 3, no. 3 (Mar. 2019): pp. 221–29; https://doi.org/10.1038/s41562-018-0522-1. But in a world where any given finding could fall to pieces after a replication attempt, complex theories could end up leading us down completely the wrong avenues. See Ian J. Deary, *Looking Down on Human Intelligence: From Psychometrics to the Brain*, Oxford Psychology Series, no. 34 (Oxford: Oxford University Press, 2000), particularly pp. 108–109. A more sublunary goal than theory-building, but still one we might want to consider, is *triangulation*: coming at a question from many different angles, using studies of different kinds with different underlying assumptions, checking whether they all converge on a single answer. See Marcus R. Munafò & George Davey Smith, 'Robust Research Needs Many Lines of Evidence', *Nature* 553, no. 7689 (25 Jan. 2018): pp. 399–401; https://doi.org/10.1038/d41586-018-01023-3; and Debbie A. Lawlor et al., 'Triangulation in Aetiological Epidemiology', *International Journal of Epidemiology* 45, no. 6 (20 Jan. 2017): pp. 1866–86; https://doi.org/10.1093/ije/dyw314. For an historical example of triangulation, see George Davey Smith, 'Smoking and Lung Cancer: Causality, Cornfield and an Early Observational Meta-Analysis', *International Journal of Epidemiology* 38, no. 5 (1 Oct. 2009): pp. 1169–71; https://doi.org/10.1093/ije/dyp317. Again, though, if we can't rely on the individual findings themselves, attempting to use them for triangulation will be a non-starter.

119   A classic example is the discovery of the green fluorescent protein (GFP), which glows brightly under ultraviolet light and is now routinely used in biology as an indicator or tag of the presence of specific proteins in the cell. It was a drastically important advance for our biological knowledge, opening up all sorts of new research questions, but it started off very humbly: Osamu Shimomura (who jointly won the Nobel Prize in Chemistry in 2008 for its discovery) discovered it in the 1960s after purifying proteins from a particular kind of bioluminescent jellyfish in a project that originally looked like a dead-end. See Osamu Shimomura, 'Biographical: NobelPrize.org', *Nobel Media* (2008); https://www.nobelprize.org/prizes/chemistry/2008/shimomura/biographical/. Other examples of blue-sky research eventually and unexpectedly leading to major advances are discussed in Jay Bhattacharya & Mikko Packalen, 'Stagnation and Scientific Incentives', *National Bureau of Economic Research* Working Paper no. 26752 (Feb. 2020); https://doi.org/10.3386/w26752

120   'not only can it be argued': Bhattacharya & Packalen, 'Stagnation and Scientific Incentives'.

121   Michèle B. Nuijten et al., 'Practical Tools and Strategies for Researchers to Increase Replicability', *Developmental Medicine & Child Neurology* 61, no. 5 (Oct. 2018): pp. 535–39; https://doi.org/10.1111/dmcn.14054

122   See e.g. Daniele Fanelli, 'Opinion: Is Science Really Facing a Reproducibility Crisis, and Do We Need It To?', *Proceedings of the National Academy of Sciences* 115, no. 11 (13 Mar. 2018): pp. 2628–31; https://doi.org/10.1073/pnas.1708272114

Epilogue

1   https://eventhorizontelescope.org

2   'severe immune deficiencies': Ewelia Mamcarz et al., 'Lentiviral Gene Therapy Combined with Low-Dose Busulfan in Infants with SCID-X1', *New England Journal of Medicine* 380, no. 16 (18 April 2019): pp. 1525–34; https://doi.org/10.1056/NEJMoa1815408; 'cystic fibrosis': Francis S. Collins, 'Realizing the Dream of Molecularly Targeted Therapies for Cystic Fibrosis', *New England Journal of Medicine* 381, no. 19 (7 Nov. 2019): pp. 1863–65; https://doi.org/10.1056/NEJMe1911602

3   Alison J. Rodger et al., 'Risk of HIV Transmission through Condomless Sex in Serodifferent Gay Couples with the HIV-Positive Partner Taking Suppressive Antiretroviral Therapy (PARTNER): Final Results of a Multicentre, Prospective, Observational Study', *Lancet* 393, no. 10189 (June 2019): pp. 2428–38; https://doi.org/10.1016/S0140-6736(19)30418-0

4   Kazuya Tsurumoto et al., 'Quantum Teleportation-Based State Transfer of Photon Polarization into a Carbon Spin in Diamond', *Communications Physics* 2, no. 1 (Dec. 2019): 74; https://doi.org/10.1038/s42005-019-0158-0

5   Yuqian Ma et al., 'Mammalian Near-Infrared Image Vision through Injectable and Self-Powered Retinal Nanoantennae', *Cell* 177, no. 2 (April 2019): pp. 243–55; https://doi.org/10.1016/j.cell.2019.01.038

6   Elizabeth A. Handley, 'Findings of Research Misconduct', *Federal Register* 84, no. 216 (7 Nov. 2019): pp. 60097–98; https://ori.hhs.gov/sites/default/files/2019-11/2019-24291.pdf. For background, see Alison McCook, '$200M Research Misconduct Case against Duke Moving Forward, as Judge Denies Motion to Dismiss', *Retraction Watch*, 28 April 2017; https://retractionwatch.com/2017/04/28/200m-research-misconduct-case-duke-moving-forward-judge-denies-motion-dismiss/

7   Ian Sample, 'Top Geneticist "Should Resign" Over His Team's Laboratory Fraud', *Guardian*, 1 Feb. 2020; https://www.theguardian.com/education/2020/feb/01/david-latchman-geneticist-should-resign-over-his-team-science-fraud

8   Original paper: D. R. Oxley et al., 'Political Attitudes Vary with Physiological Traits', *Science* 321, no. 5896 (19 Sept. 2008): pp. 1667–70; https://doi.org/10.1126/science.1157627. The replication authors wrote up the story: Kevin Arceneaux et al., 'We Tried to Publish a Replication of a Science Paper in Science. The Journal Refused', *Slate*, 20 June 2019; https://slate.com/technology/2019/06/science-replication-conservatives-liberals-reacting-to-threats.html. Their replication was eventually published as Bert N. Bakker et al., 'Conservatives and Liberals Have Similar Physiological Responses to Threats', *Nature Human Behaviour* (10 Feb. 2020); https://doi.org/10.1038/s41562-020-0823-z

9   Dalmeet Singh Chawla, 'Russian Journals Retract More than 800 Papers after "Bombshell" Investigation', *Science*, 8 Jan. 2020; https://doi.org/10.1126/science.aba8099

10   Richard Van Noorden, 'Highly Cited Researcher Banned from Journal Board for Citation Abuse', *Nature* 578, no. 7794 (Feb. 2020): pp. 200–201; https://doi.org/10.1038/d41586-020-00335-7

11   Drummond Rennie, 'Guarding the Guardians: A Conference on Editorial Peer Review', *JAMA* 256, no. 17 (7 Nov. 1986): p. 2391; https://doi.org/10.1001/jama.1986.03380170107031

12   Charles Babbage, *Reflections on the Decline of Science in England, and on Some of Its Causes* (London: B. Fellowes, 1830); https://www.gutenberg.org/files/1216/1216-h/1216-h.htm

13 *International Biographical Dictionary of Computer Pioneers*, ed. John A.N. Lee (Chicago, Ill.: Fitzroy Dearborn, 1995).

14 Simon Chaplin et al., 'Wellcome Trust Global Monitor 2018', Wellcome Trust, 19 June 2019; https://wellcome.ac.uk/reports/wellcome-global-monitor/2018, Chapter 3.

15 An Ipsos MORI poll on trust in science showed the percentage of people in the UK who trust scientists to tell the truth was 85 per cent in 2018, up 22 per cent since polling began in 1997. Gideon Skinner & Michael Clemence, 'Ipsos MORI Veracity Index 2018', Ipsos MORI, Nov. 2018; https://www.ipsos.com/sites/default/files/ct/news/documents/2018-11/veracity_index_2018_v1_161118_public.pdf

16 For a psychology-specific example, see Farid Anvari and Daniël Lakens, 'The Replicability Crisis and Public Trust in Psychological Science', *Comprehensive Results in Social Psychology* 3, no. 3 (2 Sept. 2018): pp. 266–86; https://doi.org/10.1080/23743603.2019.1684822. In another study on how people react when they're explicitly told that there have been replication failures in science, only 17 per cent of the (admittedly small) sample agreed that this is a reason not to trust science in general. See Markus Weißkopf et al., 'Wissenschaftsbarometer 2018', *Wissenschaft im Dialog*, 2018; https://www.wissenschaft-im-dialog.de/fileadmin/user_upload/Projekte/Wissenschaftsbarometer/Dokumente_18/Downloads_allgemein/Broschuere_Wissenschaftsbarometer2018_Web.pdf [German].

17 To the best of my knowledge, the original quotation is from John Diamond. Quoted, for example, in Nick Jeffery, '"There Is No Such Thing as Alternative Medicine"', *Journal of Small Animal Practice* 56, no. 12 (Dec. 2015): pp. 687–88; https://doi.org/10.1111/jsap.12427

18 Alex Csiszar, *The Scientific Journal: Authorship and the Politics of Knowledge in the Nineteenth Century* (Chicago: University of Chicago Press, 2018). pp. 262–3.

19 Ben Guarino, 'USDA Orders Scientists to Say Published Research Is "Preliminary"', *Washington Post*, 19 April 2019; https://www.washingtonpost.com/science/2019/04/19/usda-orders-scientists-say-published-research-is-preliminary/.
    The administration has also been accused of engaging in a politically-motivated kind of reverse hype: it abandoned the standard policy of publicising studies by scientists at the US Department of Agriculture when the studies in question highlighted the dangers of climate change. Helena Bottemiller Evich, 'Agriculture Department Buries Studies Showing Dangers of Climate Change', *Politico*, 23 June 2019; https://www.politico.com/story/2019/06/23/agriculture-department-climate-change-1376413

20 Ben Guarino, 'After Outcry, USDA Will No Longer Require Scientists to Label Research "Preliminary"', *Washington Post*, 10 May 2019; https://www.washingtonpost.com/science/2019/05/10/after-outcry-usda-will-no-longer-require-scientists-label-research-preliminary/

21 Adam Marcus & Ivan Oransky, 'Trump Gets Something Right about Science, Even If for the Wrong Reasons', *Washington Post*, 1 May 2019; https://www.washingtonpost.com/opinions/2019/05/01/trump-gets-something-right-about-science-even-if-wrong-reasons/

22 'Trofim Lysenko': a good brief account is given in John Grant, *Corrupted Science: Fraud, Ideology and Politics in Science* (London: Facts, Figures & Fun, 2007). 'Stalin's USSR and Mao's China': https://www.theatlantic.com/science/archive/2017/12/trofim-lysenko-soviet-union-russia/548786/. For a disturbing description of a recent resurgence in popularity of Lysenko's ideas in Russia, see Edouard I. Kolchinsky et al., 'Russia's New Lysenkoism', *Current Biology* 27, no. 19 (Oct. 2017): R1042–47; https://doi.org/10.1016/j.cub.2017.07.045

23 'creationism': Gayatri Devi, 'Creationism Isn't Just an Ideology – It's a Weapon of Political Control', *Guardian*, 22 Nov. 2015; https://www.theguardian.com/commentisfree/2015/nov/22/creationism-isnt-just-an-ideology-its-a-weapon-of-political-control; 'vaccines': perhaps ironically, one of the major anti-vaccine politicians in Italy was admitted to hospital in 2019 after having caught chickenpox: Tom Kington, 'Italian "Anti-Vax" Advocate Massimiliano Fedriga Catches Chickenpox', *The Times*, 20 March 2019; https://www.thetimes.co.uk/article/massimiliano-fedriga-no-vax-advocate-catches-chickenpox-cbnpkdbh6; 'HIV and AIDs': Pride Chigwedere et al., 'Estimating the Lost Benefits of Antiretroviral Drug Use in South Africa', *JAIDS* 49, no. 4 (Dec. 2008): pp. 410–15; https://doi.org/10.1097/QAI.0b013e31818a6cd5; 'stem-cell technology': Sohini C, 'Bowel Cleanse for Better DNA: The Nonsense Science of Modi's India', *South China Morning Post*, 13 Jan. 2019; https://www.scmp.com/week-asia/society/article/2181752/bowel-cleanse-better-dna-nonsense-science-modis-india

24 'clean and green': Scottish National Party, 'Why Have the Scottish Government Banned GM Crops?', n.d.; https://www.snp.org/policies/pb-why-have-the-scottish-government-banned-gm-crops/; 'cheap populism': Euan McColm, 'Ban on GM crops is embarrassing', *The Scotsman*, 18 Aug. 2015; https://www.scotsman.com/news/opinion/euan-mccolm-ban-on-gm-crops-is-embarrassing-1-3862228; 'extremely concerning': Erik Stokstad, 'Scientists Protest Scotland's Ban of GM Crops', *Science*, 17 Aug. 2015; https://doi.org/10.1126/science.aad1632

25 Émile Zola, *Proudhon et Courbet I*, quoted and translated in Dorra, *Symbolist Art Theories: A Critical Anthology* (Berkeley: University of California Press, 1994).

Appendix: How to Read a Scientific Paper

1 Daniel S. Himmelstein, 'Sci-Hub Provides Access to Nearly All Scholarly Literature', *eLife* 7 (1 Mar. 2018): e32822; https://doi.org/10.7554/eLife.32822

2 You can also consult one of the online lists of predatory journals, such as https://beallslist.weebly.com/

3 For clinical trials, links to trial registries for many countries and regions can be found at the following URL: https://www.hhs.gov/ohrp/international/clinical-trial-registries/index.html. For other fields, I'd suggest checking websites like https://arxiv.org/, https://www.biorxiv.org/ and https://osf.io/. Many pre-registered papers will link to their pre-registration, and clinical trials will have a registration ID number you can use.

4 This is often found in a section at the end of a published paper. Some journals now also award colourful 'badges' to papers that have open data, open methods, or were pre-registered. A list of such journals can be found here: https://cos.io/our-services/open-science-badges/

5 Credit for this idea goes to my friend Saloni Dattani.

6 Google Scholar helpfully provides a 'cited by' function under its entry for each paper, which is useful for this purpose.

7 https://www.sciencemediacentre.org/; there are also Science Media Centres based in other countries, such as the German version: https://www.sciencemediacenter.de/. For more see Ewan Callaway, 'Science Media: Centre of Attention', *Nature* 499, no. 7457 (July 2013): pp. 142–44; https://doi.org/10.1038/499142a

8 https://pubpeer.com/

9 One tip is to simply paste the URL of the journal article into Twitter's search bar: that way you'll see every tweet that links to it, and any comments that have

been made. Many scientists go out of their way to critique papers from their fields in accessible ways on Twitter, and it's an underused source of commentary for the general public.

10 An interesting new tool to get better insights into *how* a study is being cited is scite (https://scite.ai/): its algorithm (which has been trained on judgments from actual scientists) analyses the context in which any paper citing a given study mentions that study, and classifies it as 'supporting', 'contradicting', or simply 'mentioning' (meaning neutral). Although the algorithm is still under development and doesn't do a perfect job, it presents the text excerpt from each citing paper that contains the reference to the study in question, allowing readers to judge for themselves. It's just one example of the technological tools that are constantly appearing to make life easier for – and prevent mistakes by – scientists.

Index